Application of Polymers in Food Sciences

Editors
César Leyva-Porras
National Nanotechnology
Laboratory (Nanotech)
Advanced Materials Research
Center S.C. (CIMAV)
Chihuahua
Mexico

Zenaida Saavedra-Leos
Multidisciplinary Academic
Unit in the Altiplano Region
(COARA)
Autonomous University of
San Luis Potosí (UASLP)
Matehuala
Mexico

Editorial Office
MDPI AG
Grosspeteranlage 5
4052 Basel, Switzerland

This is a reprint of articles from the Special Issue published online in the open access journal *Polymers* (ISSN 2073-4360) (available at: www.mdpi.com/journal/polymers/special_issues/application_poly_food_science).

For citation purposes, cite each article independently as indicated on the article page online and as indicated below:

Lastname, A.A.; Lastname, B.B. Article Title. *Journal Name* **Year**, *Volume Number*, Page Range.

ISBN 978-3-7258-2194-5 (Hbk)
ISBN 978-3-7258-2193-8 (PDF)
doi.org/10.3390/books978-3-7258-2193-8

Application of Polymers in Food Sciences

César Leyva-Porras
Zenaida Saavedra-Leos

MDPI • Basel • Beijing • Wuhan • Barcelona • Belgrade • Novi Sad • Cluj • Manchester

Contents

About the Editors . . . . **vii**

Preface . . . . **ix**

Sherif S. Hindi, Uthman M. Dawoud, Iqbal M. Ismail, Khalid A. Asiry, Omer H. Ibrahim and Mohammed A. Al-Harthi et al.
A Novel Microwave Hot Pressing Machine for Production of Fixed Oils from Different Biopolymeric Structured Tissues
Reprinted from: *Polymers* **2023**, *15*, 2254, doi:10.3390/polym15102254 . . . . **1**

Ma. Cristine Concepcion D. Ignacio, Khairun N. Tumu, Mita Munshi, Keith L. Vorst and Greg W. Curtzwiler
Suitability of MRF Recovered Post-Consumer Polypropylene Applications in Extrusion Blow Molded Bottle Food Packaging
Reprinted from: *Polymers* **2023**, *15*, 3471, doi:10.3390/polym15163471 . . . . **27**

Brandon Van Rooyen, Maryna De Wit, Gernot Osthoff, Johan Van Niekerk and Arno Hugo
Microstructural and Mechanical Properties of Calcium-Treated Cactus Pear Mucilage (*Opuntia* spp.), Pectin and Alginate Single-Biopolymer Films
Reprinted from: *Polymers* **2023**, *15*, 4295, doi:10.3390/polym15214295 . . . . **48**

Brandon Van Rooyen, Maryna De Wit, Gernot Osthoff, Johan Van Niekerk and Arno Hugo
Effect of pH on the Mechanical Properties of Single-Biopolymer Mucilage (*Opuntia ficus-indica*), Pectin and Alginate Films: Development and Mechanical Characterisation
Reprinted from: *Polymers* **2023**, *15*, 4640, doi:10.3390/polym15244640 . . . . **61**

Lissage Pierre, Julio Elías Bruna Bugueño, Patricio Alejandro Leyton Bongiorno, Alejandra Torres Mediano and Francisco Javier Rodríguez-Mercado
Colorimetric Indicator Based on Gold Nanoparticles and Sodium Alginate for Monitoring Fish Spoilage
Reprinted from: *Polymers* **2024**, *16*, 829, doi:10.3390/polym16060829 . . . . **72**

Brandon Van Rooyen, Maryna De Wit, Gernot Osthoff and Johan Van Niekerk
Cactus Pear Mucilage (*Opuntia* spp.) as a Novel Functional Biopolymer: Mucilage Extraction, Rheology and Biofilm Development
Reprinted from: *Polymers* **2024**, *16*, 1993, doi:10.3390/polym16141993 . . . . **93**

Vicente Espinosa-Solis, Yunia Verónica García-Tejeda, Oscar Manuel Portilla-Rivera, Carolina Estefania Chávez-Murillo and Víctor Barrera-Figueroa
Effect of Mixed Particulate Emulsifiers on Spray-Dried Avocado Oil-in-Water Pickering Emulsions
Reprinted from: *Polymers* **2022**, *14*, 3064, doi:10.3390/polym14153064 . . . . **108**

Hector Alfonso Enciso-Huerta, Miguel Angel Ruiz-Cabrera, Laura Araceli Lopez-Martinez, Raul Gonzalez-Garcia, Fidel Martinez-Gutierrez and Maria Zenaida Saavedra-Leos
Evaluation of Two Active System Encapsulant Matrices with Quercetin and *Bacillus clausii* for Functional Foods
Reprinted from: *Polymers* **2022**, *14*, 5225, doi:10.3390/polym14235225 . . . . **127**

César Leyva-Porras, Zenaida Saavedra-Leos, Manuel Román-Aguirre, Carlos Arzate-Quintana, Alva R. Castillo-González and Andrés I. González-Jácquez et al.
An Equilibrium State Diagram for Storage Stability and Conservation of Active Ingredients in a Functional Food Based on Polysaccharides Blends
Reprinted from: *Polymers* **2023**, *15*, 367, doi:10.3390/polym15020367 . . . . **140**

Barbara Johana González-Moreno, Sergio Arturo Galindo-Rodríguez, Verónica Mayela Rivas-Galindo, Luis Alejandro Pérez-López, Graciela Granados-Guzmán and Rocío Álvarez-Román
Enhancement of Strawberry Shelf Life via a Multisystem Coating Based on *Lippia graveolens* Essential Oil Loaded in Polymeric Nanocapsules
Reprinted from: *Polymers* **2024**, *16*, 335, doi:10.3390/polym16030335 . . . . **156**

Daniela Soto-Madrid, Florencia Arrau, Rommy N. Zúñiga, Marlén Gutiérrez-Cutiño and Silvia Matiacevich
Development and Characterization of a Natural Antioxidant Additive in Powder Based on Polyphenols Extracted from Agro-Industrial Wastes (Walnut Green Husk): Effect of Chickpea Protein Concentration as an Encapsulating Agent during Storage
Reprinted from: *Polymers* **2024**, *16*, 777, doi:10.3390/polym16060777 . . . . **176**

Elibet Moscoso-Moscoso, Carlos A. Ligarda-Samanez, David Choque-Quispe, Mary L. Huamán-Carrión, José C. Arévalo-Quijano and Germán De la Cruz et al.
Preliminary Assessment of Tara Gum as a Wall Material: Physicochemical, Structural, Thermal, and Rheological Analyses of Different Drying Methods
Reprinted from: *Polymers* **2024**, *16*, 838, doi:10.3390/polym16060838 . . . . **195**

Angela Monasterio, Emerson Núñez, Valeria Verdugo and Fernando A. Osorio
Stability and Biaxial Behavior of Fresh Cheese Coated with Nanoliposomes Encapsulating Grape Seed Tannins and Polysaccharides Using Immersion and Spray Methods
Reprinted from: *Polymers* **2024**, *16*, 1559, doi:10.3390/polym16111559 . . . . **214**

Zenaida Saavedra-Leos, Anthony Carrizales-Loera, Daniel Lardizábal-Gutiérrez, Laura Araceli López-Martínez and César Leyva-Porras
Exploring the Equilibrium State Diagram of Maltodextrins across Diverse Dextrose Equivalents
Reprinted from: *Polymers* **2024**, *16*, 2014, doi:10.3390/polym16142014 . . . . **236**

About the Editors

César Leyva-Porras

Dr. César C. Leyva Porras holds a Bachelor's degree in Chemical Engineering from the National Technological Institute of Mexico, Chihuahua campus (ITCH, 1998). He has earned two Master's degrees in Materials Science, one from the Advanced Materials Research Center (CIMAV, 2004) and the other from Arizona State University (ASU, 2010), as well as a PhD from the University of Guadalajara (UdG, 2018). He has extensive experience in the operation and handling of transmission electron microscopes (TEMs) and scanning electron microscopes (SEMs), as well as in the synthesis and characterization of materials. His expertise has been instrumental in supporting students, academics, and industry professionals with sample analysis and result interpretation. He collaborates with professors and researchers from various institutions across Mexico on materials-related projects.

Since 2016, he has actively participated as an organizer and instructor in the School of Microscopy. He is currently in charge of the field emission scanning electron microscopy laboratory at the National Nanotechnology Laboratory (NANOTECH). Dr. Leyva Porras is also a member of the Mexican Society of Materials (SMM) and the Mexican Association of Microscopy and Microanalysis (AMMM), and he holds Level I membership in the National System of Researchers (SNI). He frequently serves as a peer reviewer for scientific articles and is a Guest Editor for the journal Polymers by MDPI.

Zenaida Saavedra-Leos

Dr. María Zenaida Saavedra Leos holds a Bachelor's degree in Food Chemical Engineering (2005), a Master's degree in Food Science and Technology (2007), and a PhD in Bioprocess Sciences (2011) from the Autonomous University of San Luis Potosí (UASLP). In 2013, she completed her postdoctoral research stay at the Advanced Materials Research Center (CIMAV) in Monterrey. Since 2014, she has been a full-time Research Professor at UASLP's Multidisciplinary Academic Unit in the Altiplano Region (COARA).

Her research areas include the chemical, physical, and thermal characterization of food products, biopolymer characterization as excipients, pharmacokinetics of active compounds, and the use of chemistry as a tool in biomedical sciences. Dr. Saavedra Leos currently leads the social program "Children in Science in Mexico" and is the founder and leader of the academic group "Chemistry and Technology Consolidation" (UASLP-CA-259). She also founded the collaborative network "Research in Pharmaceutical, Food, Environmental, and Health Biotechnology" (BioFAAS) and is a Level II member of Mexico's National System of Researchers (SNI-II).

Dr. Saavedra Leos has received several national and international recognitions, including the "25 Women in Science in Latin America" award (3rd edition, 2023), the Potosino Award for Science, Technology, and Innovation (2021) in the "Engineering Sciences" category, and the 2024 "Women Who Inspire" distinction by El Sol de San Luis Potosí. At an institutional level, she has been responsible for securing funding through industry-linked projects, overseeing partnership agreements with national and international institutions, and contributing to the creation and design of Undergraduate, Master's, and PhD academic programs.

Preface

The application of polymers in food science has become a transformative force, driving the development of novel functional foods, preserving organoleptic properties, and extending the shelf life of food products. As the global demand for safe, nutritious, and durable food products grows, polymers—both natural and synthetic—play a critical role across various stages of the food production chain, offering innovative solutions that enhance the quality and safety of our food supply.

This Special Issue, titled "Polymer Applications in Food Science", gathers cutting-edge research and review articles that explore the diverse roles of polymers in this field. Contributions were invited from leading academics and researchers worldwide, aiming to provide a comprehensive overview of how polymers can improve food products—from synthesis and characterization to packaging and shelf-life testing.

The topics covered include the following: food chemistry, synthesis, characterization, synthesis and characterization of materials in food development, packing polymers as protecting barriers, microencapsulation of active ingredients, and development of novel food products.

This reprint, based on the Special Issue "Applications of Polymers in Food Sciences", serves as a platform for the exchange of knowledge, highlighting the significant impact of polymer science on food technology. Through these contributions, we aim to advance the understanding of the vital role polymers play in creating safer, more sustainable, and higher-quality food products for consumers worldwide. The reprint is organized into two chapters: "Synthesis, Characterization, and Development of Novel Polymer-Based Foods" (articles 1-6) and "Microencapsulation of Novel and Functional Polymer-Based Foods" (articles 7-15). Each chapter presents manuscripts in chronological order, reflecting the evolution of research in this dynamic field.

We extend our heartfelt gratitude to all the authors who contributed their valuable research to this Special Issue. Their collaborative efforts have been instrumental in the successful publication of this reprint. We also express our sincere appreciation to Mr. Paul Bian, Section Managing Editor of Polymers, for his unwavering support and insightful guidance throughout this process. His expertise was pivotal in achieving the objectives of this Special Issue. Lastly, we are deeply grateful for the support provided by the public institutions in Mexico, where the Guest Editors of this Special Issue, Dr. María Zenaida Saavedra-Leos from the Autonomous University of San Luis Potosí (UASLP) and Dr. César Leyva-Porras from the Advanced Materials Research Center S.C. (CIMAV), are affiliated. Their commitment to research excellence was crucial in the successful completion of this endeavor.

César Leyva-Porras and Zenaida Saavedra-Leos
Editors

polymers

MDPI

Article

A Novel Microwave Hot Pressing Machine for Production of Fixed Oils from Different Biopolymeric Structured Tissues

Sherif S. Hindi [1,*], Uthman M. Dawoud [2], Iqbal M. Ismail [3], Khalid A. Asiry [1], Omer H. Ibrahim [1], Mohammed A. Al-Harthi [1], Zohair M. Mirdad [1], Ahmad I. Al-Qubaie [1], Mohamed H. Shiboob [4], Najeeb M. Almasoudi [1] and Rakan A. Alanazi [1]

1 Department of Agriculture, Faculty of Environmental Sciences, King Abdullaziz University (KAU), Jeddah 21589, Saudi Arabia
2 Department of Chemical and Materials Engineering, Faculty of Engineering, King Abdullaziz University (KAU), Jeddah 21589, Saudi Arabia
3 Department of Chemistry, Faculty of Science, Center of Excellence in Environmental Studies, King Abdullaziz University (KAU), P.O. Box 80216, Jeddah 21589, Saudi Arabia
4 Department of Environment, Faculty of Environmental Sciences, King Abdullaziz University (KAU), Jeddah 21589, Saudi Arabia
* Correspondence: shindi@kau.edu.sa; Tel.: +966-56-676-0086

Abstract: A microwave hot pressing machine (MHPM) was used to heat the colander to produce fixed oils from each of castor, sunflower, rapeseed, and moringa seed and compared them to those obtained using an ordinary electric hot pressing machine (EHPM). The physical properties, namely the moisture content of seed (MCs), the seed content of fixed oil (S_{cfo}), the yield of the main fixed oil (Y_{mfo}), the yield of recovered fixed oil (Y_{rfo}), extraction loss (EL), six Efficiency of fixed oil extraction (E_{foe}), specific gravity (SG_{fo}), refractive index (RI) as well as chemical properties, namely iodine number (IN), saponification value (SV), acid value (AV), and the yield of fatty acid (Y_{fa}) of the four oils extracted by the MHPM and EHPM were determined. Chemical constituents of the resultant oil were identified using GC/MS after saponification and methylation processes. The Ymfo and SV obtained using the MHPM were higher than those for the EHPM for all four fixed oils studied. On the other hand, each of the SG_{fo}, RI, IN, AV, and pH of the fixed oils did not alter statistically due to changing the heating tool from electric band heaters into a microwave beam. The qualities of the four fixed oils extracted by the MHPM were very encouraging as a pivot of the industrial fixed oil projects compared to the EHPM. The prominent fatty acid of the castor fixed oil was found to be ricinoleic acid, making up 76.41% and 71.99% contents of oils extracted using the MHPM and EHPM, respectively. In addition, the oleic acid was the prominent fatty acid in each of the fixed oils of sunflower, rapeseed, and moringa species, and its yield by using the MHPM was higher than that for the EHPM. The role of microwave irradiation in facilitating fixed oil extrusion from the biopolymeric structured organelles (lipid bodies) was protruded. Since it was confirmed by the present study that using microwave irradiation is simple, facile, more eco-friendly, cost-effective, retains parent quality of oils, and allows for the warming of bigger machines and spaces, we think it will make an industrial revolution in oil extraction field.

Keywords: fixed oil; hot pressing machine; microwave irradiation; methylation; GC/MS

Citation: Hindi, S.S.; Dawoud, U.M.; Ismail, I.M.; Asiry, K.A.; Ibrahim, O.H.; Al-Harthi, M.A.; Mirdad, Z.M.; Al-Qubaie, A.I.; Shiboob, M.H.; Almasoudi, N.M.; et al. A Novel Microwave Hot Pressing Machine for Production of Fixed Oils from Different Biopolymeric Structured Tissues. *Polymers* **2023**, *15*, 2254. https://doi.org/10.3390/polym15102254

Academic Editor: Alfonso Jimenez

Received: 6 February 2023
Revised: 27 April 2023
Accepted: 1 May 2023
Published: 10 May 2023

1. Introduction

Oil-producing crops have been attracting more attention due to their high profit compared to other crops. Many species are suitable to be a source of fixed oils, for example, but not limited to castor bean, sunflower, rapeseed, moringa, oil palm, macauba palm, babassu palm, buriti palm, pequi, oiticia, coconut, avocado, Brazil nut, macadamia nut, jatrova, jojoba, pecan, bacuri, ghoper plant, pissava, olive tree, opium poppy, peanut, cocoa, linseed, sesame, and tung oil tree [1].

Fixed oils are synthesized in plants in special biopolymeric tissues of the oily plants and accumulate in their seeds. These seeds differ anatomically (histologically) from each other [2–6]. Different studies have been undertaken to know the morphological characteristics of oily seeds, showing that the structure (macro or micro) of raw materials is a key parameter to understanding the effect of processing and the final yield and quality of products [3,5].

Lipids are generally stored as triacylglycerols in oil bodies. Although there are quantitative differences in the accumulation of storage reserves in seeds, it was not clear whether this would also qualitatively affect the fatty acid profiles of triacylglycerols in these seeds [5].

It was reported by Hu et al. [5] that lipids and protein are major storage reserves in mature Brassica seeds. An inverse relationship between oil and protein accumulation in seeds has been reported for some plant species, including rapeseed. In addition, seeds of many plant species, including rapeseed, accumulate lipids to supply the energy requirements for germination and seedling growth. Such lipids are generally stored as triacylglycerols in small, spherical, and discrete intracellular organelles called oil bodies. Accordingly, the greater the number of oil bodies, the higher the oil content of a seed [5]. In *moringa oleifera*, lipid bodies surrounded the protein bodies and filled most of the remaining space in all healthy cells [6].

The castor bean (*Ricinus communis* L.), also called the castor oil plant, is an annual flowering plant in the Ricinus genus. There are large populations of this plant throughout tropical and subtropical regions, and it has also been found in agricultural sites in Saudi Arabia [7]. Due to its unique ability to withstand thermal fluctuations, castor oil is a valuable raw material in the pharmaceutical, medical, food, chemicals, agricultural, textile, paper, plastic, rubber, cosmetics, perfumeries, electronics, paints, inks, additives, lubricants, and biofuel industries [8–13]. According to [14–16], castor oil has different chemical compositions and physical–chemical properties according to its origin. Malaysian castor seeds, for instance, yield castor oil with a saponification value of 182.96 mg KOH/g as well as a 43.3% dry weight [14], while Nigerian castor seeds yield castor oil with a saponification value of 178.00 mg KOH/g and a dry weight of 48%. In addition, its viscosity and specific gravity are high, making it soluble in alcohols with very low solubility in petroleum solvents [14–17]. Seed oil content ranges from 30% to 60% by weight with variations according to variety, geographic location, and extraction method [11,18,19]. The castor oil's high boiling and low melting points make it superior to petroleum-based oils particularly at high and low temperatures [17–19].

Castor oil consists primarily of ricinoleic acid which has three functional groups. These functional groups are responsible for regulating the chemical composition of the oil which is used to produce resins, waxes, polymers, and plasticizers. It has been demonstrated that carboxylic groups can give rise to a variety of esterification products, while the point of unsaturation can be modified by hydrogenation, epoxidation, or vulcanization [20,21]. Additionally, this oil is important because of its hydroxyl groups and double bonds in the ricinoleic acid component [16]. As a result of the presence of the hydroxyl group in RA and its derivatives, the oil is oxidatively stable with a number of facilitated chemical reactions, including halogenation, dehydration, alkoxylation, esterification, and sulfation [9,22]. In addition to comprising esters (triglycerides molecules) comprised of 3-carbon alcohols (glycerol) and 3-carbon fatty acids (18 carbon atoms), such as other vegetable oils [11,15,23], castor oil is also unique among vegetable oils due to the presence of the hydroxylated fatty acid ricinoleic acid which is the only source of this fatty acid commercially available [11]. It is well known that castor oil contains a wide range of beneficial compounds for health, including fatty acids, flavonoids, phenolic compounds, amino acids, terpenoids, and phytosterols [24]. There are a number of important fatty acids detected in castor oil, including linoleic, oleic, stearic, and linolenic fatty acids [11,22]. It can be used as a bio-based lubricant obtained from the fatty acids of castor oil [25].

The sunflower (*Helianthus annuus* L.) belongs to the family Asteraceae and originated in eastern North America. It is grown throughout the world as one of the most widely cultivated oil crops due to its relatively short growing season [1] and its ability to grow in large semi-arid regions without irrigation [26,27]. It has been found that the oleic acid content of commercial hybrids of sunflower varies between 10 and 50% depending on the climatic conditions of the field and the temperature at which the seeds are growing. Several studies have demonstrated a strong negative correlation between linoleic acid and oleic acid [1,28]. On average, this variety contains 75% oleic acid, although individual plants range between 50 and 80% [29] and individual seeds vary between 19 and 94% [30]. There were significant increases in oleic content in offspring of this variety even when temperatures varied with values exceeding 83% under various conditions [30,31].

Rapeseed is an essential crop for many different purposes, including edible oil, biodiesel, lubricant, and feed. The oil of this plant is high in oleic acid which gives it a competitive advantage over other cooking oils [32]. In terms of oil content, rapeseeds varied between 30.6% and 48.3% of their dry weight which was found to be a higher percentage compared with canola seed oils. The most abundant fatty acids are oleic (56.80–64.92%) and linoleic (17.11–20.92%) as indicated by Matthaus et al. [33] with an omega-6 to omega-3 ratio of 2:1 which offers cardiovascular health benefits [34]. High levels of tocopherols, polyphenols, phytosterols, and carotenoids have been reported in cold-pressed rapeseed oil. Due to its rich content of antioxidants and nutraceuticals, this oil possesses anti-hypercholesterolemic and anti-inflammatory properties. As a result of the high nutritional value of cold-pressed rapeseed oil, it can be used in foods, cosmetics, and pharmaceuticals [34].

Moringa oleifera Lam. is a widely distributed species of the Moringaceae family and is a fast growing, hardwood tree native to the sub-Himalayan area of Northern India [35,36]. In addition to its nutritional and medicinal value, the plant contains phytochemicals that are effective antioxidants and antimicrobials [37]. In addition to their high oil content (up to 40%), seed kernels also contain high levels of fatty acids (oleic acid > 70%) as well as exceptional resistance to oxidative damage. In fact, this rich oil profile makes moringa seeds ideal for both human consumption and commercial use [38,39]. Seeds, on the other hand, have attracted scientific attention due to their high fatty acid content and after refining, significant resistance to oxidative degradation [39]. About 76.73 percent of the oil consists of monounsaturated fatty acids, mainly oleic acid, as well as gadoleic acid and palmitoleic acid. It contains 21.18% saturated fatty acids of which palmitic acid is the predominant followed by behenic, stearic, and arachidic acids. In addition, small quantities of cerotic, lignoceric, myristic, margaric, and caprylic acids have been reported. As a whole, the content of polyunsaturated fatty acids is very low, on average 1.18%, while that of linoleic and linolenic acids is 0.76% and 0.46%, respectively. Moreover, the extraction method does not appear to significantly impact the oil's fatty acid composition. Only one study reported a modest increase in myristic acid and stearic acid content in solvent-extracted oil compared to cold-pressed oil [40].

There are two methods for extracting fixed oils from seeds: mechanical pressing (cold or hot), solvent extraction, or a combination of the two [41]. The mechanical pressing method has the disadvantage of extracting less oil from seeds (about 45%), making it necessary to re-extract the remaining oil from the seed cake. On the other hand, the high-temperature hydraulic pressing method was found to be effective in extracting about 80% of the oil [42]. Moreover, cold-pressed castor oils have low acid and iodine values and have a slightly higher saponification value than solvent-extracted oils [11] obtained a maximum yield of 50.16% of castor oil by pressing for 60 min at 108 °C, using isopropanol/hexane (50:50 v/v) as a solvent system.

Using electric heating tools, such as coils or bands require thick cords, complicated circuits, consumes more electric energy, and can only heat the space covered by the coil or band. Contrarily, microwave offers a thermally homogeneous spaces in any desired size [43–49]. In addition, the ordinary available electric hot presser machines depend for their heat transferring, mainly on the conduction phenomenon that increases the oil extraction durations and reduces the extraction efficiency itself.

This study was initiated to evaluate the yield and properties of each of the four fixed oils squeezed using two machines that differed in their heating tools (EHPM and MHPM) to ensure that the novel microwave procedure used did not distort the parent quality of fixed oils.

2. Materials and Methods

2.1. Management Plan

As shown in Figure S1 presented in the supplementary material, testing the efficiency of the novel microwave hot pressing machine in extracting high yield and quality of fixed oils was applied according to the following steps:

Selection of the species.

1. Specifying the seeds amounts required for the extraction processes.
2. Preparation of the electric hot pressing machine by inserting the electric heating coil around the extruder head of the machine.
3. Preparation of the microwave hot pressing machine by releasing the electric heating coil from the extruder head of the machine.
4. Pressing the seeds of each of the four oil crops (castor, sunflower, rapeseed, or moringa).
5. Dehydration, settlement, and filtration of the crude fixed oils.
6. Characterization of the fixed oils and comparing the oil extracted by microwave to that obtained by electric band heater.

2.2. Raw Materials

Four different species were used to be the source of fixed oils, namely castor bean (*Ricinus communis* L., Family: Euphorbiaceae), common sunflower (*Helianthus annuus* L., Family: Asteraceae), rapeseed (*Brassica napus* L., Family: Brassicaceae), and moringa (*Moringa oleifera* Lam., Family: Moringaceae) as shown in Figure S2a–d. A random sample of three shrubs/plants of each species was collected during the 2021–2022 season from the Agricultural Research Center of King Abdulaziz University at Hada Al-Sham which is located about 120 km from Jeddah, Saudi Arabia. The shrubs/plants were grown in sandy soil at a latitude of 21° 46′.839N and a longitude of 39° 39′.911E which was 206 m above sea level.

2.2.1. Preparation of the Seeds

Figure S2a–d illustrates the steps involved in the extraction of each of the four fixed oils from the air-dried seeds of castor bean (Figure S2a), common sunflower (Figure S2b), rapeseed (Figure S2c), and moringa (Figure S2d). A bean de-huller machine (Figure S2a–d) was used to separate the seeds from their hulls/pods after being collected from the selected shrubs/plants, and the seeds were, then, cleaned to remove any foreign materials [7].

2.2.2. The Mechanical Extraction of the Four Fixed Oils

For each castor bean, common sunflower, rapeseed, and moringa, about one kg of the clean seeds were air dried to reduce the moisture of the seeds [48,50,51]. The air-dry seeds were allowed to be oven dried at about 65–70 °C in an electric furnace. To reduce the oil viscosity within seeds tissues that can maximize the oil yield [52,53], the seeds were exposed to a microwave beam for about ten minutes within the seed hopper. Then, the seeds were hot-pressed mechanically using each of the two hot pressing machines (MHPM and EHPM) as shown in Figure 1. After settling for one hour, the oily supernatants were collected individually and filtered through vacuum filters to remove contaminants from the oil. Hydraulic pressure was used to release the remaining oil from the settled cake [50].

2.2.3. The Hot Pressing Machine (HPM)

Microwave radiation is generated using the magnetron device, the essential component of the microwave generator unit (MGU), by converting the input alternate electric current (AC) into microwave beam that is responsible for warming the pressing machine. The HPM (Figure 1) consists of a motor, extruder, colander, and a heating tool (electric heater or microwave beam). It is worth mentioning that two devices were used in the present investigation for squeezing each of the four oily seeds, namely electric hot pressing machine (EHPM) and microwave hot pressing machine (MHPM). Both of them use the same hot pressing machine but they differ in the heating tool necessary to warm the extruder's colander of the machine. For more illustration, the EHPM uses a dc-electric band heater (Figure 2a–c) for this purpose, while the MHPM was constructed to be dependent on a microwave generator unit that emitters microwave beams responsible to release the colander temperature.

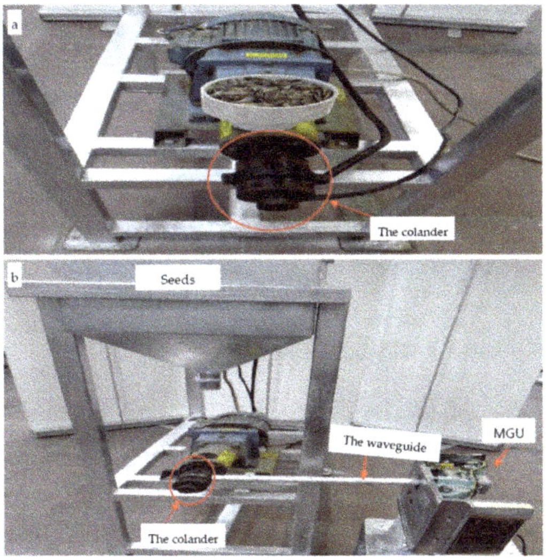

Figure 1. The microwave hot pressing machine (MHPM) used for extraction of the four fixed oils: (**a**) An overall front image for the HPM before connecting the microwave source, (**b**) An overall front image for the MHPM showing the colander, the waveguide that used for directing microwave beam to the extrusion head, and the microwave generator unit (MGU).

The main components of the HPM are the Taby motor (380 Volt 3 phase 10 amp 50/60 Hz, Sweden) that rotates the extruder shaft (Figure 2a,b) of the HPM. The extruder shaft is used to deliver the oily seeds from the seed hopper to the colander, in which the seeds are compressed mechanically, allowing the fixed oil to excrete from biopolymeric tissues of seeds through the oil channels of the colander as clear from Figures 1 and 2. The power screw geometry of its straight shaft is apparent in Figure 2b. Heating the oily seed tissue found within its inner central channel allowing for extraction of their fixed oil and excreted through its oil capillary channels where they were collected in a suitable reservoir. In addition, the colander is the terminal part of the pressing machine coupled directly to the extruder (Figures 1a,b, 2a,c, 3a,c and 4e). It has a hollow central tunnel (to feed the extracted seed tissues known as the cake) as well as an oil channel that accumulates the final oil yield in the production reservoir. To soften seeds tissues and enhance fixed oils yield, a heating source must be used to worm the colander of the HPM. In this investigation, two heating techniques were performed and compared to each other, namely electric band heater used for EHPM and MGU used for MHPM.

To prepare the EHPM, the electric heating coil must be installed, while for installing the microwave hot pressing, the electric band heater must be released from its colander and replaced with the MGU in which the microwave beam was directed to the extruder's colander by using an ideal waveguide, a short metallic waveguide (Figures 1b and 4e).

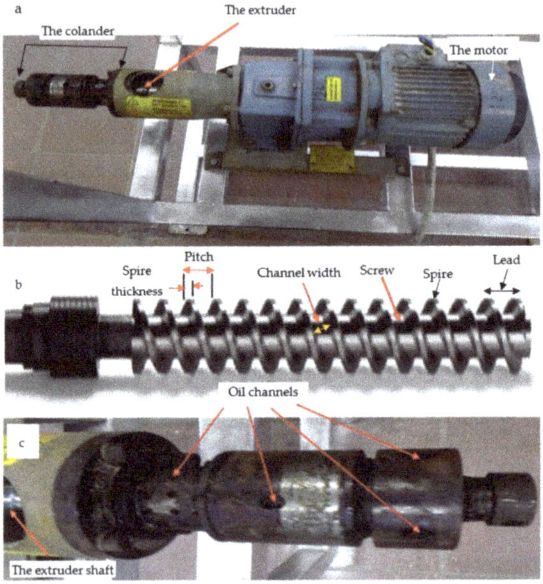

Figure 2. The hot pressing machine (HPM): (**a**) Upper view, (**b**) The straight shaft and its power screw geometry, and (**c**) The colander of the hot pressing machine.

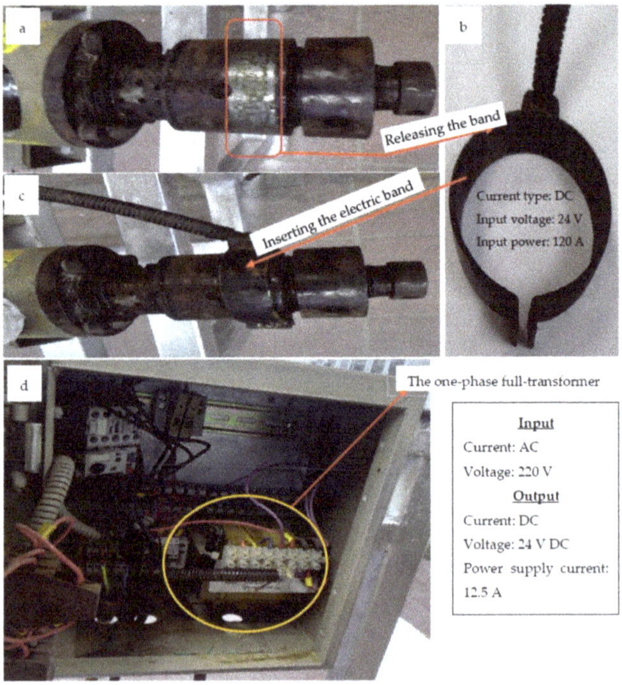

Figure 3. Removing the dc-electric band from the EHPM to be exchanged to the MHPM: (**a**) The DC-electric band heater used to heat the colander of the extruder, (**b**) The thermal band, (**c**) Fixing the band tighten around the colander, and (**d**) The electric box feeding the HPM.

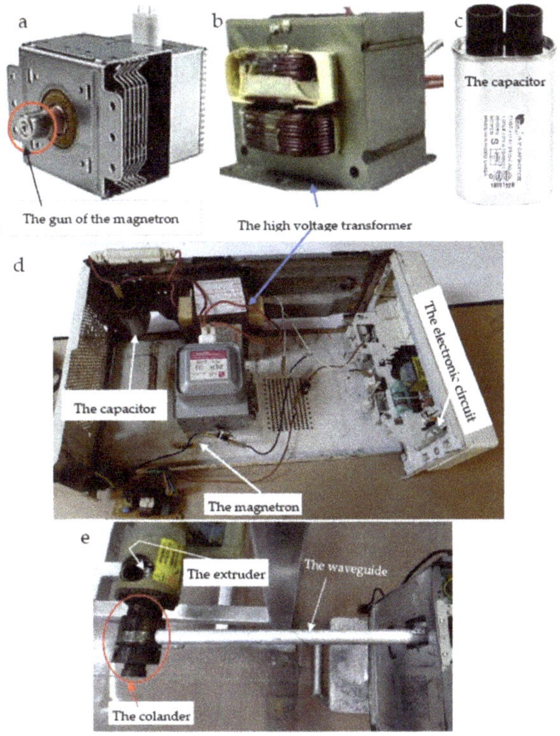

Figure 4. The microwave generator unit (MGU) used for heating the head of the microwave hot pressing machine: (**a**–**e**): (**a**) The high-voltage magnetron, (**b**) The high-voltage transformer, (**c**) The high-voltage capacitor, (**d**) Overall image of the MGU, and (**e**) Directing microwave beam to the colander of the MHPM through the waveguide.

The band heater (Figure 3b,c) was used to heat the colander that is coupled to the extruder front. It is also suitable for heating any metal pipes, nozzles, barrels, and other cylindrical parts. These fast-heat band heaters electrically warm the external surface of a cylindrical part to provide indirect heating. It is worth mentioning that the heating process is an essential facility to prevent plugging the fine holes that exert the oil, reducing the oil viscosity for ease of handling within the extruder as well as increasing the oil yield released from the seed tissues.

In the MGU, the magnetron is the primary and only source of microwave energy (Figure 4a,d). Originally, it was taken from a domestic microwave oven. In terms of technical specifications, the magnetron has a power of 900 W, an anode voltage of 4.20 kVp, and a frequency of 2.46 GHz. The magnetron, 2M214 39F (06B), code of 2B71732E for LG microwave ovens (900 W, 4.20 kVp anode voltage, 2460 MHz frequency) was used in this study as a self-excited microwave oscillator, cavity magnetrons convert high-voltage electric energy into microwave beams (Figure 4a,d). It is worth mentioning that magnetrons produce high-power outputs by crossing electron and magnetic fields. In this study, a magnetron emits a constant frequency of 2.46 GHz, since magnetrons are usually designed on a constructively fixed frequency as illustrated by Hindi et al. [45–48].

As shown in Figures 1b and 4e, microwave energy is radiated into the air through a stainless-steel pipe with a diameter of 1 inch, then is conducted into a metal cavity. In order to melt the fixed oil and force it to excrete from seed tissues, temperature must be adjusted to be approximately 75–80 °C.

The high voltage transformer (HVT) is used to convert alternate current (AC) into direct current (DC) required for working the electric band heater. In addition, the high voltage transformer (Teknik, Amal, Sweden) was used in the present investigation as shown in Figure 4b,d. It has been reported that large industrial and commercial ovens can use 915 megahertz magnetrons to excite the larger cavities within the ovens [49].

The high voltage capacitor (2100V.AC, 1%, 50/60 Hz, and 10 MΩ). An electrical connection was made between the capacitor and the magnetron, the transformer, as well as the outlet of the waveguide via a diode (Figure 4c).

Other accessories were used in the MGU, such as an electronic diode, the double-terminal component in which the current conducts primarily in one direction (asymmetric conductance). Furthermore, its resistance is low in one direction (ideally zero) and high in the other direction (ideally infinite). In addition, a small electric fan was constructed to distribute air around the transformer and magnetron to cool them.

2.3. *Characterization of the Four Species*

After collecting the crude oil, it was weighed and stored until it was subjected to different characterizations. A set of oil specifications were determined using methods reported in the standard methods or applied by researchers working in the same or related fields [50–54].

The values of the different physical and chemical properties of the seeds and fixed oils of the four crops species were calculated as presented in Table 1.

Table 1. Calculation of different chemical and physical properties of the seeds and the extracted fixed oils [7,42,48,51–53].

Equation	Definitions
[1] $MC, \% = [(W_{ads} - W_{ods})/W_{ods}] \times 100$ [2] $S_{cfo}, \% = (W_{mfo} + W_{rfo})/W_{ods}$ [3] $Y_{mfo}, \% = (W_{mfo}/W_{ods}) \times 100$ [4] $Y_{rfo}, \% = (W_{ods} - W_{rc})/W_{ods}] \times 100 = (W_{rfo}/W_{ods}) \times 100$ [5] $EL, \% = [\{W_{ods} - (W_{rfo} + W_{rc})\}/W_{ods}] \times 100$ [6] $E_{foe}, \% = [W_{rfo}/(S_{cfo} \times W_{ods})] \times 100$	W_{ads}: Weight of air-dried seeds. W_{ods}: Weight of oven-dried seeds, g. W_{mfo}: Weight of the main fixed oil (g). W_{rfo}: Weight of recovered fixed oil (from seed'cake), g W_{rc} = Weight of residual cake after extraction. S_{cfo}: Seed content of fixed oil
[7] $SG_{fo} = (W_1 - W_2)/(W_2 - W_0)$	W_0: Weight of an empty bottle. W_1: Weight of the bottle filled with fixed oil. W_2: Weight of the bottle filled with deionized water.
[8] $IN = 12.69 \times C \times (V_1 - V_2)/W$	C, V_1, V_2: Parameters of sodium thiosulphate: C: Concentration. V_1: The volume used for the blank test. V_2: The volume used for the fixed oil. W: The fixed oil weight.
[9] $SV = 56.1N \times (V_1 - V_2)/W$	V_1: Volume of the solution used for the blank test. V_2: Volume of the solution used for fixed oil. N: The actual normality of the HCl used. W: The fixed oil weight.
[10] $AV = 5.61(V \times N)/W$	V: Volume of KOH, IN mL. N: Normality of KOH. W: Fixed oil weight.
[11] $Y_{fa} = (A_1/A_2) \times 100$	A_1: Peak area (mm^2) of a certain fatty acid detected at a certain retention time (min.) A_2: The overall peak's areas (mm^2) of all fatty acids in the fixed oil.

[1] Moisture content of seed, [2] Seed content of fixed oil, [3] Yield of the main fixed oil, [4] Yield of recovered fixed oil, [5] Extraction loss, [6] Efficiency of fixed oil extraction, [7] Specific gravity, [8] Iodine number, [9] Saponification value, [10] Acid value, [11] and Yield of fatty acid.

2.3.1. Physical Characterization of the Seeds and Oils

The physical properties of fixed oils are known to be influenced by chemical composition as well as operating parameters, such as pressure and temperature. These properties were determined and compared with the cited standard [7,51].

Concerning seed qualifications related to their suitability for producing fixed oils, moisture content (MC) as well as seed content of fixed oil (S_{cfo}) were determined. For the MC investigation, about 100 g of cleaned intact seeds were dried in an oven at 90 °C, weighed at 1 h intervals until obtaining a constant weight after about 6 h [42]. It is worth mentioning that the S_{cfo} refers to the amount of oil that can be derived from an oil seed. It is usually represented as a percentage based on the parent seed weight as shown in Table 1 [52].

Regarding fixed oil extraction parameters, four properties were determined, namely yield of the main fixed oil (Y_{mfo}), yield of recovered fixed oil (Y_{rfo}), extraction loss (EL), and efficiency of fixed oil extraction (E_{foe}).

The Y_{mfo} refers to the quantity of oil that can be extracted from an oil seed. It was determined as the ratio of the weight of main oil extracted to the weight of the ground seed specimen before extraction. Furthermore, it was determined mechanically using each of the MHPM and EHPM, while both Y_{rfo} and EL were investigated by solvent extraction procedures using analytical grade n-hexane anhydrous (95% purity) purchased from Sigma-Aldrich, Germany in a Soxhlet extractor.

Prior to the chemical extraction performed, only for the purpose of estimating both Y_{rfo} and EL, the air dry seeds were oven dried at about 65–70 °C in a heating furnace. After that, the oven-dried seeds were ground using a rotary grinding machine and sieved using standard sieves to be passed through a standard sieve of 60 mesh and retained on that of 80 mesh (60/80 mesh). This range of the seeds' particle size facilitates the diffusion of the soluble compounds and the release of the oil [52,53].

The suitable particle-sized seed powder was put in a cotton thimble (2.5 × 7 cm) and centered in one siphon of the Soxhlet extractor fixed with a round flask (250 mL) containing about 150 mL of n-hexane. After that, a rotary evaporator with a vacuum pump was also used to separate the oil and the solvents after oil extraction. The recovered fixed oil was calculated in the manner indicated in Table 1.

Extraction loss (EL) was calculated as the difference between the weight of the ground seeds specimen before extraction and the sum overall weights of both the oil recovered and residual cake after extraction by hexane divided by the parent weight of the ground seed sample before extraction as clear in Table 1 [52].

The E_{foe}, an important parameter in oilseed processing, is the percentage of oil extracted in relation to the amount of oil present in the seed. It was computed as the ratio of the weight of oil recovered to the product of the seed oil content and the weight of the crushed seed sample before extraction [52]. The E_{foe} was computed as a ratio of the weight of oil recovered to the product of the seed oil content and the weight of the crushed seed sample before extraction [52].

Regarding the specific gravity of fixed oil (SG_{fo}), in a clean 25 mL oven dried, stoppered bottle, the fixed oil was pyrchonometrically determined at 25 °C. The weight of the bottle was measured before and after it was filled with the fixed oil. Following the washing and oven drying of the bottle, the fixed oil was substituted with deionized water and the bottle was weighed (W_2). The SG_{fo} is calculated based on the equation shown in Table 1 [7,51,55,56].

Moreover, the refractive Index (RI) provides a rapid indicator of its purity and quality. A refractometer model No. 922313 (Bellingham and Stanley Ltd., London, UK) was used to determine the RF of the castor oil at 30 °C. For each fixed oil, three repetitions were performed and recorded [51,57,58].

2.3.2. Chemical Characterization of the Fixed Oils

Determination of Iodine Number (IN)

An IN represents the amount of unsaturationality (the presence of double bonds) in a fixed oil determined by the amount of iodine that reacts with 100 g of oil. Oils with a higher IN contain more double bonds [59]. The iodine value is counted as g I_2/100 g of oil.

The IN was analyzed using the method described by AL-Hamdany and Jihad [60] and Omari et al. [51]. After dissolving the fixed oil in 10 mL of chloroform, the solution was mixed with 30 mL of Hanus iodine solution, kept in a tightly closed flask, and vigorously swirled for 30 min in the dark. Thereafter, approximately 10 mL of potassium iodide (15%) and 100 mL of distilled water were added under shaking. Titration was carried out with the iodine solution against sodium thiosulfate (0.1N) until a yellowish color was achieved. Approximately two to three drops of the starch solution were then added, turning the solution color blue. Titration continued until the blue color faded, and the volume of $Na_2S_2O_3$ at this point was calculated. A blank test (without oil) was conducted in the same manner. In order to calculate the iodine number (IN), the formula presented in Table 1 was used [7,51,53].

Determination of Saponification Value (SV)

By measuring the amount of potassium hydroxide required to saponify one gram of fixed oil, SV provides an indication of the chain length of the fatty acids contained in the oil. Based on the standard procedure described by Akpan et al. [57], the SV was determined. About 2 g of a fixed oil was added to 25 mL of 0.1 N ethanolic-KOH with constant stirring was allowed to be boiled gently for 60 min under refluxing. A few drops of phenolphthalein indicator was added to the warm mixture and then titrated with HCl (0.5 M) up to the end point in which the pink color of the indicator just disappeared. Based on the equation presented in Table 1, we were able to determine the saponification value.

Determination of Acid Value (AV)

The AV of an oil is defined as the milligram amount of potassium hydroxide required to neutralize one gram of the oil. The term is used to indicate how many carboxylic acid groups (-COOH) a fatty acid contains. In order to determine the AV of fixed oil, 0.5 g of the oil was dissolved in 25 mL of absolute ethanol phenolphthalein indicator. A shaking water bath was used to heat the mixture for five minutes. Afterward, a drop-wise titration with KOH (0.1 N) was conducted while the free acid was still hot. The solution was continuously agitated until a pink color appeared which remained for a minimum of 10 s. Upon approaching the endpoint, vigorous shaking was performed to ensure thorough mixing. During the reaction with the oil, KOH (0.1N) was measured in terms of volume based on the formula shown in Table 1 [51,53].

Determination of pH Value

In a beaker with a slow stirring, about 2 g of each fixed oil was mixed with 13 mL of hot distilled water. Afterward, the emulsion was cooled in a cold-water bath until it reached 25 °C. Using suitable buffer solutions, an electrode was standardized and immersed in oil to determine its pH value [51,55].

Determination of the Fatty Acids of the Fixed Oils

Two samples were analyzed for each crop, whereby one of them was extracted by the electric hot pressing machine (EHPM), and the other was produced by the microwave–hot pressing machine (MHPM). Statistically, three repetitions were performed to detect the experimental error.

To ensure high-accuracy chemical analyses, the fixed oil samples were pretreated as the following scheme before characterization by using the GC/MS (Figure S3).

1. Preparation of the Methyl Esters

Using silica gel columns, each fixed oil was eluted with hexane [7,61,62]. Under reduced pressure, hexane was evaporated, resulting in the so-called hexane fraction (Figure S3).

The procedure includes the following steps: saponification, esterification, fractionation on a column chromatography filled with silica gel, and analysis by GC/MS. To remove the free lipids, about 2 g of fixed oil was extracted with 10 mL mixture of $CHCl_3$/MeOH solution (2:1 v/v) and sonicated for about 15 min, each. To release "bound" components, the free lipids-fixed oil sample was saponified with 0.5 M NaOH (MeOH/H_2O solution, 9:1 *v*/*v*, 5 mL, 70 °C, 1 h).

After centrifugation of the cold mixture at 2500 rpm for 15 min, the supernatant was discarded, the precipitate was collected, and its pH was readjusted to 3 by adding 3 M HCl. About 1 mL of solvent-extracted water and 3 mL of $CHCl_3$ were used to extract the hydrolyzed lipids. After the complete elimination of the solvent by evaporation under a steady flowing helium atmosphere, the hydrolyzed acid components were methylated using 100 L of BF_3/MeOH (14% *w*/*v*) at 70 °C for 1 h). Then, the methyl esters were extracted with about 2 mL of $CHCl_3$ (3 times) from the solvent-extracted H_2O.

Using a glass column filled with activated silica gel preheated at 120 °C for 24 h, the methylated extracts were fractionated and eluted using n-hexane.

The extracts were separated using the elutropic series of the following four solvent systems, including n-hexane, dichloromethane (DCM), and methanol (MeOH). The four systems were: 6 mL of n-hexane, 2 mL of a mixture of n-hexane/dichloromethane (9:1 *v*/*v*), 6 mL of dichloromethane, and 5 mL DCM/MeOH (1:1 *v*/*v*) to yield an elution rate of 15 mL min1. The DCM/MeOH method allowed the dihydroxy-fatty acids to be eluted. Under a mild steady stream of helium gas, all fractions were collected, and the solvent was removed.

2. Gas Chromatography-Mass spectrometer (GC–MS)

Since saponification/methylation processes (Figure S3) are helpful for identifying fatty acids [62,63], the major constituents of the oil were detected as methyl or methyl esters derivatives. GC-Trace ultra-system (Thermo Co. USA) was used to analyze each of the studied fixed oils. Fractionation and separation were conducted using a fused-silica capillary column (Elite- 5MS column, Thermo Scientific GC Column, 30 m × 0.25 mm × 0.25 µm) with chemically bonded phases DB1 (J&W Scientific). The initial temperature (IT) was adjusted to 40 °C for 4 min as residence time (RT) and raised to a maximum final temperature (MFT) of 220 °C at a rate of 4 °C/min for 15 min. The temperature was set at 250 °C for the injector, where the injection volume was 1 µL of the tested sample at a transfer temperature of 280 °C. Helium was used as a carrier gas at a flow rate of 20 mL/min. The EI mass spectra were collected in full scan mode (*m*/*z* 50–600) at 180 °C ion source temperature and 70 eV ionization voltage. Based on a library search using NIST, USA, and Wiley Registry 8-edition as well as a comparison of retention indices among the peaks, tentative identifications were made. A peak area normalization method was to estimate the yield of fatty acids (Y_{fa}) in the fixed oil [7,64] by calculating the peak area (in mm^2) of a certain fatty acid detected at a certain retention time (in minutes) based on the overall areas in mm^2 representing chemical constituents of the fixed oil that arisen at different retention times (in minutes) at the chromatogram as clear in Table 1. Moreover, their chromatograms are presented in Figure S4. The percentage allocations of the fatty acids in the fixed oils are presented in Figure 5.

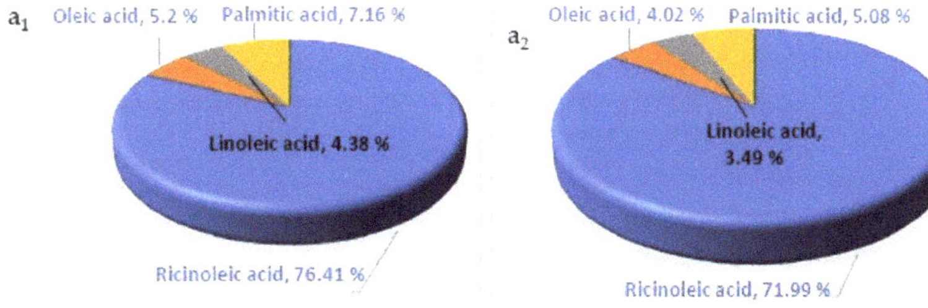

Figure 5. *Cont.*

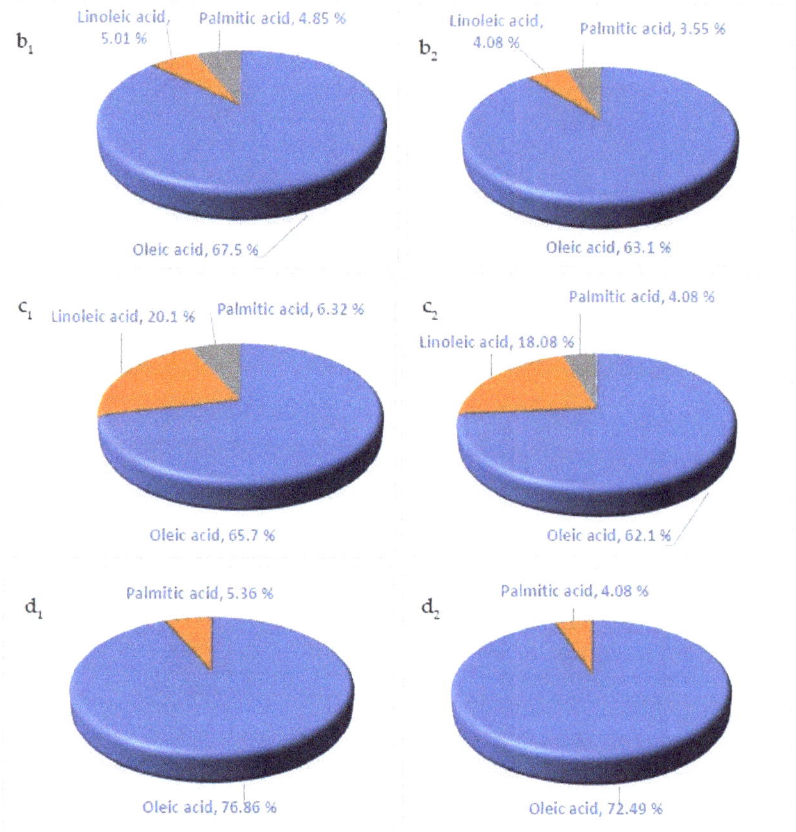

Figure 5. The main fatty acids and their contents of (**a_1**,**a_2**) Castor oil: (**a_1**) Microwave-assisted extracted oil (MAEO), (**a_2**) Electric hot pressed oil (EHPO), (**b_1**,**b_2**) Sunflower oil: (**b_1**) For MAEO, (**b_2**) EHPO, (**c_1**,**c_2**) Rapeseed oil: (**c_1**) MAEO, (**c_2**) EHPO, and (**d_1**,**d_2**) Moringa: (**d_1**) MAEO, (**d_2**) EHPO.

2.4. Scanning Electron Microscopy (SEM)

SEM imaging was used to study the surface morphology and anatomical features in biopolymeric structured tissues. The SEM technique and image analysis were efficient tools for the characterization of the microstructure of the four oily seeds [3–5].

Sample preparation was conducted by isolating a small central block of cotyledon (about 3 mm in length) from an air-dried seed.

For the SEM examination, the sample was placed on a double side carbon tape on Al-stub and dried in air. It is worth mentioning that all samples were examined without any pretreatments, such as embedding, fixation and sputtering processes [3]. The samples were examined with an SEM Quanta FEG 450, FEI, Amsterdam, The Netherlands. The microscope was tried at an accelerating voltage ranging from 5–20 kV.

2.5. Statistical Design and Analysis

The present experiment was statistically designed to be a split plot design in three replicates (blocks) with multiple observations (five repetitions) per sub-plot per block for each property of each fixed oil. This was conducted to detect the differences between the four fixed oils produced from castor, sunflower, rapeseed, and moringa seeds squeezed by the hot pressing machine heated with two different heating tools, namely electric band heater (EHPM) and microwave irradiation (MHPM). In addition, the least significant difference at a 95% level of confidence ($LSD_{0.05}$) method was used to compare the differences between species means for all the properties studied [65].

3. Results

The physical and chemical properties of each of the four fixed oils were determined. The comparisons between the mean values of these properties were extended to those within species as well as between species. Accordingly, the comparisons were conducted between the fixed oils extracted using the microwave –hot pressing machine (MHPM) and the electric hot pressing machine (EHPM) as well as between the oily species at the same level of machinery used.

3.1. Characterization of the Fixed Oils

Eleven physical properties of the four fixed oils, namely the moisture content of the seed, [2] seed content of the fixed oil, [3] yield of the main fixed oil, [4] yield of recovered fixed oil, [5] extraction loss, [6] efficiency of fixed oil extraction, [7] specific gravity, [8] iodine number, [9] saponification value, [10] acid value, [11] and yield of fatty acid were determined and are presented at Table 2. Furthermore, the chemical properties of these oils, namely iodine number (IN), saponification value (SV), acid value (AV), and pH were also investigated and included in Table 2.

Table 2. Mean values * of physical and chemical properties of the four fixed oils, namely castor (C), sunflower (S), rapeseed (R), and moringa (M) extracted by microwave –hot pressing machine (MHPM) and electric hot pressing machine (EHPM).

Properties		Extraction Method [3]								ASTM [4]
		MHPM				EHPM				
		C	S	R	M	C	S	R	M	
MC, %	AD	±	±	±	±	±	±	±	±	
	OD	3.09 ± 0.15	±	±	±	4.47 ± 0.18	±	±	±	N.d. [12]
OY [5], %		44.8 ± 1.03	42.5 ± 0.98	42.2 ± 1.02	39.6 ± 0.89	38.3 ± 1.09	36.4 ± 1.25	35.2 ± 1.33	34.6 ± 0.93	N.d. [12]
SG [6]		0.973 ± 0.08	0.914 ± 0.07	0.906 ± 0.1	0.946 ± 0.093	0.968 ± 0.09	0.918 ± 0.07	0.914 ± 0.09	0.957 ± 0.088	0.96–0.97
RI [7]		1.48 ± 0.092	1.473 ± 0.08	1.465 ± 0.069	1.464 ± 0.13	1.43 ± 0.07	1.469 ± 0.086	1.467 ± 0.102	1.563 ± 0.17	1.48–1.49
IN [8] ($g\ I_2/100\ g$ oil)		83.7 ± 1.09	126.3 ± 1.23	114.5 ± 1.48	69.7 ± 1.44	84.8 ± 1.05	127.1 ± 1.19	115.1 ± 1.29	71.4 ± 1.38	82–88
SV [9] (mg KOH/g oil)		181.5 ± 3.79	188.4 ± 2.34	168.3 ± 1.98	180.7 ± 2.18	182.1 ± 3.07	194.6 ± 2.33	181.2 ± 2.56	190.4 ± 1.97	175–187
AV [10] (mg KOH/g oil)		0.869 ± 0.004	0.914 ± 0.008	2.107 ± 0.08	1.433 ± 0.005	0.882 ± 0.008	1.139 ± 0.001	2.306 ± 0.006	1.344 ± 0.007	0.4–4.0
pH [11]		6.48 ± 0.359	6.46 ± 0.45	6.43 ± 0.77	6.45 ± 0.69	6.53 ± 0.48	6.38 ± 0.39	6.49 ± 0.78	6.37 ± 0.59	N.d. [12]

* Each value is an average of 5 samples ± standard deviation, [3] Yield of the main fixed oil, [4] Yield of recovered fixed oil, [5] Extraction loss, [6] Efficiency of fixed oil extraction, [7] Specific gravity, [8] Iodine number, [9] Saponification value, [10] Acid value, [11] Yield of fatty acid, and [12].

3.1.1. Physical Properties of Seeds

Investigation of the moisture content of seeds (MC_s) revealed that the castor beans had the lowest MC_s (3.6%), while moringa had the highest MC_s value (5.34%) among the four oily species as is clear in Table 2. The obtained mean values approach those determined by Muzenda et al. [42] and differed from those reported in the literature findings of 4.15% [66–68]. Moreover, comparing the two heating techniques (MHPM and EHPM) revealed that the MHPM dried all four seeds to lower levels compared to the EHPM.

For the seed content of the fixed oil (S_{cfo}), comparing the S_{cfo} values between species indicate that the castor bean seeds contained the highest S_{cfo} value (49.87%) than the other species. Moreover, the MHPM extracted more fixed oils than those obtained by the EHPM (Table 2).

3.1.2. Physical Properties of the Fixed Oils

For the yield of the main fixed oil (Y_{mfo}), it is clear from Table 2 that at the same level of heating technique (MHPM and EHPM), comparing the Y_{mfo} values within species revealed that castor seeds gave the highest yield (44.8 % and 38.3 %, respectively) comparing to those obtained from sunflower (42.5 % and 36.4 %, respectively), rapeseed (42.2 % and 35.2 %, respectively), and moringa (35.6 % and 34.6 %, respectively) as presented in Table 2. At the same level of oily species, the MHPM produces higher Y_{mfo} values than those obtained by the EHPM (Table 2). This is clear for all the four species examined in the present study.

Regarding the yield of the recovered fixed oil (Y_{rfo}), it can be seen from Table 2 that the Y_{rfo} obtained by the MHPM (6.93–8.85%) that was higher than those for the EHPM (3.34–5.06%).

Studying the extraction loss (EL) indicates that the results of MHPM (1.02–2.03%) were lower than those obtained from the EHPM (2.35–2.75%). This finding confirms the superiority of the invented MHPM. In addition, comparing species at the same level of the heating tool revealed that there are no significant differences between them.

For the efficiency of the fixed oil extraction (E_{foe}), the MHPM had the highest efficiency (90.22–92.8%) compared to those for the EMHPM. Moreover, comparing species at the same level of the hot pressing procedure revealed that there are no significant differences between them.

Specific Gravity of Fixed Oil (SG_{fo})

For the SG_{fo} property, there are statistical differences between the SG_{fo} values among the four species examined, and the comparisons made within or between species. For more illustration, at the same level of heating tool (MHPM and EHPM), comparing the SG_{fo} values within species revealed that castor seeds gave the highest yield (0.973 and 0.968, respectively) comparing to those obtained from sunflower (0.914–0.918, respectively), rapeseed (0.906 and 0.914, respectively), and moringa (0.946 and 0.957, respectively) as presented in Table 1. The mean SG_{fo} of castor oil was found to be 0.973. However, the mean value of castor oil SG_{fo} determined represents a slight increase over the ASTM specification range (0.957–0.968).

At the same level of oily species, the MHPM produces nearly identical values for those obtained by the EHPM (Table 2). This indicates that microwave irradiation did not alter the SG_{fo} of the resultant fixed oils protruding the economic importance of this novel investigation.

The refractive index (RI) of the fixed oil was also examined in this study. It was found that there are no statistical differences between the RI values among the four species examined as well as the comparisons made within each of the four oily species. At the same level of the hot pressing tool (MHPM and EHPM), statistical comparisons of the RI values within species revealed that castor seeds gave nearly the same index value (1.48 and 1.43, respectively) compared to those obtained from sunflower (1.473 and 1.469, respectively), rapeseed (1.465 and 1.467, respectively), and moringa (1.464 and 1.563, respectively) as presented in Table 2. Moreover, at the same level of oily species, the MHPM produces nearly identical values to those obtained by the EHPM (Table 1). The RI for castor oil (1.48), is within the range of values cited [51]. This indicates that microwave beams did not cause any difference in the RI of the resultant fixed oils. This result confirms the suitability of the present study for the present industrial application.

3.1.3. Chemical Properties of the Fixed Oils

Concerning the iodine number (IN), it was found that at the same level of pressing machines (MHPM), sunflower (126.3) and rapeseed (114.5) seeds gave higher INs than those for castor oil (83.7) and moringa (69.7). Furthermore, for the IN values obtained using EHPM, the same trend was repeated with an identical sequence as follows: Sunflower (127.1) and rapeseed (115.1) seeds gave higher INs than those for castor oil (84.8) and moringa (71.4). In addition, within species, there are no significant differences between using MHPM and EHPM among all the species studied (Table 2).

As shown in Table 2, although the IN values for castor oil (83.7 and 84.8, respectively) were lower than 100 IN-unit for both MHPM and EHPM; these values are located at the ASTM-specification limit. On the other hand, those values for moringa fixed oil for MHPM and EHPM (69.7 and 71.4, respectively) were lower than the ASTM specification range.

For the saponification value (SV) of the fixed oil, examining Table 2 revealed that sunflower oil had the highest SV value for both MHPM and EHPM (188.4 and 194.6, respectively) compared to the other resources examined, namely castor bean (181.5 and 182.1, respectively), rapeseed (168.3 and 181.2, respectively), and moringa (180.7 and 190.4). In addition, microwave beams produced a slight reduction in the SV compared to the action of the electric band heater.

Moreover, speculating the acid value (AV) revealed that at the same level of the heating techniques (MHPM and EHPM) comparing the AV values within species revealed that castor seeds gave the highest AV (44.8% and 38.3%, respectively) compared to those obtained from sunflower (42.5% and 36.4%, respectively), rapeseed (42.2% and 35.2%, respectively), and moringa (35.6% and 34.6%, respectively) as presented in Table 2.

At the same level of oily species, the MHPM produces higher AV values than those obtained by the EHPM (Table 2). This is clear for all the four species examined in the present study.

In comparison to the literature [42,51], the AV of castor oil (0.869 mg KOH/g oil) lies within ASTM-specification guidelines (0.4–4.0 mg KOH/g oil).

The pH value of the fixed was achieved to estimate the oil's dissociation state since fixed oils may contain free fatty acids. As shown in Table 2, there are no significant differences between the pH values, and for both comparisons made within or between species. The pH values lie within the normal range although it is slightly higher than those found by Omari et al. [51], especially for castor oil.

Through the GC/MS analysis of the fixed oil, the main fatty acids constituting castor bean, sunflower, rapeseed, and moringa oils examined in this study were identified (Figures 5 and S4). The mass spectra and fragmentation patterns of the predominant fatty acids constituents were determined by comparison with the mass spectra stored in different available databases [69].

Examining Figure 5 concerning the four fixed oils resulting from the microwave–hot pressing machine (MHPM) and the electric hot pressing machines (EHPM) revealed that four fatty acids were detected in all four fixed oils but in different concentrations.

1. For castor oil (Figure 5a_1,a_2), ricinoleic acid was the most abundant fatty acid (76.41 and 71.99%) among the four fixed oils resulting from MHPM (Figure 5a_1) and EHPM (Figure 5a_2), respectively. However, castor oil had the lowest fatty acid content due to lower contributions from oleic acid (3.31%), linoleic acid (4.32%), and palmitic acid (4.32%).

It is worth noting that the high concentration of ricinoleic acid (76.41%) in the present study corresponds with the range (70–90%) cited by numerous researchers [11,22,70].

The amount of oleic acid in MHPM and EHPM was approximately 5.2% and 4.02%, respectively (Figure 5a_1,a_2). In addition, linoleic acid accounted for approximately 4.38% and 3.49% of the total fatty acids in the MHPM and the EHPM, respectively (Figure 5a_1,a_2). Furthermore, the amount of palmitic acid in the castor oil was approximately 7.16% and 5.08%, respectively. There may be an economic value to its sodium salt which is used in the soap and cosmetic industries [58,71].

2. For the sunflower fixed oil (Figure 5b_1,b_2), the oleic acid was the prominent fatty acid with high contents of 67.5% and 61.1% for the MHPM and EHPM, respectively. In addition, linoleic and palmitic acid had lower contents in the sunflower oil.
3. For the rapeseed fixed oil (Figure 5c_1,c_2), oleic acid, linoleic acid, and palmitic acid were the major fatty acids detected in the rapeseed fixed oil (65.7%, 20.1%, and 6.32%,

respectively) produced by the MHPM, while their yields were 62.1%, 18.08%, and 4.08%, respectively, by using the EHPM.

4. For the moringa fixed oil (Figure 5d_1,d_2), oleic acid and palmitic acid were found in the moringa fixed oil in concentrations of 76.86% and 5.36%, respectively, using the MHPM, while their contents were 72.49 and 4.08% for the EHPM.

Based on the data presented in Figure 2, it can be concluded that the prominent fatty acid of the castor oil is ricinoleic acid, making up 76.41% and 71.99% of the MHPM and EHPM, respectively. In addition, the oleic acid was the prominent fatty acid in each of the fixed oils of sunflower, rapeseed, and moringa species (65.7–76.86% for the MHPM and 62.1–72.49% for the EHPM).

The chemical structure of the compound along with the mass spectra-chromatogram of the four fatty acids detected at the four fixed oils extracted are shown in Figure S4 patterns [7,72–74].

To confirm the chemical structures of the prominent fatty acid (ricinoleic acid) in the castor oil, its molecular mass and fragmentation pattern were studied and presented in Figure S7a. Based on the mass spectrum, the prominent peaks were observed at 28, 29, 41, 43, 55, 67, 69, 74, 82, 83, 84, 87, 96, 97, 98, 124, 166, and 198 m/z. The presence of the compound was confirmed by comparing its mass spectrum with that of the standard 9-octadecenoic acid, 12-hydroxy-, methyl ester, and [R-(Z)]-profile (Figure S4a) as both compounds exhibit similar fragmentation.

In addition, the molecular mass and fragmentation pattern was studied to confirm the chemical structure of the prominent fatty acid (oleic acid) detected in the fixed oils of sunflower, rapeseed, and moringa species and presented in Figure S4b. According to the mass spectrum, prominent peaks were observed at 27, 29, 41, 43, 55, 83, and 97 m/z. The existence of the compound was confirmed by comparing its electron ionization chromatogram with the standard (9Z)-octadec-9-enoic profile.

Furthermore, the chemical structure of linoleic acid was confirmed using its mass spectra chromatogram (Figure S4c). It was sorted as the second prominent fatty acid in the sunflower fixed oil, the prominent peaks were clearly present at 29, 41, 55, 67, 68, 81, 96, 110, and 124 m/z. Comparing its mass spectrum with its standard profile, the compound existence was confirmed where they have the same fragmentation patterns. All four fixed oils contain palmitic acid (Figure S4d) in minor amounts.

A mass spectrum of the sample revealed prominent peaks at 41, 43, 56, 74, 87, and 143 m/z. The existence of this compound was confirmed by comparing its mass spectrum with the standard hexadecanoic acid and methyl ester profile.

3.2. Effect of Microwave Irradiation on Biopolymeric Structured-Tissues (BST)

Technical information relevant to the microwave irradiation device that replaced ordinary electric heating coils to warm the hot pressing machine (HPM) essential for extracting fixed oils from the BST is presented in Figures 1–4 and 6. This figure shows the microwave generator unit (MGU), termed as a high-voltage magnetron, the high voltage-transformer, the high voltage-capacitor, and the manner for directing the microwave beam to the colander of the microwave hot pressing machine (MHPM) through a special waveguide.

Furthermore, the theoretical behavior of the microwave beam (Figure 4) illustrates the sinusoidal wave curve (SWC), energy level along with the SWC, and the proportionality between square amplitude and the energy carried by the wave [46–48].

For studying the biopolymeric structured tissues as affected by microwave irradiation, the four oily seed precursors were anatomically examined by SEM and are presented in Figure 7. These micrographs show the cellular structure of each castor bean (*Ricinus communis*, common sunflower (*Helianthus annuus*), rapeseed (*Brassica napus*), and moringa (*Moringa oleifera*). Furthermore, Figure 7 shows the endosperm cells (EC), cell walls (CW), cytoplasm (CY), compound lipid bodies (CLBs), and subcellular organelles of singular lipid bodies (SLBs).

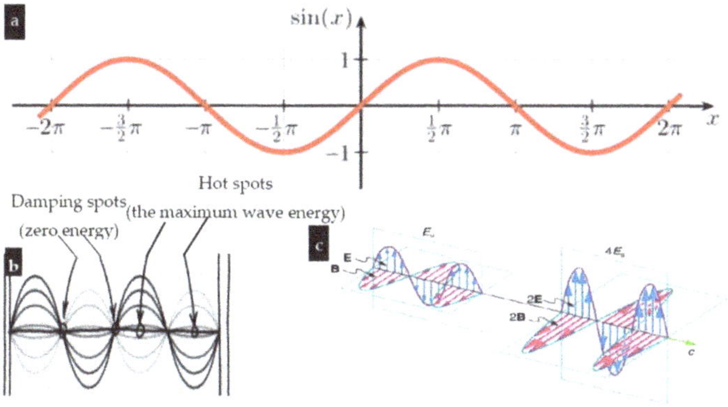

Figure 6. The microwave beam: (**a**) Sinusoidal wave curve (SWC), (**b**) Energy level along with the SWC, and (**c**) The proportionality between square amplitude and the energy carried by the wave.

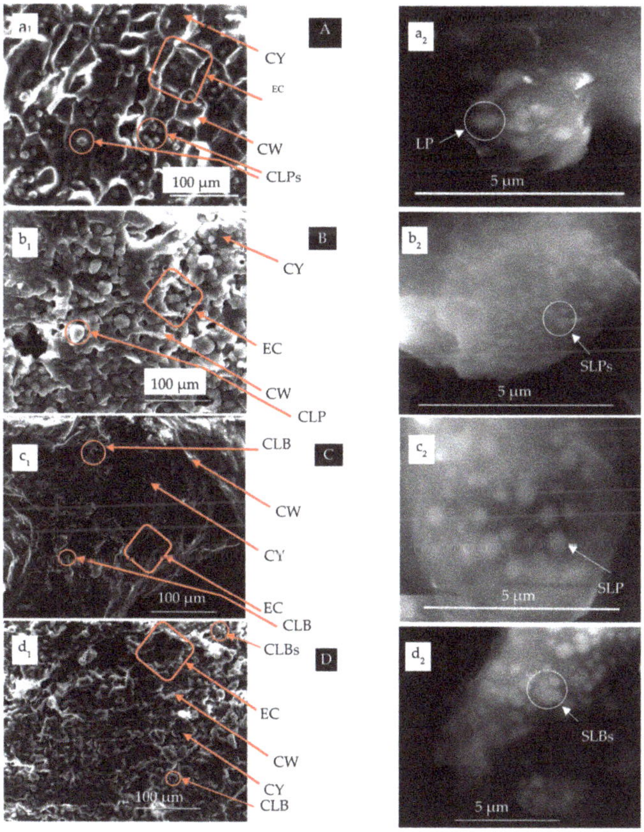

Figure 7. SEM micrograph of endosperm sections of the four oily seeds (**A**–**D**): (**A**) Castor bean (*Ricinus communis* L), (**B**) Common sunflower (*Helianthus annuus* L.), (**C**) Rapeseed (*Brassica napus* L.) and (**D**) Moringa (*Moringa oleifera* Lam.) showing cellular structure ($\mathbf{a_1}$–$\mathbf{d_1}$) with endosperm cells (EC), cell walls (CW), cytoplasm (CY), and compound lipid bodies (CLBs) as well as subcellular organelles ($\mathbf{a_2}$–$\mathbf{d_2}$) of singular lipid bodies (SLBs).

For comparing the oily bodies inserted in the four structured tissues, their estimated results were obtained using image analysis and presented at Table 3.

Table 3. Mean values [1] of lipid body (LB) diameters of the four oily seeds for each of the compound lipid body (CLP) and the single LB (SLB).

Species	Body Diameter (μm)	
	CLB	SLB
Castor bean	6.67 [0.58]–24.9 [3.25]	0.44 [0.39]–0.94 [0.03]
Common sunflower	8.7 [0.48]–22.09 [2.74]	0.19 [0.12]–0.47 [0.05]
Rapeseed	7.27 [0.68]–14.55 [2.06]	0.97 [0.14]–0.59 [0.08]
Moringa	6.52 [0.69]–17.39 [1.83]	0.24 [0.09]–0.71 [0.09]

[1] Each value is an average of three observations ± the standard deviations.

4. Discussion

4.1. Physical Characterization of Physical Characterization of the Fixed Oils

1. The moisture content of seed (MC_s) values were found to be varied in seeds that ranged from 3.6% to about 7% [42,67]. The variation could be attributed to the difference in the nature of beans from different locations [42] and/or seed macro-structure, including hull-to-kernel weight ratio, hull thickness, and their oil content [66].
2. Focusing on the seed content of fixed oil (S_{cfo}) showed that the recovery of the remained traces in the seeds' cake by using solvent extraction procedure allowed the yield of recovered fixed oil (Y_{rfo}) to be maximized as well as reducing the extraction loss (EL) for the fixed oils. Some remaining traces of fixed oils in the cakes can be attributed to incomplete cell lysis within the seeds which possibly trapped and retained some amount of oil [52].

The efficiency of fixed oil extraction (E_{foe}) for the MHPM was higher than that for the EHPM.

A specific gravity of fixed oil (SG_{fo}) can be used as an indicator of its purity and as a tool to distinguish a variety of oily solutions as stated by Omari et al. [51]. Furthermore, they noted that the SG_{fo} of oil floats over an aqueous solution when a spill occurs where compounds are found in water as a result of the spill [51]. Due to the use of the hot pressing extraction technique in the present study, some impurities were present in the castor oil that may have contributed to the minor differences in the SG_{fo} values between literature studies [51].

It has been reported by Omari et al. [51] that minor differences between cited data in the RI are due to differences in planting and harvesting conditions of the mother shrubs as well as research conditions. A number of impurities may alter the RI values, including gums, phosphates, etc. [51]. Castor oil's RI is generally related to its degree of unsaturation or conjugation and vice versa [51]. Due to the two direct relationships between the unsaturationality and the iodine number and refractive index, its moderate RI value for the castor oil matches its moderate iodine number (84.8 iodine unit).

Since the MHPM produces nearly identical values for those obtained by the EHPM at the same level of oily species (Table 2), it can be concluded that microwave beams did not cause any difference for the RI of the resultant fixed oils. This result confirms the suitability of the present study for the present industrial application.

4.2. Chemical Characterization of the Fixed Oils

Iodine number, saponification value, acid value, and yield of fatty acid of the four fixed oils were studied to ensure whether microwave irradiation may influence their chemical quality as well as fatty acids yield.

The lower IN values can be attributed to the presence of more saturated fatty acids which did not react with the Hanus iodine solution [51]. In contrast, higher IN values for sunflower and rapeseed (>100) indicate higher levels of unsaturation which increases the amount of iodine absorbed by unsaturated acids which makes them ideal as drying oils in the cosmetic, varnish, and coating industries as well as to make cosmetics.

Castor and moringa oils, however, had INs lower than 100 iodine units and could be classified as non-drying oils, incompatible with paint industries but suitable for soap industries [75].

According to Omari et al. [51], higher SV indicates lower molecular weights (MW) of triglycerides and vice versa. As a result, castor oil was expected to contain a lower MW of triglycerides than other fixed oils with a lower SV. It was shown in Table 2 that castor oil had a high SV, which was within the ASTM specification range of 175–187 SV-units and comparable to those reported in the literature [51,76,77]. It is possible to attribute the variations seen between the cited data to the small differences in the fatty acid composition of castor oil as these values are characteristic of castor oil.

The castor oil had low free fatty acid content since the AV is an indicator of the number of carboxylic acid groups in fatty acids constituting an oil and it is also a measure of the amount of free fatty acids. This study obtained a lower AV for castor oil than that of Omari et al. [51], as the seeds had been collected from the ground following curing for a period of time. Thus, the seeds had been exposed to sufficient amounts of lipase enzyme to hydrolyze their triglycerides into free fatty acids, thus improving their acid content.

Among the four fixed oils examined, the higher pH value of castor oil is expected to be attributed to its high free fatty acid content which is correlated with its acid value [51].

These fatty acids are represented as methyl or methyl esters due to pretreatments of the fixed oils using the saponification/methylation route. It has been demonstrated that methylation of oil causes the oil to become more volatile which subsequently adapts the Programme Temperature Volume of the injector of the GC-MS device [71,74]. There is no difference in fragmentation patterns between methyl esters derivatives of free fatty acids and their parent free fatty acids except that methyl esters have a higher molecular weight than their parent free fatty acids by 14 g/moles [74].

Castor oil is a unique source of ricinoleic acid, a hydroxylated fatty acid (Figure 5), it can cause the oil to be converted from a non-drying oil to a drying oil by removing the hydroxyl groups in its chain. This is useful in the manufacture of alkyd resin coating [74] and adds to its industrial advantages.

4.3. *Effect of Microwave Irradiation on Biopolymeric Structured-Seed Tissues*

Several researchers have been trying to use microwave irradiation, a sustainable non-contact heating source, for oil extraction from aromatic and oily plants [48,78–83]. The magnetron receives energy from a high-voltage transformer (HVT). Regarding technical specifications, the HVT is 1000E-1E, 220 V and 60 Hz [48]. In order to achieve the required 2.45 GHz, the magnetron converts the high voltage alternate current (AC). It has been reported that large industrial and commercial ovens can use 915 megahertz magnetrons to excite the larger cavities within the ovens [46,48]. It is worth mentioning that wavelengths of microwaves range from approximately one meter to one millimeter with frequencies between 300 MHz (1 m) and 300 GHz (1 mm). While electromagnetic waves clearly exhibit wave characteristics, they also exhibit particle characteristics at high frequencies [48,75].

As shown in Figure 6a–c, the microwave beams can pass through oily seed tissues via the microwave generator. There is usually a frequency of 2.46 GHz (with a wavelength of 12.24 cm) for microwave beams. In microwave heating, microwaves penetrate the interior of biopolymeric structured –tissues, including but not limited to seeds as well as foods affecting their dipolar molecules of water, fats, or sugars. Accordingly, heat is generated because dipolar molecules absorb electromagnetic radiation which leads to their intensive vibration producing friction that leads to rapid temperature rising and, consequently, efficient water evaporation and/or melting fats. This results in a greatly increased vapor pressure differential between the center and surface of the biopolymeric tissue, allowing fast transfer of moisture and or essential oil out of the tissue. Hence, microwave drying as well as the microwave hot pressing technique is rapid, more uniform, and energy efficient compared to ordinary ones [48,84,85]. Through their electric and magnetic fields, electromagnetic (sinusoidal) waves (Figure 6a) are able to bring energy into a system [46–48]. In addition to exerting force and moving charges in the system, these fields may also perform work on the charges. It is much more efficient to transfer energy when the electromagnetic wave frequency matches the natural frequency of the system (such as microwaves at the resonant frequency of water molecules). The energy of a wave is proportional to the square root of its amplitude (Figure 6c). A larger E-field and a larger ß-field exert greater forces and are capable of performing more work with electromagnetic waves.

Furthermore, microwave beams have hot and cold spots which makes them unsuitable for heating the extruder's colander (Figure 6b). By facing a microwave beam with damp thermal paper, this phenomenon can be detected. It is apparent from examining the propagation line (the baseline) of the microwave sinusoidal curve that cold (damping) spots are analogous to the intersection points of both the magnetic and electric wave curves. With our invention, the hot spots have been considered and eliminated since the extruder rotates continuously, alternating the damping spots.

For the EHPM, heat energy and entropy are transferred mainly by conduction according to Fourier's law. On the other hand, for the MHPM, heat is transferred mainly via conduction (according to the Fourier's law), convection (according to Newton's law of cooling), and some amount of heat is transferred via radiation (Kirchhoff's law of thermal radiation).

Conduction heat is responsible for moving heat from the surficial regions of the oily seed tissues to their cores, while convection combines conduction heat transfer and circulation to force molecules in the air to move from the hottest zones to cooler ones. Moreover, radiation is the process where heat and light waves strike and penetrate your food. As such, there is no direct contact between the heat source and the cooking food.

Accordingly, the microwave hot pressing machine is an ideal tool for heat transfer and, subsequently, heating the oily seed tissues increases the fixed oil yield [45,84].

4.4. *Effect of Ultrastructure of Seeds on Fixed Oils Extraction*

Understanding the microstructure of different oily seed organelles would be helpful to isolate and characterize their lipids more efficiently. In particular, the knowledge of the seed microstructure may be important in industrial processing [3], especially for detecting the suitability of exchanging microwave irradiation with convenient heating techniques in the oil extraction process.

Histological features of the four oil species are shown in Figure 5 for castor bean (*Ricinus communis* L.), (b) Common sunflower (*Helianthus annuus* L.), (c) Rapeseed (*Brassica napus* L.), and (d) Moringa (*Moringa oleifera* Lam.).

It is clear from Figure 7 that there are two spherical-shaped biopolymeric organelles spreading in the endoplasm termed novelly as compound LB (CLB) and singular LB (SLB) as clear in Figure 7A–D. The difference between the CLB and SLB is that the former contains huge internal spheres of LB, while the SLB is a single individual oily body.

The difference between the four species was estimated by analyzing their images for CLB (Figure 7a_1–d_1) and SLB (Figure 7a_2–d_2).

For the castor bean, the CLB's diameter was found to be ranged from 6.67 to 22.2 µm, while the SLB's diameter varied from 0.44 to 0.92 µm. Furthermore, regarding the sunflower biopolymeric tissue, CLB's diameter ranges from 8.7 µm to 22.09 µm and that for SLB differed from 0.19 µm to 0.47 µm. In addition, examining the diameters of the lipid organelles of the rapeseed revealed the CLB's diameter ranged from 7.27 µm to 14.55 µm, while the SLB's diameter differed between 0.97 µm to 0.59 µm. For the lipid body found immersed in the endosperm of the moringa seed, it was found that the diameter of the CLB (6.52 µm–17.39 µm) was higher than that for the SLB (0.24 µm to 0.71 µm) as shown in Table 3 and Figure 7.

Lipids are generally stored as triacylglycerols in oil bodies. Although seeds differed quantitatively in the accumulation of storage reserves, it was not clear whether this would also qualitatively affect the fatty acid profiles of triacylglycerols in these seeds [5].

In addition, examining Figure 7a_1–d_1, the cells are configured in a rectangular shape which can be better packaged than spherical cells in the endosperm [3]. These results were from the approach found by other researchers [3–6].

It can be seen from Table 3 and Figure 7a_1–d_1 that castor bean has the highest SLB diameter (0.44 μm–0.94 μm) followed by those moringa seeds. This finding is agreed with that found and reported by 3. Perea-Flores et al. [3] for the castor oil crop as well as other species [3,5–7]. Although our presentations (Table 3 and Figure 7) covered the differences between the four examined biopolymeric structured tissues [7,48,51,74,86], they did not give impressions concerning the overall fixed oil content of them. This explains the highest fixed oil yield obtained from such species.

Compared with conventional methods, microwave treatment for oil extraction has many advantages, e.g., improvement of extracted oil yield and quality, direct extraction capability, lower energy consumption, faster processing time, and reduced solvent contents [7,80]. These results may be attributed to microwave irradiation which provides a potential alternative to induce stress reactions in ultrastructured tissues within the oil seeds. By using microwave radiation in oil seeds, a higher extraction yield and an increase in mass transfer coefficients can be obtained because the cell membrane is more severely ruptured. Apart from this, permanent pores were generated accordingly, this enables the oil to move through permeable cell walls [42,44].

5. Conclusions

1. A microwave beam was used to heat the extruder's colander of a hot pressing machine instead of the ordinary electric one.
2. The invented microwave–hot pressing machine was used to produce, individually, four different fixed oils extracted from seeds of castor, sunflower, rapeseed, and moringa species and compared to those obtained using the ordinary electric–hot pressing machine.
3. The physical properties, namely moisture content, oil content, oil yield, oil extraction efficiency, specific gravity, and refractive index, were determined for the four fixed oils.
4. The chemical properties, namely iodine number, saponification value, acid value, pH, and chemical constituents, using gas chromatography coupled with a mass spectrometer of the four oils extracted by using both heating tools were evaluated based on those in the literature for the four fixed oils.
5. The higher oil extraction efficiency indicated that using microwave irradiation enhanced the oil yield with retaining the parent quality of the fixed oils that is very encouraging for candidating such an invention as a pivot of the industrial fixed oil projects.
6. Studying the histological features of the biopolymeric structured tissues revealed that castor bean species has the highest singular lipid body diameter (0.44 μm–0.94 μm) followed by those for moringa seeds. This explains the highest fixed oil yield obtained from such species.
7. Comparing with conventional methods, microwave treatment for oil extraction has many advantages, e.g., improvement of extracted oil yield and quality, direct extraction capability, lower energy consumption, faster processing time, and reduced solvent contents [7,75].

Patent: Hindi, S.S.; Dawoud, U.M.; Ismail, I.M.; Asiry, K.A.; Ibrahim, O.H.; Al-Harthi, M.A.; Mirdad, Z.M.; Al-Qubaie, A.I.; Shiboob, M.H.; Almasoudi, N.M.; Alanazi, R.A. 2023. Microwave-assisted extraction of fixed oils from seeds. US Patent, application no. 63445512, filling date: 14 February 2023.

Supplementary Materials: The following supporting information can be downloaded at: https://www.mdpi.com/article/10.3390/polym15102254/s1, Figure S1: The management plan applied to investigate the efficiency of the microwave-hot pressing machine for fixed oil extraction; Figure S2. The extraction steps of the four fixed oils from seeds: (**a**) Castor (*Ricinus communis* L.), (**b**) Sunflower (*Helianthus annuus* L.), (**c**) Rapeseed (*Brassica napus* L.), and (**d**) Moringa (*Moringa oleifera* Lam.); Figure S3. Preparation of fatty acids methyl esters of the four fixed oils to be analyzed by GC-MS; Figure S4. Chemical structure and the mass spectra-chromatogram of: (**a**) Ricinoleic acid methyl ester detected, the main fatty acid of the castor oil, (**b**) Oleic acid, the main fatty acid in the fixed oils of sunflower, rapeseed, and moringa species, (**c**) Linoleic acid detected as the second main fatty acids in the sunflower fixed oil, and (**d**) Palmitic acid detected at the four fixed oils.

Author Contributions: Supervision, conceptualization, and methodology, S.S.H.; formal analysis, and writing—review and editing, U.M.D.; software and assistance with writing—original draft preparation, K.A.A.; statistical design, validation and review, I.M.I.; assistance with practical work and correcting—original draft preparation, O.H.I.; preparation and assistance with practical and theoretical works, M.A.A.-H.; preparation and assistance with fixed oil sampling and statistical analysis Z.M.M.; methodology and review and validation of the results, A.I.A.-Q.; chemical characterization of fixed oils and data validation, M.H.S.; writing, editing and statistical analysis, N.M.A.; practical participating for sampling, laboratorial examinations, R.A.A. All authors have read and agreed to the published version of the manuscript.

Funding: This work was funded by the Deanship of Scientific Research (DSR), KAU, Jeddah under grant no. DF-116-155-1441.

Institutional Review Board Statement: Not applicable.

Informed Consent Statement: Informed consent was obtained from all subjects involved in the study.

Data Availability Statement: Not applicable.

Acknowledgments: The authors are deeply thankful to the DSR, KAU, Jeddah for funding this research work.

Conflicts of Interest: The authors declare no conflict of interest.

Nomenclature

Symbol	Definition	Symbol	Definition
AC	Alternate current	S	Sunflower (*Helianthus annuus* L.)
ACS	The American Chemical Society	MAEO	Microwave-assisted extracted oil
ADB	Air-dried membranes	MFT	Maximum final temperature
AFM	Atomic force microscopy	MGU	Microwave generator unit
ASTM	American Society for Testing and Materials	MHPM	Microwave hot pressing machine
AV	Acid value	NDB	Nanodehydrated-bioplastic membrane
BS	Basic extraction	NIST	The National Institute of Standards and Technology
BST	Biopolymeric Structured-Tissues	NPS	Nanometric particle Size
C	Castor bean (*Ricinus communis* L.)	PD	Pore diameter
CI	Crystallinity index	pH	The acidity or basicity number
CLB	Compound lipid bodies	PS	Particle size
CW	Cell walls	PubChem	An open chemistry database managedby the National Institutes of Health (NHI)
CY	Cytoplasm	R	Rapeseed (*Brassica napus* L.)
DC	Direct current	SD	Standard deviation
DSC	Differential scanning calorimetry	SE	Secondary extraction
DTA	Differential thermal analysis	RI	Refractive index
EC	endosperm cells	SEP	Self-electrostatic peeling
EHPO	Electric hot pressed-oil	SG_{fo}	Specific gravity of fixed oil
EHPM	Electric-hot pressing machine	SLB	Singular lipid bodies
FEG	Field emission gun in the SEM	SP	Statistical parameters
FEI	Field electron and ion US-company	SV	Saponification value

EL	Extraction loss, %	SWC	Sinusoidal wave curve
E_{foe}	Efficiency of fixed oil extraction, %	TGA	Thermogravimetric analysis
FTIR	Fourier transform infrared spectroscopy	TR	Temperature range
GC-MS	Gas chromatography-mass spectrometer	VFHF	Vibrated-free horizontal flow
GHz	Frequency	VV	Void volume
HC	Heat change in μVs/mg	W_{gvs}	Weight of ground virgin seeds, g
HPM	Hot pressing machine	W_{mfo}	Weight of main fixed oil, g
HVT	High voltage transformer	W_{rfo}	Weight of recovered fixed oil, g
IN	Iodine number	XRD	X-ray diffraction
LSD	Least significant differencce	Y_{fa}	Yield of fatty acids
M	Moringa (*Moringa oleifera* Lam.)	Y_{mfo}	Yield of main fixed oil, %
RI	Refractive index	Y_{rfo}	Yield of recovered fixed oil, %

References

1. Fernández-Luqueño, F.; López-Valdez, F.; Miranda-Arámbula, M.; Rosas-Morales, M.; Pariona, N.; Espinoza-Zapata, R. An introduction to the sunflower crop. In *Sunflowers*; Arribas, J.I., Ed.; Nova Science Publishers, Inc.: Hauppauge, NY, USA, 2014.
2. Boesewinkel, F.D.; Bouman, F. The Seed: Structure. In *Embryology of angiosperms*; Johri, B.M., Ed.; Springer: Berlin/Heidelberg, Germany, 1984. [CrossRef]
3. Perea-Flores, M.J.; Chanona-Perez, J.J.; Garibay-Febles, V.; Calderon-Dominguez, G.; Terrés-Rojas, E.; Mendoza-Perez, J.A.; Herrera-Bucio, R. Microscopy techniques and image analysis for evaluation of some chemical and physical properties and morphological features for seeds of the castor oil plant (*Ricinus communis*). *Ind. Crops. Prod.* **2011**, *34*, 1057–1065. [CrossRef]
4. Walters, C.; Pierre Landre, P.; Hill, L.; Corbineau, F.; Bailly, C. Organization of lipid reserves in cotyledons of primed and aged sunflower seeds. *Planta* **2005**, *222*, 397–407. [CrossRef] [PubMed]
5. Hu, Z.-Y.; Hua, W.; Zhang, L.; Deng, L.-B.; Wang, X.-F.; Liu, G.H.; Hao, W.J.; Wang, H.Z. Seed structure characteristics to form ultrahigh oil content in rapeseed. *PLoS ONE* **2013**, *8*, e62099. [CrossRef] [PubMed]
6. Fotouo, M.H.; du Toit, E.S.; Robbertse, P.J. Germination and ultrastructural studies of seeds produced by a fast-growing, drought-resistant tree: Implications for its domestication and seed storage. *AoB PLANTS* **2015**, *7*, plv016. [CrossRef] [PubMed]
7. Hindi, S.S.; Dawoud, U.M. Characterization of hot pressed-fixed oil extracted from *Ricinus communis* L. seeds. *Int. J. Innov. Res. Technol. Sci. Eng.* **2019**, *8*, 10172–10191.
8. Mascolo, N.; Izzo, A.A.; Autore, G.; Barbato, F.; Capasso, F. Nitric oxide and castor oil-induced diarrhea. *J. Pharmacol. Exp. Ther.* **1994**, *268*, 291–295.
9. Trevino, A.S.; Trumbo, D.L. Acetoacetylated castor oil in coatings applications. *Prog. Org. Coat.* **2002**, *44*, 49–54. [CrossRef]
10. Girard, P.; Pansart, Y.; Lorette, I.; Gillardin, J.M. Dose–response relationship and mechanism of action of *Saccharomyces boulardii* in castor oil-induced diarrhea in rats. *Dig. Dis. Sci.* **2003**, *48*, 770–774. [CrossRef]
11. Ogunniyi, D.S. Castor oil: A vital industrial raw material. *Bioresour. Technol.* **2006**, *97*, 1086–1089. [CrossRef]
12. Saraf, S.; Sahu, S.; Kaur, C.D.; Saraf, S. Comparative measurement of hydration effects of herbal moisturizers. *Pharmacognosy Res.* **2010**, *2*, 146–151. [CrossRef]
13. Berman, P.; Nizri, S.; Wiesman, Z. Castor oil biodiesel and its blends as alternative fuel. *Biomass Bioenergy.* **2011**, *35*, 2861–2866. [CrossRef]
14. Salimon, J.; Noor, D.A.M.; Nazrizawati, A.T.; Firdaus, M.M.; Noraishah, A. Fatty acid composition and physicochemical properties of Malaysian castor bean *Ricinus communis* L. seed oil. *Sains Malays.* **2010**, *39*, 761–764.
15. Nangbes, J.G.; Nvau, J.B.; Buba, W.M.; Zukdimma, A.N. Extraction and characterization of castor (*Ricinus communis*) seed oil. *IJES* **2013**, *2*, 105–109.
16. Salihu, B.; Gana, A.K.; Apuyor, B.O. Castor oil plant (*Ricinus communis* L.): Botany, ecology and uses. *IJSR* **2014**, *3*, 1333–1341.
17. Yeganeh, H.; Hojati-Talemi, P. Preparation and properties of novel biodegradable polyurethane networks based on castor oil and poly (ethylene glycol). *Polym. Degrad. Stab.* **2007**, *92*, 480–489. [CrossRef]
18. Mukherjea, R.N.; Saha, K.K.; Sanya, S.K. Plasticizing effect of acetylated castor oil on castor oil-based, moisture-cured polyurethane film. *J. Am. Oil. Chem. Soc.* **1978**, *55*, 653–656. [CrossRef]
19. Marc, R.L.; Furst, M.R.L.; Le Goff, R.; Quinzler, D.; Mecking, S.; Botting, C.H.; Cole-Hamilton, D.J.C. Polymer precursors from catalytic reactions of natural oils. *Green. Chem. J.* **2012**, *14*, 472–477.
20. Imankulov, N. Preparation and research on properties of castor oil as a diesel fuel additive. *Appl. Tech. Innov. J.* **2012**, *6*, 30–37. [CrossRef]
21. Mubofu, E.B. Castor oil as a potential renewable resource for the production of functional materials. *Sustain. Chem. Process.* **2016**, *4*, 11. [CrossRef]
22. Dunford, N.T. *Food and Industrial Bioproducts and Bioprocessing*, 1st ed.; Dunford, N.T., Ed.; John Wiley & Sons Inc.: Hoboken, NJ, USA, 2012; pp. 115–143.
23. Alwaseem, H.; Donahue, C.J.; Marincean, S. Catalytic transfer hydrogenation of castor oil. *J. Chem. Educ.* **2014**, *91*, 575–578. [CrossRef]

24. Marwat, S.K.; Rehman, F.; Khan, E.A.; Baloch, M.S.; Sadiq, M.; Ullah, I.; Javaria, S.; Shaheen, S. *Ricinus communis*: Ethnomedicinal uses and pharmacological activities. *Pak. J. Pharm. Sci.* **2017**, *30*, 1815–1827. [PubMed]
25. Campos Flexa Ribeiro Filho, P.R.; Rocha do Nascimento, M.; Otaviano da Silva, S.S.; Tavares de Luna, F.M.; Rodríguez-Castellón, E.; Loureiro Cavalcante, C., Jr. Synthesis and frictional characteristics of bio-based lubricants obtained from fatty acids of castor oil. *Lubricants* **2023**, *11*, 57. [CrossRef]
26. Osorio, J.; Fernandez-Martínez, J.M.; Mancha, M.; Garces, R. Mutant sunflower with high concentration of saturated fatty acids in the oil. *Cropl. Sci.* **1995**, *35*, 739–742. [CrossRef]
27. Piva, G.; Bouniols, A.; Mondies, G. Effect of cultural conditions on yield, oil content and fatty acid composition of sunflower kernel. In Proceedings of the 15th International Sunflower Conference, Toulouse, France, 12–16 June 2000; pp. 61–66.
28. Vrânceanu, A.V.; Soare, G.; Craiciu, D.S. Breeding sunflower for high oleic acid content. *An. ICCPT-Fundulea* **1995**, *LXII*, 96–97.
29. Meller, J.F.; Zimmerman, D.C. Inheritance of high oleic fatty acid content in sunflower. In *Proc. Sunflower Research workshop, Fargo, N.D. 26 January*; National Sunflower Association: Bismark, ND, USA, 1983; p. 10.
30. Urie, L.A. Inheritance of high oleic acid in sunflower. *Crop. Sci.* **1985**, *25*, 986–989. [CrossRef]
31. Fernandez-Martinez, J.; Munoz, J.; Jimenez-Ramirez, A.; Dominguez Jimenez, J.; Alcantara, A. Temperature effect on the oleic and linoleic acid of three genotypes in sunflower. *Grasas Aceites.* **1986**, *37*, 327–333.
32. Woźniak, E.; Waszkowska, E.; Zimny, T.; Sowa, S.; Twardowski, T. The rapeseed potential in Poland and Germany in the context of production, legislation, and intellectual property rights. *Front. Plant Sci.* **2019**, *10*, 1423. [CrossRef]
33. Matthaus, B.; Özcan, M.M.; Al Juhaimi, F. Some rape/canola seed oils: Fatty acid composition and tocopherols. *Z. Naturforsch C. J. Biosci.* **2016**, *71*, 73–77. [CrossRef]
34. Chew, S.C. Cold pressed rapeseed (*Brassica napus*) oil. In *Cold Pressed Oils*; Ramadan, M.F., Ed.; Academic Press: Cambridge, MA, USA, 2020; pp. 65–80.
35. Bhatnagar, A.S.; Gopala Krishna, A.G. Natural antioxidants of the Jaffna variety of Moringa oleífera seed oil of Indian origin as compared to other vegetable oils. *Grasasy Aceites* **2013**, *64*, 537–545.
36. Ramachandran, C.; Peter, K.V.; Gopalakrishnan, P.K. Drumstick (*Moringa oleifera*): A multipurpose Indian vegetable. *Econ. Bot.* **1980**, *34*, 276–283. [CrossRef]
37. Kayode, R.M.O.; Afolayan, A.J. Cytotoxicity and effect of extraction methods on the chemical composition of essential oils of *Moringa oleifera* seeds. *J. Zhejiang Univ. Sci. B.* **2015**, *16*, 680–689. [CrossRef]
38. Anwar, F.; Ashraf, M.; Bhanger, M.I. Interprovenance variation in the composition of Moringa oleifera oil seeds from Pakistan. *J. Am. Oil Chem. Soc.* **2005**, *82*, 45–51. [CrossRef]
39. Leone, A.; Spada, A.; Battezzati, A.; Schiraldi, A.; Aristil, J.; Bertoli, S. Moringa oleifera seeds and oil: Characteristics and uses for human health. *Int. J. Mol. Sci.* **2016**, *17*, 2141. [CrossRef]
40. Ogunsina, B.S.; Indira, T.N.; Bhatnagar, A.S.; Radha, C.; Debnath, S.; Gopala Krishna, A.G. Quality characteristics and stability of *Moringa oleifera* seed oil of Indian origin. *J. Food Sci. Technol.* **2014**, *5*, 503–510. [CrossRef]
41. Dasari, S.R.; Goud, V.V. Comparative extraction of castor seed oil using polar and nonpolar solvents. *IJCET* **2013**, *1*, 121–123.
42. Muzenda, E.; Member, I.; Kabuba, J.; Mdletye, P.; Belaid, M. Optimization of process parameters for castor oil production. In Proceedings of the World Congress on Engineering, London, UK, 3–5 July 2012.
43. Kittiphoom, S.; Sutasinee, S. Effect of microwaves pretreatments on extraction yield and quality of mango seed kernel oil. *Int. Food Res. J.* **2015**, *22*, 960–964.
44. Azadmard-Damirchi, S.; Alirezalu, K.; Fathi Achachlouei, B. Microwave pretreatment of seeds to extract high quality vegetable oil. *World Acad. Sci. Eng. Technol.* **2011**, *57*, 72–75.
45. Watanabe, S.; Karakawa, M.; Hashimoto, O. Temperature of a heated material in a microwave oven considering change of complex relative permittivity. In Proceedings of the 2009 European Microwave Conference (EuMC), Rome, Italy, 29 September –1 October 2009; pp. 798–801.
46. Hindi, S.S.; Dawoud, U.M.; Assiry, K. System and Method for Manufacturing Shellac Floss. U.S. Patent 11060208, 22 March 2021.
47. Hindi, S.S.; Dawoud, U.M.; Asiry, K.A. Bioplastic floss of a novel microwave-thermospun shellac: Synthesis and bleaching for some dental applications. *Polymers J.* **2023**, *15*, 142. [CrossRef]
48. Hindi, S.S.; Dawoud, U.M.; Ismail, I.M.; Asiry, K.A.; Ibrahim, O.H.; Alharthi, M.A.; Mirdad, Z.M.; Al-Qubaie, A.I.; Shiboob, M.H.; Alanazi, R.A.; et al. Microwave-Assisted Extraction of Fixed Oils from Seeds. U.S. Patent Application No. 63445512, 14 February 2023.
49. Whelan, M.; Dehn, R.; Kornrumpf, W.; Microwave Ovens. Edison Tech. *Center.* 2012. Available online: https://edisontechcenter.org/Microwaves.html (accessed on 28 July 2022).
50. Abitogun, A.S.; Alademeyin, O.J.; Oloye, D.A. Extraction and characterization of castor seed oil. *Internet. J. Nutr. Wellness.* **2009**, *8*, 1–8.
51. Omari, A.; Mgani, Q.A.; Mubofu, E.B. Fatty acid profile and physico-chemical parameters of castor oils in Tanzania. *Green Sustain. Chem.* **2015**, *16*, 154–163. [CrossRef]
52. Alenyorege, E.A.; Hussein, Y.A.; Adongo, T.A. Extraction yield, efficiency and loss of the traditional hot water floatation (HWF) method of oil extraction from the seeds of *Allanblackia floribunda*. *Int. J. Sci. Technol. Res.* **2015**, *4*, 92–95.
53. Keneni, Y.G.; Bahiru, L.A.; Marchetti, J.M. Effects of different extraction solvents on oil extracted from jatropha seeds and the potential of seed residues as a heat Provider. *Bio. Energy Res.* **2021**, *14*, 1207–1222. [CrossRef]
54. Lavenburg, V.M.; Rosentrater, K.A.; Jung, S. Extraction methods of oils and phytochemicals from seeds and their environmental and economic impacts. *Processes* **2021**, *9*, 1839. [CrossRef]

55. Isah, A.G. Production of detergent from castor oil. *Leonardo El J. Pract. Technol.* **2006**, *9*, 153–160.
56. Auta, M. Extraction and characterization of drilling oil from castor oil. *Int. J. Innov. Appl. Stud.* **2013**, *3*, 382–387.
57. Akpan, U.G.A.; Jimoh, A.; Mohamed, A.D. Extraction, characterization and modification of castor seed oil. *Leonardo. J. Sci.* **2006**, *5*, 43–52.
58. Warra, A.A.; Wawata, I.G.; Gunu, S.Y.; Aujara, K.M. Extraction and physicochemical analysis of some selected Northern Nigerian industrial oils. *Arch. Appl. Sci. Res.* **2011**, *3*, 536–554.
59. Thomas, A.; Matthaus, B.; Fiebeg, H.-J. Fats and fatty oils. In *Ullmann's Encyclopedia of Industrial Chemistry*; Wiley-VCH Verlag GmbH & Co.: KgaA, NJ, USA, 2015.
60. AL-Hamdany, A.J.; Jihad, T.W. Oxidation of some primary and secondary alcohols using pyridinium chlorochromate. *Tikrit J. Pure Sci.* **2012**, *17*, 72–76.
61. Hansel, F.A.; Bull, I.D.; Evershed, R.P. Gas chromatographic mass spectrometric detection of dihydroxy fatty acids preserved in the 'bound' phase of organic residues of archaeological pottery vessels. *Rapid Commun. Mass Spectrom.* **2011**, *25*, 1893–1898. [CrossRef]
62. Silva, R.A.C.D.; de Lamos, L.G.; Ferreira, D.A.; Monte, F.J.Q. Chemical study of the seeds of *Ximenia americana*: Analysis of methyl esters by gas chromatography coupled to mass spectrometry. *J. Anal. Pharm. Res.* **2018**, *7*, 79–83.
63. Bhukya, G.; Kaki, S.S. Design and synthesis of sebacic acid from castor oil by new alternate route. *Eur. J. Lipid Sci. Technol.* **2022**, *124*, e2100244. [CrossRef]
64. Gaul, J.A. Quantitative calculation of gas chromatographic peaks in pesticide residue analyses. *J. Assoc. Off. Anal. Chem.* **1966**, *49*, 389–399. [CrossRef]
65. El-Nakhlawy, F.S. *Experimental Design and Analysis in Scientific Research*; Sci. Pub. Center, King Abdulaziz University: Jeddah, Saudi Arabia, 2009.
66. Chapman, G.W.; Robertson, J.A. Moisture content/relative humidity equilibrium of high-oil and confectionery type sunflower seed. *J. Stored Prod. Res.* **1987**, *23*, 115–118. [CrossRef]
67. Akaranta, O.; Anusiem, A.C.I. A boiresource solvent for extraction of castor oil. *Ind. Crops. Prod.* **1996**, *5*, 273–277. [CrossRef]
68. Doan, L.G. Ricin: Mechanism of toxicity, clinical manifestations, and vaccine development. *A Review. J. Toxicol.* **2004**, *42*, 201–208. [CrossRef]
69. Jans-Joachim, H. Handbook of GC/MS. In *Fundamental and Applications, 2nd eds.*; WILEY-VCH, Verlag Gbmh: Weinheim, Germany, 2009.
70. Foglia, T.A.; Jones, K.C.; Sonnet, P.E. Selectivity of lipases: Isolation of fatty acids from castor, coriander and Meadowfoam oils. *Eur. J. Lipid Sci. Techn.* **2000**, *102*, 612–617. [CrossRef]
71. Gunstone, F.D.; Harwood, J.L.; Dijkstra, A.D. *The Lipid Handbook*; CRC Press: Cambridge, MA, USA, 2007; p. 1472.
72. Anonymous. National Institute Standards and Technology, NIST Chemistry WebBook, SRD69. 2012. Available online: https://webbook.nist.gov/chemistry/ (accessed on 28 July 2022).
73. Choi, G.H.; Kim, L.; Lee, D.Y.; Jin, C.L.; Lim, S.-J.; Park, B.J.; Cho, N.J.; Kim, J.-H. Quantitative analyses of ricinoleic acid and ricinine in *Ricinus communis* extracts and its biopesticides. *J. Appl. Biol. Chem.* **2016**, *59*, 165–169. [CrossRef]
74. Omowanle, J.; Ayo, R.J.; Habila, J.; Ilekhaize, J.; Adegbe, E.A. Physico-chemical and GC/MS analysis of some selected plant seed oils; castor, neem and rubber seed oils. *FUW Trends Sci. Tech. J.* **2018**, *3*, 644–651.
75. Odoom, W.; Edusei, V.O. Evaluation of saponification value, iodine value and insoluble impurities in coconut oils from Jomoro District in the western region of Ghana. *J. Agric. Food Sci.* **2015**, *3*, 494–499.
76. Saribiyik, O.Y.; Ozcanli, M.; Serin, H.; Serin, S.; Aydin, K. Biodiesel production from *Ricinus communis* oil and its blends with soybean biodiesel. *J. Mech. Eng.* **2010**, *56*, 811–816.
77. Uzoh, C.F.; Nwanbanne, J. Investigating the effect of catalyst type and concentration on the functional group conversion in castor seed oil alkyd resin production. *Adv. Chem. Eng. Sci.* **2016**, *6*, 190–200. [CrossRef]
78. Farag, R.S.; Hewedi, F.M.; Abu-Raiia, S.H.; Elbaroty, G.S. Comparative study on the deterioration of oils by microwave and conventional heating. *J. Food Prot.* **1992**, *55*, 722–727. [CrossRef] [PubMed]
79. Takagi, S.; Ienaga, H.; Tsuchiya, C.; Yoshida, H. Microwave roasting effects on the composition of tocopherols and acyl lipids within each structural part and section of a soya bean. *J. Sci. Food Agric.* **1999**, *79*, 1155–1162. [CrossRef]
80. Takagi, S.; Yoshida, H. Microwave heating influences on fatty acid distribution of triacylglycerols and phospholipids in hypocotyls of soybeans (*Glycine max* L.). *Food Chem.* **1999**, *66*, 345–351. [CrossRef]
81. Ramanadhan, B. Microwave Extraction of Essential Oils (from Black Pepper and Coriander) at 2.46 Ghz. Master's Thesis, University of Saskatchewan, Saskatoon, SK, Canada, 2005.
82. Uquiche, E.; Jerez, M.; Ortiz, J. Effect of pretreatment with microwaves on mechanical extraction yield and quality of vegetable oil from Chilean hazelnuts (*Gevuina avellana* Mol). *J. Innov. Food Sci. Emerg. Technol.* **2008**, *9*, 495–500. [CrossRef]
83. Yoshida, H.; Hirakawa, Y.; Tomiyama, Y.; Mizushina, Y. Effects of microwave treatment on the oxidative stability of peanut (*Arachis hypogaea*) oils and the molecular species of their triacylglycerols. *Eur. J. Lipid Sci. Technol.* **2003**, *105*, 351–358. [CrossRef]
84. Anonymous. Microwave Assisted Drying. 1993. Available online: http://ecoursesonline.iasri.res.in/mod/page/view.php?id=882 (accessed on 4 April 2023).

85. Rakesh, V.; Seo, Y.; Datta, A.K.; McCarthy, K.L.; McCarthy, M.J. Heat transfer during microwave combination heating: Computational modeling and MRI experiments. *Bioeng. Food Nat. Prod.* **2010**, *56*, 2468–2478. [CrossRef]
86. Patel, V.R.; Dumancas, G.G.; Viswanath, L.C.K.; Maples, R.; Subong, B.J. Castor oil: Properties, uses, and optimization of processing parameters in commercial production. *PMC J.* **2016**, *9*, LPI-S40233. [CrossRef]

polymers

MDPI

Article

Suitability of MRF Recovered Post-Consumer Polypropylene Applications in Extrusion Blow Molded Bottle Food Packaging

Ma. Cristine Concepcion D. Ignacio [1,2], Khairun N. Tumu [1,3], Mita Munshi [1,3], Keith L. Vorst [1,3] and Greg W. Curtzwiler [1,3,*]

1 Polymer and Food Protection Consortium, Iowa State University, Ames, IA 50011, USA; mignacio@iastate.edu (M.C.C.D.I.); kntumu@iastate.edu (K.N.T.); mmunshi@iastate.edu (M.M.); kvorst@iastate.edu (K.L.V.)
2 Department of Agricultural and Biosystems Engineering, Iowa State University, Ames, IA 50011, USA
3 Department of Food Science and Human Nutrition, Iowa State University, Ames, IA 50011, USA
* Correspondence: gregc@iastate.edu

Abstract: Polypropylene (PP) is one of the most abundant plastics used due to its low price, moldability, temperature and chemical resistance, and outstanding mechanical properties. Consequently, waste from plastic materials is anticipated to rapidly increase with continually increasing demand. When addressing the global problem of solid waste generation, post-consumer recycled materials are encouraged for use in new consumer and industrial products. As a result, the demand is projected to grow in the next several years. In this study, material recovery facility (MRF)-recovered post-consumer PP was utilized to determine its suitability for extrusion blow molded bottle food packaging. PP was sorted and removed from mixed-polymer MRF-recovered bales, ground, trommel-washed, then washed following the Association of Plastics Recyclers' protocols. The washed PCR-PP flake was pelletized then manually blended with virgin PP resin at 25%, 50%, 75, and 100% PCR-PP concentrations and fed into the extrusion blow molding (EBM) machine. The EBM bottles were then tested for physical performance and regulatory compliance (limits of TPCH: 100 µg/g). The results showed an increased crystallization temperature but no practical difference in crystallinity as a function of PCR-PP concentrations. Barrier properties (oxygen and water vapor) remained relatively constant except for 100% MRF-recovered PCR-PP, which was higher for both gas types. Stiffness significantly improved in bottles with PCR-PP (p-value < 0.05). In addition, a wider range of N/IAS was detected in PCR-PP due to plastic additives, food additives, and degradation byproducts. Lastly, targeted phthalates did not exceed the limits of TPCH, and trace levels of BPA were detected in the MRF PCR-PP. Furthermore, the study's results provide critical information on the use of MRF recovered in food packaging applications without compromising performance integrity.

Keywords: extrusion blow molding; municipal recovery facility; polypropylene; polymer recycling; post-consumer recycling; polymer processing

Citation: Ignacio, M.C.C.D.; Tumu, K.N.; Munshi, M.; Vorst, K.L.; Curtzwiler, G.W. Suitability of MRF Recovered Post-Consumer Polypropylene Applications in Extrusion Blow Molded Bottle Food Packaging. *Polymers* **2023**, *15*, 3471. https://doi.org/10.3390/polym15163471

Academic Editors: Zenaida Saavedra-Leos and César Leyva-Porras

Received: 25 July 2023
Revised: 9 August 2023
Accepted: 17 August 2023
Published: 19 August 2023

1. Introduction

Increasing plastic production continues to amplify the negative impact of global solid waste management on the environment and human health. Plastic production reached 390.7 million tons in 2021, consisting of 90% fossil-based plastics, 8.3% post-consumer recycled plastics, and 1.5% bio-based plastics [1]. Among the four major plastics (polyethylene (PE), polypropylene (PP), polystyrene (PS), and polyethylene terephthalate (PET)) used in packaging, PP is considered to be one of the most abundant plastics used worldwide [2,3]. In 2015, the global PP production was about 68 million metric tons, generating about 55 million metric tons of plastic waste [3]. Polypropylene is used in a wide range of applications due to its low price, moldability, temperature and chemical resistance, and outstanding mechanical properties [2,4]. Thus, PP production is anticipated to grow rapidly over the next decade to USD 108.57 billion at a CAGR of 5.2% from 2021 to 2028 [5]. Consequently, global solid-waste generation will continue to increase, along with its environmental impacts, which affect soil and water resources because of plastic landfill disposal and accumulation.

The recycling process is one notable solution to minimizing plastic waste and significantly contributes to promoting circular economy approaches, which include designing products for reuse, finding innovative techniques for recycling plastics, and incorporating PCR products [1]. Worldwide efforts and programs have been initiated and implemented to encourage the recycling and upcycling of plastic products. In the United States (U.S.), the Environmental Protection Agency (EPA) coordinated the development of the National Recycling Strategy in 2021. The goal is to increase the U.S. recycling rate to 50% by 2030 by improving markets for recycling commodities, improving infrastructure, reducing contamination, enhancing policies to support circularity, and standardizing measurement and data collection [6]. On the other hand, the findings of the "Circular Economy for Plastics 2020 EU27 + 3" (Norway, Switzerland, and the United Kingdom) study showed that the recycling rate increased to nearly 35%, while an increase of 1.3 percentage points was obtained for plastic parts and products with post-consumer recycled (PCR) content [7].

Furthermore, the benefits of and perspectives on plastic recycling have been established in published studies. Plastics recycling has challenges not limited to the presence of contaminants in direct food-contact applications [8], inadequate recycling infrastructure and sorting systems for certain types of plastics, such as polypropylene [2,9], high costs and economic viability, and limited demand for recycled plastics. In the United States, only 9% of the majority of plastics wastes (PE, PP, PS, and PET) was mechanically recycled in 2018 [3], and this fell to a rate of 5–6% in 2020 [10]. Despite all developments worldwide, more effort must be made to improve collection, sorting, and recycling technologies to sufficiently meet the targets [6,7]. Due to PP's wide range of applications, its waste can have very different properties and contaminants [2]; thus, the recycling process can be challenging. In addition, PP and PE are immiscible plastics that are difficult to separate with current recycling practices [9], which raised the question of how to efficiently recover high-value post-consumer plastics with high purity from single-stream collection. To address this matter and improve the recycling rate and systems, the European Union (EU) has enforced Extended Producer Responsibility (EPR) laws on packaging products since 1994, a recognized efficient waste-management policy that has become more popular in the U.S. Research studies have reported different techniques and processes to recover PP wastes [2], multilayer packaging films [11], plastic waste from material recovery facilities [12], and mixed wastes [13], aiming to achieve climate change mitigation and a circular economy.

PCR materials are known to play a vital role in sustainable packaging strategies. Studies have predicted a high market potential for recycled PP, with demands for increases in post-consumer polypropylene (PCR-PP) and global brands showing more interest in the inclusion of PCR-PP in their products [14,15]. With the enforcement of new laws and regulations, awareness of sustainable plastic waste management and reductions in environmental impacts through plastic recycling are increasing. One such initiative is the Polypropylene Recycling Coalition, which, in two years, granted 24 MRFs to acquire upgrades in sorting technology and produce high-quality recycled PP for reuse in packaging [16]. However, one of the major issues in using recycled materials is their suitability for direct food-contact applications and impact on the packaging integrity and performance. Single-stream and material recovery facility (MRF)-recovered plastics with food- and nonfood-grade materials can introduce unapproved additives and contamination to food-grade materials. Thus, pre-sorting becomes necessary to reduce potential human exposure risk to PCR in food packaging or to obtain single-sourced materials from known food-grade applications. The Toxics in Packaging Clearinghouse Model legislation prohibits the intentional use of four metals (cadmium, lead mercury, and hexavalent chromium) in any finished package, with a combined limit of 100 ppm [17]. In 2021, the legislation added the class of perfluoroalkyl and poly-fluoroalkyl substances (PFAS) and ortho-phthalate as regulated chemicals in the packaging [18]. Thus, an effective risk assessment is important to identify undesired substances and their sources in PCR plastics to limit contamination and ensure compliance. In addition, the main requirements for a material to be used in packaging often include barriers to oxygen and water vapor and proper mechanical properties to ensure package performance [19]. Lastly, an evaluation of the mechanical, thermal, and barrier performance of plastic packaging, incorporated with PCR plastics, must be accurately measured and documented.

This research aimed to determine the compliance and physical performance of extrusion blow-molded MRF-recovered PCR-PP bottles. The results of this research provide baseline information on the viability of using MRF-recovered PCR-PP while maintaining properties and ensuring compliance for direct food-contact applications.

2. Materials and Methods

2.1. Materials and Bottle Manufacturing Process

Post-consumer polypropylene was collected from a sorted #3–7 (resin identification codes) bale that could include polyvinyl chloride (PVC; #3), low-density polyethylene (LDPE; #4), polypropylene (PP; #5), polystyrene (PS; #6), and other (#7), sourced from a material recovery facility (MRF) located in Iowa, USA. Unwashed post-consumer PP flakes were washed following a typical wash procedure in a commercial recycling facility [20]. The PP flakes were washed in a laboratory-sized trommel separator with water to remove contaminants such as glass, woods, paper, and metals using the APR detergent wash solution (0.5% wt. of NaOH and 0.3% wt. of Triton X-100) in a five-gallon stainless-steel tank with four baffles [20]. The washed flake was then rinsed with water and dried.

Figure 1 shows the MRF bale-to-washing-to-bottle manufacturing process conducted in this study. The dried flakes from the washing process were then pelletized using a micro 18GL 18 mm twin-screw-co-rotating extruder (Leistritz, Sommerville, NJ, USA) and lab-scale pelletizer BT 25 (Bay Plastics Machinery, Bay City, MI, USA).

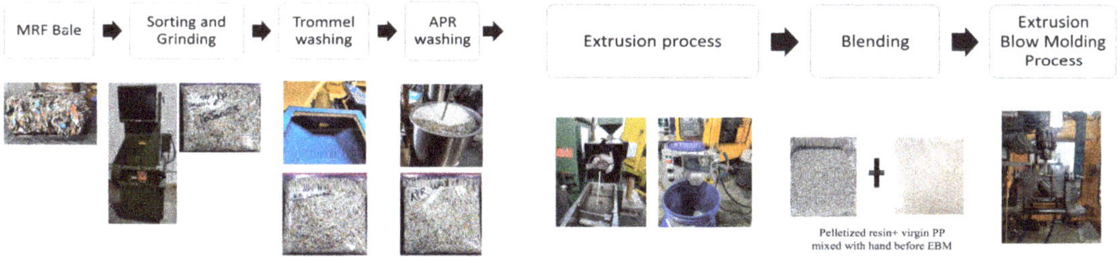

Figure 1. MRF-recovered PCR-PP to bottle manufacturing process.

Virgin PP resin was a non-food-grade polypropylene copolymer manufactured by Huntsman (Huntsman Copolymer PP RX PP18S07A-G Lot#A103HPP3048). The pelletized MRF-recovered PCR-PP pellets were manually mixed with virgin PP resin at different weight percentages (25%, 50%, 75%, and 100%); then, at least 20 bottles were manufactured using an Extrusion Blow Molder Model CS1 (Rocheleau Tool and Die Company, Fitchburg, MA, USA). Figure 2 shows the PCR-PP EBM bottles produced and analyzed in this study. The optimized EBM parameters used for PCR-PP bottles are listed in Table S1 (Supplementary Data).

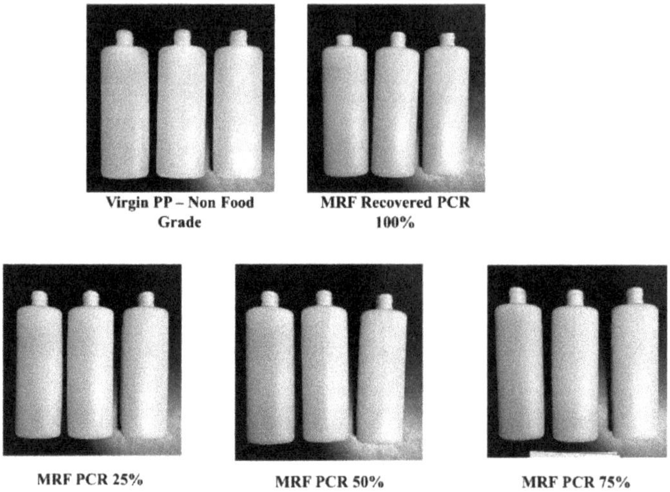

Figure 2. Optical images of the PCR-PP extrusion blow model bottles used in the study.

A complete randomized design (CRD) was used in this study (Table 1). Five treatments correspond to different weight concentrations of MRF-recovered PCR-PP mixed with PP virgin resin. For each treatment, at least three EBM bottles, as repeated measures, were used for physical performance tests (viscosity, molecular structure, thermal, barrier and tensile properties) and suitability analyses (heavy metals and CFR analysis) for direct food-contact applications.

Table 1. Experimental design.

Treatments	% (wt) MRF-Recovered Post-Consumer PP
1	0 (virgin PP)
2	25
3	50
4	75
5	100

2.2. *Characterizing MRF-Recovered Post-Consumer PP EBM Bottles*

2.2.1. Melt Flow Index

The melt flow rate of MRF-recovered post-consumer PP EBM bottles and virgin PP bottle was obtained in accordance with Procedure A of ASTM D1238–20 [21] in three repeated measures. The test utilized the specified parameters for polypropylene (230 °C, 2.16 kg) using a D4004 Melt Flow Indexer (Dynisco, Morgantown, PA, USA). Approximately 3.0–8.0 g of sample was loaded into the cylinder and the material was pre-heated for 6 min. The cut-off time was different for each of the samples depending on the flow rate. By applying the appropriate factor, the mass was then converted into grams per 10 min [22].

2.2.2. Infrared Spectroscopy

Attenuated total reflectance Fourier-Transform Infrared Spectroscopy (ATR-FTIR) was used to determine changes in chemical bonding and interactions of post-consumer PP bottles and virgin PP bottles in three repeated measures. A Nicolet 6700 infrared spectrometer (Thermo Scientific, Waltham, MA, USA) with a DTGS detector was used with 32 scans per run and a 2 cm^{-1} resolution at ambient temperature (22 °C). All spectra were baseline-corrected with OMINIC ™ 8.3 software (Thermo Fisher, Waltham, MA, USA) [22].

The diamond window of the ATR assembly was cleaned with an isopropanol wipe after each measurement to ensure no cross-contamination [23].

2.2.3. Differential Scanning Calorimetry

A Q2000 differential scanning calorimeter (TA Instruments, New Castle, DE, USA) was used to investigate the thermal transition properties for each PCR-PP and virgin PP bottle blend using a heat/cool/heat cycle between −10 °C and 310 °C, at a rate of 10 °C/min, under an N_2 atmosphere. Each treatment was tested with three repeated measurements with a sample mass of from 3 to 6 mg in a hermetically sealed aluminum DSC pan, according to ASTM D3418–15 [24]. An empty pan was used as a reference.

2.2.4. Oxygen Induction Time

The oxidation induction time (OIT) was used to evaluate the thermo-oxidative stability of the polymer blends. The OIT tests were carried out according to ASTM D3895 [25] using a Q2000 differential scanning calorimeter (TA Instruments, New Castle, DE, USA). A mass of 3–6 mg was placed in an open standard aluminum DSC pan, while an empty pan was used as reference. The sample was initially heated from 50 °C to 200 °C at a heating rate of 10 °C/min, under a nitrogen flow of 50 mL/min. Once the temperature reached 200 °C, the sample was isothermally conditioned for 5 min; then, the atmosphere switched from nitrogen to air. The sample was held at 200 °C until the sample went through oxidative degradation, where an exothermic peak appeared in the DSC curve. The time interval between the switch from nitrogen to air and the onset of the thermo-oxidation exotherm was reported as the OIT time [26].

2.2.5. Barrier Properties

Representative PCR-PP EBM bottles were selected as being free of defects, serving as representative samples of each treatment in three repeated measures. The oxygen transmission rate (OTR) values were determined using ASTM D3985-17 [27] using a Mocon Ox-Tran Model 2/21 (AMETEK MOCON Inc., Brooklyn Park, MN, USA). The OTR was measured in cm^3/pkg-day at atmospheric O_2 and 23 °C. The water vapor transmission rate (WVTR) was measured as g/pkg–day at 90% RH, 37.8 °C, and 10 standard cubic centimeters per minute (sccm), according to ASTM F1249-20 [28] using a Permatran W Model 3/33 (AMETEK MOCON Inc., MN, USA).

2.2.6. Mechanical Properties

The tensile properties (modulus of elasticity and yield stress) of virgin PP and post-consumer PP EBM bottles were examined using a Shimadzu Autograph AGS-J Tensile Tester (Shimadzu Instruments Manufacturing, Co., Ltd.; Analytical & Measuring Instruments Division, International Operations Department 1-3, Kanda Nishiki-cho, Chiyoda-ku, Tokyo 101-8448, Japan). The method used to determine mechanical properties was ASTM D638-14 (standard test method for tensile properties of plastics) [29]. The test specimens sectioned from bottles as strips, using a calibrated 2.54 cm wide shear, were then attached to the grips with 50 mm separation and separated at a rate of 50 mm/min. Each treatment has five repeated measures.

2.3. Suitability of Post-Consumer PP EBM Bottles for Direct Food Contact

2.3.1. Heavy Metal Analysis

Three samples of 0.1500 ± 0.0005 g were collected from pelletized MRF-recovered PCR-PP and virgin PP resins with three repeated measures. Samples were digested using microwave digestion (milestone UltraWave) in 5 mL HNO_3 (Thermo Scientific 67% *v*/*v* Trace Metal Grade) and 1 mL HCl (Thermo Scientific 34% *v*/*v* Trace Metal Grade). Digested samples were evaluated for the presence of Al (aluminum), Cd (cadmium), Cr (chromium), Fe (iron), Pb (lead), Sb (antinomy), and Ti (titanium) using Inductively Coupled Plasma—Optical Emission Spectrometry (ICP-OES, Thermo Scientific iCap-7400 Duo).

2.3.2. CFR Analysis and Cramer Classification

Di-ethyl phthalate (DEP), di-isobutyl phthalate (DIBP), di-butyl phthalate (DBP), Dipentyl phthalate (DPENP), dihexyl phthalate (DHEXP), di-cyclohexyl phthalate (DCHP), di-(2-ethylhexyl) phthalate (DEHP), Diisononyl phthalate (DINP), diisodecyl phthalate (DIDP), Bisphenol-F (BPF), Bisphenol-A (BPA), Bisphenol-B (BPB), Bisphenol-S (BPS), Bisphenol A diglycidyl ether (BADGE), benzophenone and Ethyl-P-Toluate were purchased from Sigma-Aldrich (St. Louis, MO, USA), with purity higher than 99%. Acetone (Fisher Scientific Inc., Fair Lawn, NJ, USA), xylene (Fisher Scientific, Hanover Park, IL, USA) were HPLC grade. The water used was purified using a Milli-Q gradient A10 system (Billerica, MA, USA).

All the materials (spatula, scissors, glass materials) in direct contact with the sample were washed with a precision detergent (Alconox, Inc, New York, NY, USA), deionized water and acetone (HPLC-grade), then dried in the oven at 150 °C and covered with aluminum foil until use to avoid cross-contamination. The use of plastic materials for extraction and material handling was strictly avoided. The samples were shredded into small pieces (2–5 mm) using a grinder. To extract the polypropylene samples, the Code of Federal Regulations (CFR) 21 part 177.1520 was followed [30]. Each sample (5 g) was dissolved in xylene (1 L) at 120 °C for 2 hr, then cooled to room temperature. The precipitated polymer was vacuum-filtered at room temperature and the filtrate was distilled at 140 °C until approximately 100 mL remained. The concentrated filtrate was added to a pre-weighted petri dish for extractable measurement. The sample containing petri dish was transferred to a hot plate for residual solvent evaporation and then stored in desiccators for 24 h. The extractable content was calculated based on Equation (1). A similar CFR extraction approach was followed (Figure 3) until the distillation and solvent reduced to 100 mL; then, it was collected for the gas chromatographic analysis. All samples were extracted in triplicate.

$$\% \, w/w = \frac{\text{weight of extractables}}{\text{initial weight}} \quad (1)$$

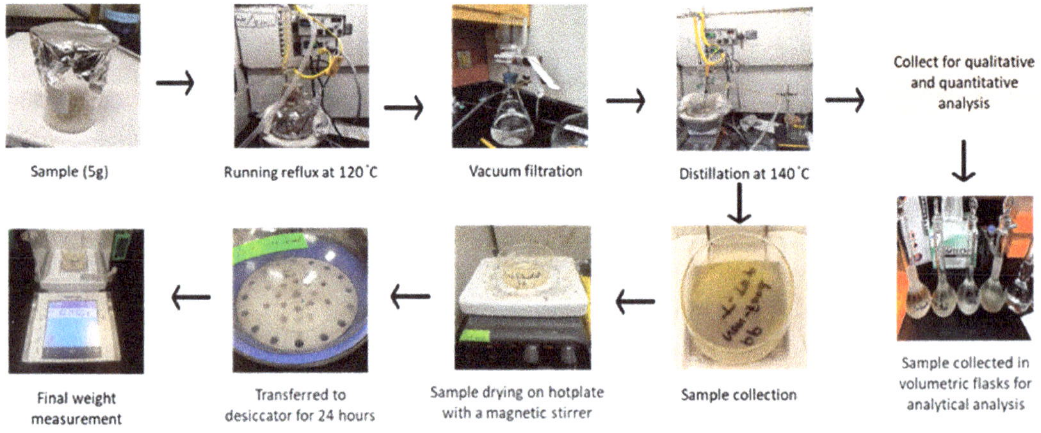

Figure 3. CFR and analytical analysis of polypropylene sample.

Both virgin and PCR PP bottle extracts were analyzed by GC-MSD in full scan mode for the unknown identification using gas chromatography and mass spectrometry (HP 6890 series GC system with an auto-sampler and HP Agilent 5973 mass selective detector). The detailed parameters are provided in Table S2 (Supplemental Materials). Compound identification was carried out using a mass spectral library (National Institute of Standards and Technology (NIST) version 2021). A threshold of reverse fit >600 and forward fit >600 from the NIST library was utilized for the identification of unknowns as per NIST recommendations [31]. The Cramer decision tree [32] was used to generate a structure-based Threshold of Toxicological Concern and classify organic chemicals into three classes of hazard level: low (I) toxicity, intermediate (II) toxicity and high (III) toxicity. Cramer class examines the presence of specific functional groups associated with known toxic effects and determines the hazard level [33].

To quantify the presence of the 9 phthalates and 5 bisphenols, stock standards were prepared in a cocktail at 1000 μg/mL. The internal calibration curve was made by serial dilution with dilution factor and followed by addition of the internal standard Ethyl p-Toluate (EPT). This was maintained at a constant 5 μg/g concentration, then analyzed by GC–QqQ- GC-MS/MS; Agilent 7000, Triple Quad, GC 7890A using multiple reaction monitoring (MRM) conditions (Table S2). The standards for the calibration curve and the samples were analyzed with the method mentioned above. BSTFA (N, O bis(trimethylsilyl)trifluoroacetamide) was used as the derivatization agent. A total of 50 μL of BSTFA and 100 μL of acetone (to accelerate the reaction rate) was added to each blank, standard, and sample before analysis. The LOD and LOQ were determined following the [34] internal calibration method. This process was performed using the standard deviation (SD) of the response and the slope of the calibration curve. The LOD was calculated as 3.3 SD/slope (μg/g) and the LOQ was 10 SD/slope (μg/g).

2.4. Statistical Analysis

Data were analyzed using one-way ANOVA considering a 95% confidence level (α = 0.05) on JMP® 16 Pro (SAS Institute., Cary, NC, USA) to determine significant differences between treatment means. Pairwise comparison was carried out using the Tukey's honestly significant difference (HSD) test.

3. Results and Discussions

3.1. Physical Performance of MRF-Recovered PCR-PP EBM Bottles

3.1.1. Polymer Viscosity

The melt flow rate is a rapid and convenient measurement technique commonly used during manufacturing for raw material quality control and qualification before processing. MFR is strongly dependent on the polymer molecular weight [35]. The MFR is associated with polymer's molecular weight, viscosity, and flow of the polymer, providing valuable information on polymer processability and its intended application [36]. Previous research by Curtzwiler et al. determined that the addition of recycled content increased measured MFR, indicating a lower viscosity [22]. This could be attributed to the increases in molecular weight distribution and lower chain molecular weight caused by reprocessing and prior utilization. The data collected herein (Figure 4) suggest that the changes in MFR depend more on the type of virgin and PCR-PP. Bottles with 50% and 75% PCR-PP did not show significant differences between the different ratios of PCR-PP added to the blend; however, they were significantly different than 100% PCR-PP. The virgin PP presented an MFR value almost 12 times lower than 100% PCR-PP. The MFR increased with PCR-PP concentration. Eriksen et al. explained MFR values between 0 and 1 g/10 min are often recommended for extrusion applications, whereas blow-molding requires 0.3 and 5 g/10 min and injection-molding values must be higher than 5 g/10 min. MFR values between 5 and 50 g/10 min can be applied to mold thicker-walled products [37].

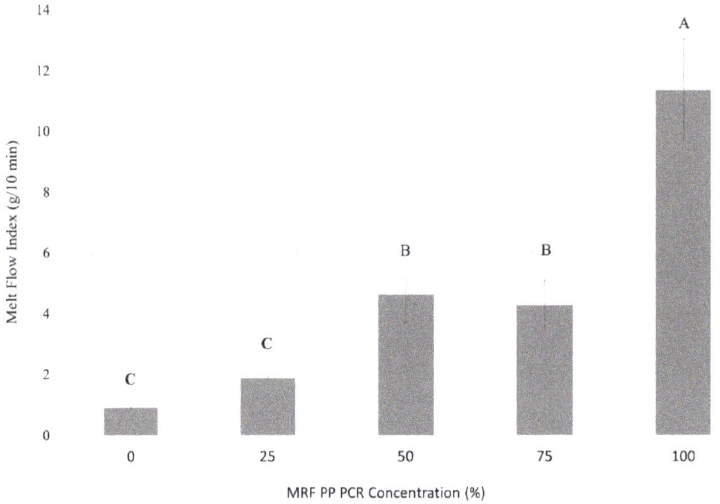

Figure 4. Melt flow rate of PP EBM bottles with varying weight percentages of MRF-recovered PCR-PP. There is evidence of a difference in measured average MFI among treatments (p-value < 0.05). Average values (±standard deviation) with the same letters (A, B, C) between MRF PP PCR concentrations are not significantly different (p-value < 0.05, Tukey test N = 15).

3.1.2. Polymer Molecular Structure

FTIR was used to compare virgin PP and PCR-PP blends to understand changes in molecular interactions and molecular composition. Figure 5 shows an overlay of representative examples of the collected spectra. Characteristic bands at 2949 cm^{-1} and 2916 cm^{-1} indicate C-H bond stretching (aliphatic hydrocarbon) [38] and bands at 1450 cm^{-1} and 1375 cm^{-1} are representative of asymmetric and symmetric C-H deformations [39]. All spectra possess bands that are characteristic of PP with no noticeable blue or red shifting in the characteristic bands, indicating similar chemical environments [38].

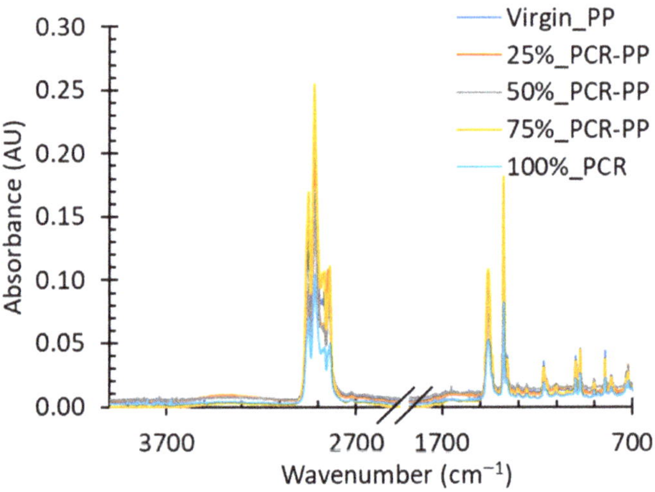

Figure 5. FTIR spectra of selected PP EBM bottles with varying weight percentages of MRF recovered PCR-PP.

3.1.3. Thermal Properties

Representative heat flow thermograms of all samples are shown in Figures 6 and 7. The Tm (melting temperature), Tc (crystallization temperature), and Xc (crystallinity; %) values of virgin PP and PCR-PP at different percentages are presented in Table 2. The Tc increased with the addition of PCR-PP to the virgin PP; this is beneficial, as higher crystallization temperatures can result in reduced cycle times and increased production rates. This observation can be attributed to residual nucleating agents in the post-consumer material and suggests that fewer additives may be required in PCR blends to achieve the same effect [40–44].

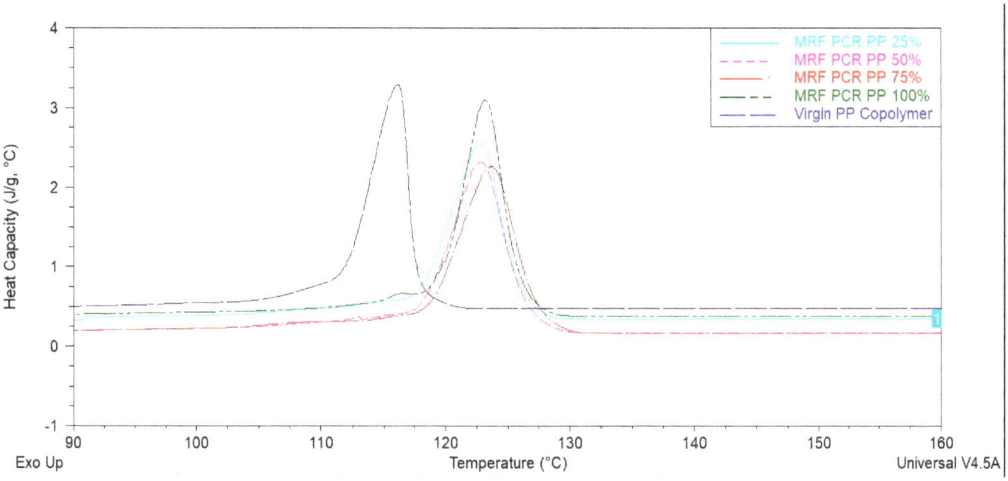

Figure 6. Heat capacity of PP EBM bottles with varying weight percentages of MRF-recovered PCR-PP.

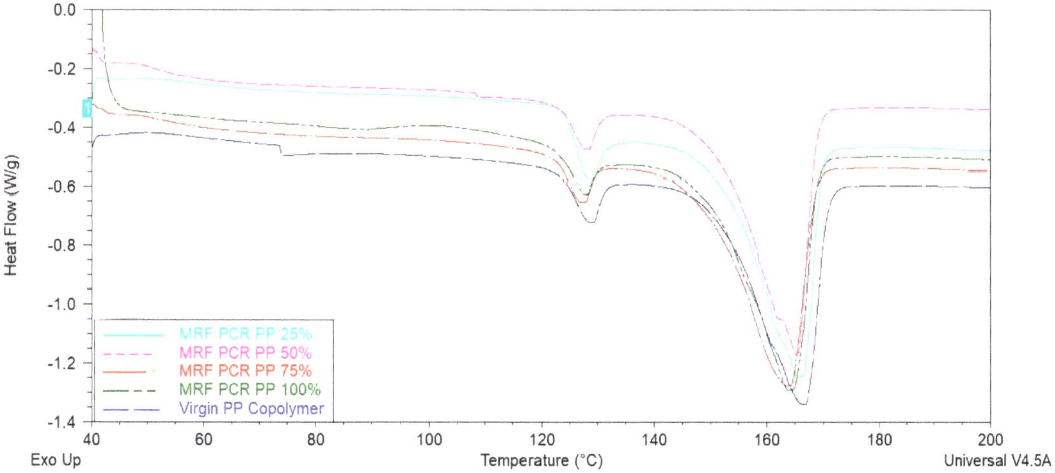

Figure 7. Heat flow of PP EBM bottles with varying weight percentages of MRF-recovered PCR-PP.

Table 2. Average thermal properties of PP EBM bottles with varying weight percentages of MRF-recovered post-consumer PP.

Treatment	Tm1 (°C)	Tm2 (°C)	Tc (°C)	Hc (J/g)	Hm1 (J/g)	Hm2 (J/g)	Crystal 1 (%)	Crystal 2 (%)
1	165.5 [a]	162.1 [a]	115.9 [b]	66.2 [ab]	49.1 [ab]	49.7 [a]	23.7 [ab]	24 [a]
2	165.9 [a]	162.3 [a]	122.8 [a]	62.6 [b]	48 [b]	51.3 [a]	23.2 [b]	24.8 [a]
3	165.3 [a]	161.2 [a]	123.2 [a]	72.6 [ab]	51.2 [ab]	51.4 [a]	24.7 [ab]	24.9 [a]
4	164.0 [a]	161.6 [a]	123.4 [a]	76.2 [a]	54.1 [ab]	54.8 [a]	26.1 [ab]	26.5 [a]
5	164.8 [a]	159.3 [a]	123.1 [a]	73.6 [ab]	54.8 [a]	54.9 [a]	26.5 [a]	26.5 [a]

Note: values with the same letter in a column are statistically the same ($\alpha = 0.05$).

The statistical analysis of the Tm values of the virgin PP and PCR-PP samples did not possess significant differences in comparison to either the first or second heat cycle, indicating that crystal quality was not influenced by the presence of the MRF-recovered PCR-PP. However, the ΔHm1 (melting enthalpy) values increased as the percentage of PCR-PP in the material increases and the 100% PCR-PP was statistically the highest; however, the melting enthalpy of the second heat cycle was statistically the same for all blends. These results suggest faster crystallization rates during the bottle manufacturing process, which are supported by the higher crystallization temperatures of PCR-PP-containing blends. The melting enthalpy values for virgin PP (49.1 J/g for Hm1 and 49.7 J/g for Hm2) are lower compared to the highest PCR-PP percentages. Due to the increase in melting enthalpy values with PCR-PP-loading into virgin PP, the degree of crystallinity of blends gradationally increased in the blends as manufactured.

3.1.4. Thermooxidative Stability of Post-Consumer PP EBM Bottles

The oxygen induction time is often used to investigate polyolefin stability, which degrades when heated in an oxidizing atmosphere such as air. The relative antioxidant efficiency in a material is verified by measuring the OIT [35]. The DSC-based OIT thermograms for virgin PP and PCR-PP blends (containing 0, 25, 50, 75 and 100 wt%, respectively) are shown in Figure S1 (Supplementary Materials). The exotherm onset is considered the oxidation initiation [45]. As anticipated, the virgin PP requires the most amount of time for the onset of oxidation to occur, as it would possess the highest concentration of unused antioxidant since it was not exposed to the environment in its first service life and was only subjected to one melt processing cycle. There appears to be sufficient antioxidant in the virgin material and enough remaining in the PCR-PP in the 25% PCR-PP blend, as the OIT of this blend was statistically the same as virgin PP (Figure 8) [26]. These results, coupled with the increased crystallization temperature, suggest that residual additives in post-consumer plastics, coupled with those in virgin plastic, can enable a similar performance to virgin plastics with a lower cost due to the synergistic effect of combining additives. The OIT results are also anticipated to translate into the mechanical properties, as noted by Martin and De Paoli, where the mechanical property degradation from melt-processing cycles was reduced for the stabilized formulations [35].

3.1.5. Barrier Properties

A gas barrier is defined as the resistance of packaging to the diffusion and sorption of a substance, which can be expressed in terms of permeability [46]. Two of the most important barrier properties in food packaging are the OTR and WVTR [19]. Oxygen permeation through the packaging results in an oxidation process that causes food spoilage. Thus, reducing the packaging material's oxygen permeation rate can contribute to increasing product shelf life and maintaining food quality. Virgin PP bottles had the lowest average OTR value of 0.48 $\pm$ 0.01 cc/pkg-day, and were significantly different from those with a 75 and 100% concentration of MRF-recovered PCR-PP (Figure 9). From published studies, crystallinity has a significant effect on the gas-barrier properties of polymer materials [47,48]. To improve polymer barrier properties, crystallinity must be higher; however, our materials possessed decreased barrier properties with slight increases in crystallinity due to the increased PCR-PP concentrations. This observation suggests that additional interactions in the polymer's amorphous fraction contributed to the more facile gas migration through the polymer. Among all treatments, bottles with 100% PCR-PP had the highest average OTR value of 0.65 $\pm$ 0.02 cc/pkg-day, corresponding to a 36% increase and significantly different from virgin PP. Also, bottles with 25 and 50% PCR-PP concentrations were comparable and not statistically different from virgin PP bottles' oxygen transmission rates, with values ranging from 0.48 to 0.58 cc/pkg–day. Thus, it is hypothesized that the decrease in oxygen barrier property mainly relates to pigments, fillers, and contaminants, as observed by the measured concentration of Al, Fe and Ti results from MRF-recovered PCR-PP (Table 3) which are known to significantly affect the crystallinity and performance properties of polymers.

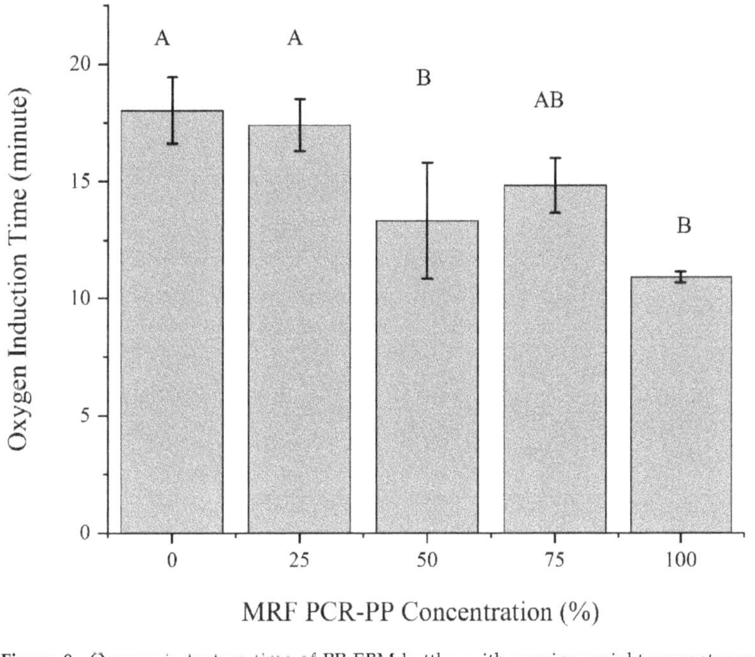

Figure 8. Oxygen induction time of PP EBM bottles with varying weight percentages of MRF-recovered PCR-PP. Note: values with the same letter in a column are statistically the same (α = 0.05).

Table 3. Measured metals compliant with CONEG.

	Element Concentration (µg/g or ppm)						
Sample	Al	Sb	Cd	Cr	Fe	Pb	Ti
PP virgin	3.45	* b	0.13	0.17	* b	4.10	* b
PP MRF	84.2	* b	* b	* b	49.02	* b	68.26
MLOD (ppm)	0.124	0.011	0.002	0.003	0.010	0.689	0.002
MLOQ (ppm)	0.41	0.04	0.01	0.01	0.03	2.30	0.01

* b—below limit of detection.

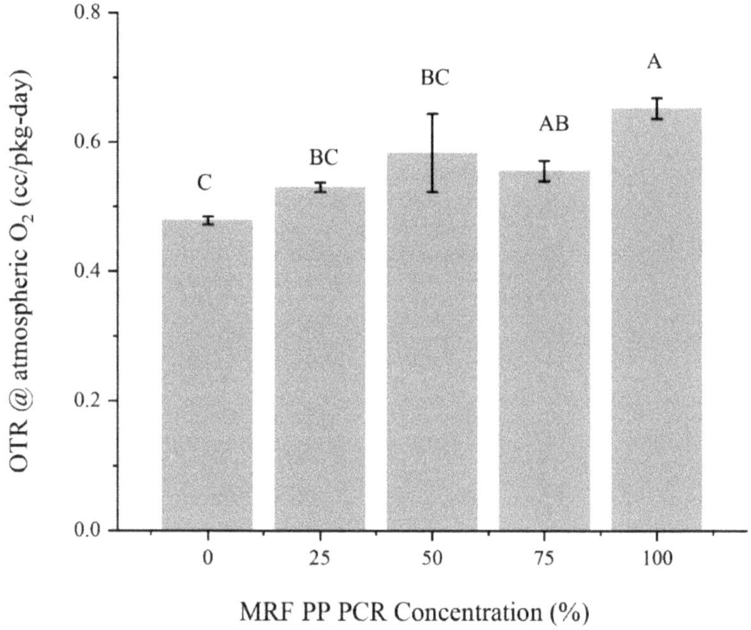

Figure 9. Oxygen transmission rates of PP EBM bottles with varying weight percentages of MRF-recovered PCR-PP. There is evidence of differences in measured average OTR among treatments (p-value < 0.05). Average values (±standard deviation) with the same letters (A, B, C) between MRF-recovered PCR-PP concentrations are not significantly different (p-value < 0.05, Tukey test N = 15).

The WVTR indicates how easily moisture vapor can permeate a packaging structure. When maintaining a safe moisture content for products during the storage period, a bottle should have a sufficiently low WVTR value. PP is known for its good-to-excellent water vapor barrier properties, as reflected in the results of this study. Mixing virgin PP resin with 25–75% concentrations of MRF-recovered PCR-PP did not significantly change the water-vapor barrier properties of virgin PP. Figure 10 shows the measured average WVTR for the five PCR-PP concentrations. The 100% MRF-recovered PCR-PP bottle had the highest value of 0.14 ± 0.03 g/pkg–day and was significantly different from all other treatments. On the other hand, bottles with 100% virgin PP and 25–75% post-consumer PP did not show significant differences, with an average WVTR ranging from 0.01 to 0.02 g/pkg-day.

The results of this study indicate that adding PCR-PP to virgin PP in EBM bottles decreases its ability to limit oxygen and water-vapor transmission, as reflected by the increasing trend of the transmission rates. However, these increases were only from 5% to 36% different, suggesting that PCR-PP can be incorporated into products without detrimental impacts on barrier performance. A previous study [19] reported the opposite trend regarding the effect of mixing PCR polyolefins in virgin PP resin, where it was reported that 100% post-consumer recycled polyolefin resulted in increased barrier properties (both for oxygen and water vapor) compared to virgin polypropylene. However, it is important to note that the barrier properties of a material can be affected by its chemical structure and processing parameters during production and the material's grade, quality, purity, filler composition, etc. [48].

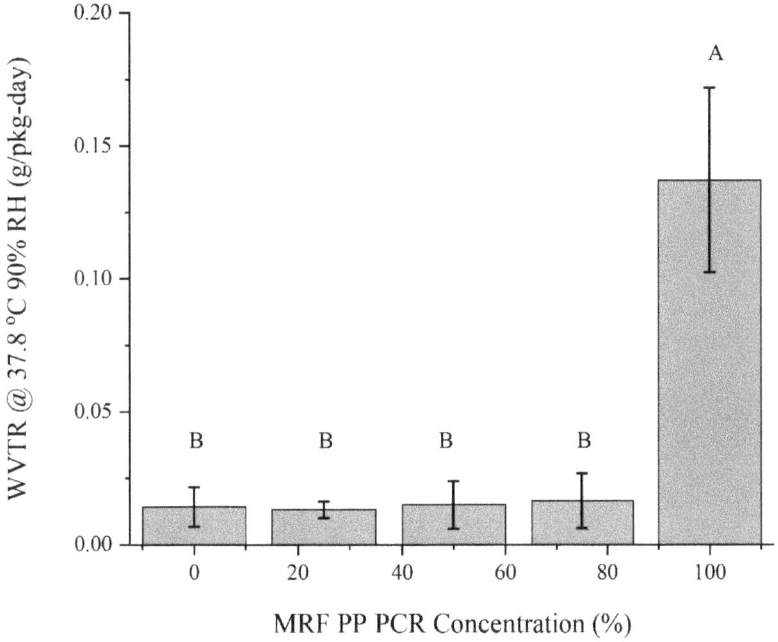

Figure 10. Water vapor transmission rates of PP EBM bottles with varying weight percentages of MRF-recovered PCR-PP. There is evidence of differences in measured average WVTR among treatments (p-value < 0.05). Average values (±standard deviation) with same letters (A, B, C) between MRF-recovered PCR-PP concentrations are not significantly different (p-value < 0.05, Tukey test N = 15).

3.1.6. Tensile Properties

Mechanical properties are a measure of materials' behavior when subjected to stress or deformations. Plastic bottles, when used, must have sufficient strength to maintain physical integrity during the handling, transport, and storage of products. These properties include the modulus of elasticity (MOE), flexibility, hardness, and yield stress. MOE is the property that describes the stiffness of the polymer, while yield stress is the point where strain increases and no change in stress and plastic deformation occurs.

Figure 11 shows that the MOE of PP EBM bottles differs significantly among the different compositions of PCR-PP (p-value < 0.05). Bottles with 75% PCR-PP had the highest MOE value of 224 ± 49 MPa, which was significantly higher than virgin PP bottles with an MOE of 140 ± 24 MPa. Variability in measured MOE could be caused by differences in the presence of impurities. Measured average yield stress ranging from 18 to 21 MPa did not show significant differences among treatments. These results indicate that introducing PCR-PP increased the stiffness, but the yield stress remained the same, providing better performance during transport and shipping.

3.2. Compliance of MRF-Recovered Post-Consumer PP EBM Bottles

3.2.1. Metals' ICP-OES

MRF-recovered PCR-PP EBM bottle samples were analyzed for common metals such as Al, Sb, Cd, Cr, Fe, Pb and Ti because these are used as catalysts during the polymerization process. Also, according to the Coalition of Northeastern Governors (CONEG), the sum of Cd, Cr^{+6}, Hg, and Pb should not exceed 100 ppm [17]. From the results, Cd, total Cr, and Pb were not detected in the MRF-recovered PCR-PP but were detected in virgin PP resin with a sum < 100 ppm. Mercury was not detected in the MRF-recovered PP using X-ray fluorescence spectroscopy (LOD = 1 ppm). These results suggest that the PP EBM bottles with post-consumer PP would comply with current CONEG metals legislation. Table 3 shows the metals measured for virgin PP resin and pelletized MRF-recovered PCR-PP. It was observed that Al, Fe, and Ti were the main materials present in both samples, which may be attributed to the additives, fillers, catalysts, or contaminants in the samples. No Sb, Cd, Cr, or Pb were detected in post-consumer PP.

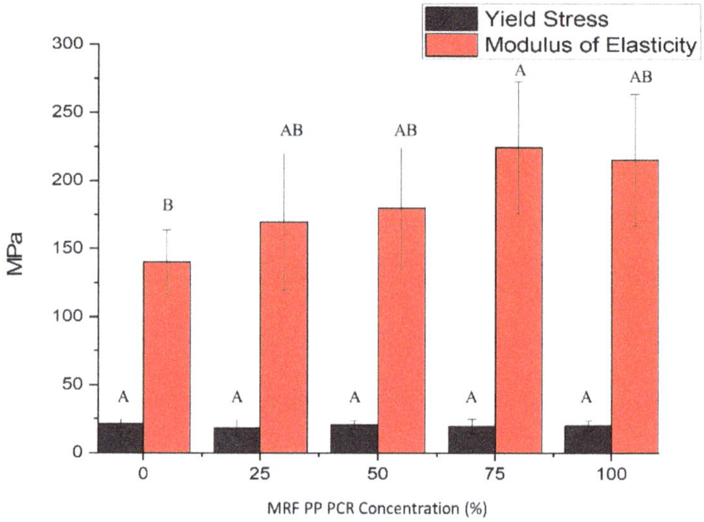

Figure 11. Tensile properties of PP EBM bottles with varying weight percentages of MRF-recovered PCR-PP. There is evidence of differences in measured average modulus of elasticity among treatments (p-value < 0.05). Average values (±standard deviation) with same letters (A, B) between MRF-recovered PCR-PP concentrations are not significantly different (p-value < 0.05, Tukey test N = 25).

3.2.2. 21 CFR 177.1520 Extraction for Extractable and Cramer Classification

The virgin and MRF-recovered PCR-PP were characterized for compliance with the Code of Federal Regulations for direct food contact via 21 CFR 177.1520. According to CFR 21, part 177.1520, the extractable fraction from polypropylene should not exceed 9.8% in weight [30]. The virgin PP comprised a higher percentage of extractable matter (13.713%) compared to 100% PCR-PP (9.164%); this was anticipated, as the virgin PP was not food-grade (Figure 12). In our experiment, the virgin PP exceeded that specification limit, while the PCR-PP was within the specification limit. In general, virgin PP has a higher molecular weight and crystallinity than the same PCR-PP polymer after its first service life and the recycling process due to the multiple heating treatments that occur during the recycling process, where a wide range of thermal and oxidative degradation occurs [48,49]. The extractables from a polymer can vary depending on various factors, for example additives, purity, oligomeric content, and the processing method (extrusion, recycling, molding, etc.) [50]. Another possibility is the alteration or degradation of the additives or fillers in the PCR sample during the recycling process, which is absent in the virgin sample, and can reduce the extractable content in the PCR-PP sample. While the virgin PP was non-food-grade PP, the MRF-recovered PP was likely composed of a mixture of food-grade and non-food-grade, which could explain the lower CFR extractables for the MRF-recovered material [30].

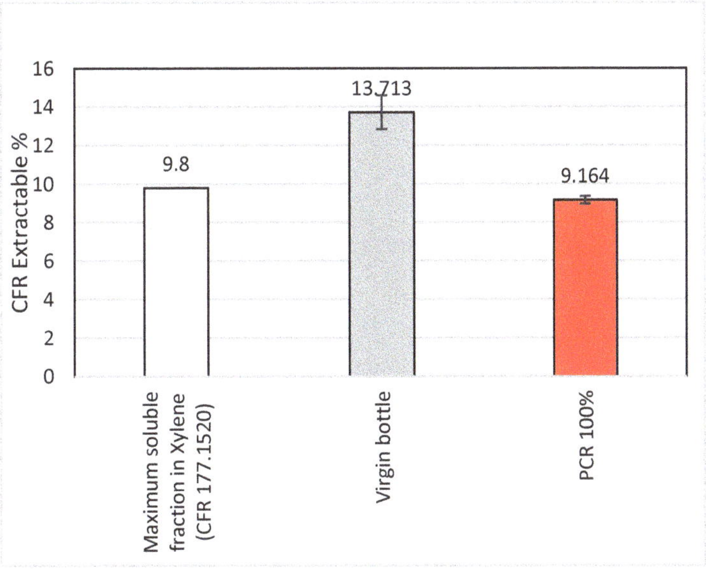

Figure 12. Extractable % virgin PP and MRF-recovered post-consumer PP samples.

The results for non-intentionally added substances (NIAS) and related Cramer classification-based toxicity levels are listed in Table 4. NIAS are substances that may be present in packaging polymers with no known application and may pose a risk to health if they migrate into food [51]. Analysis indicated that PCR samples contained more additives, pigments, and impurities than virgin PP. Substances only detected in the PCR sample, such as myristic acid, oleic acid [52], stearic [53], 1-monopalmitin, have applications as lubricants, emulsifiers, or surfactants in the industry. Glycerol monostearate is used as an antistatic additive in polypropylene and in the food sector as an emulsifier [54]. Myristic acid has diversified applications as a flow agent and emulsifier in the food sectors, as a surfactant in soaps, detergents, and as an internal or external lubricant in plastics (ACME-Hardesty). The substances found only in PCR samples may have been sourced as additives to improve flexibility and compatibility, or as contaminants from other packaging or waste during the first intended service life. Butylated hydroxytoluene (BHT) was only found in the virgin sample, which is generally used as an antioxidant in plastic packaging materials such as polyethylene and polypropylene films [55]. This prevents degradation during the service life. BHT is listed under the Registration, Evaluation, Authorisation, and Restriction of Chemicals (REACH) regulation [56], which is designed to increase safety and improve the protection of human health and the environment from hazardous chemicals.

Table 4. IAS/NIAS compounds in virgin and 100% PCR-PP bottle sample.

Virgin Sample		**100% PCR Sample**	
Compound (CAS)	Cramer Class	Compound (CAS)	Cramer Class
M-Toluic acid, TMS derivative (959296-29-8)	Class I	m-Toluic acid, TMS derivative	Class I
Butylated Hydroxytoluene (BHT) (128-37-0)	Class II	1-iodo- Decane (2050-77-3)	Class III

Table 4. *Cont.*

Virgin Sample		100% PCR Sample	
4-cyano-3-fluorophenyl 4-(4-butylcyclohexyl)benzoate (92118-83-7)	Class III	1,1′-(1,2-dimethyl-1,2-ethanediyl)bis- Benzene (4613-11-0)	-
1,1′-(1,2-dimethyl-1,2-ethanediyl)bis Benzene (5789-35-5)	Class III	1′-(1,2-ethanediyl)bis[4-methyl-Benzene	Class III
1′-(1,2-ethanediyl)bis 4-methyl-Benzene (538-39-6)	Class III	Myristic acid, TBDMS derivative (104255-79-0)	Class I
4-Allyl-2-methoxyphenyl benzoate	Class I	Tricyclo[4.4.0.0(2,7)]decan-3-one, 1-methoxy-2,6-dimethyl- (62648-63-9)	Class III
Palmitic Acid, TMS derivative (55520-89-3)	Class I	1,2-Bis(3,5-dimethylphenyl)-diazene 1-oxide (64857-67-6)	Class III
Stearic acid, TMS derivative (18748-91-9)	Class I	Palmitic Acid, TMS derivative (55520-89-3)	Class I
-	-	Linoelaidic acid, tert.-butyldimerthylsilyl ester	-
-	-	Oleic Acid, (Z)-, TMS derivative (21556-26-3)	Class I
-	-	Stearic acid, TMS derivative (18748-91-9)	Class I
-	-	1-Monopalmitin, 2TMS derivative (1188-74-5)	Class III
-	-	2-Monostearin, 2TMS derivative (53336-13-3)	Class III
-	-	4-tert-Octylphenol, TMS derivative (78721-87-6)	Class I
-	-	Glycerol monostearate (GMS), 2TMS derivative (1188-75-6)	Class I

The compounds detected (8 in virgin PP and 15 in PCR-PP) in both types of samples can often be considered intentionally added substances. For example, m-Toluic acid is used as a dye or polymer stabilizer [57] and palmitic acid, a fatty acid found in both virgin and PCR samples, has applications as a stabilizer in polymers [58]. The Cramer decision tree [32,33] classifies organic chemicals into three classes of hazard level: low (I) toxicity, intermediate (II) toxicity and high (III) toxicity. Cramer class examines the presence of specific functional groups associated with known toxic effects and determines the hazard level [58]. Of the compounds found in the virgin sample, Cramer class I consists of 50%, class II consists of 12.5%, and class III consists of 37.5%. In the PCR sample, 15 chemicals were detected: class I consists of 46.67%, class II consists of 0%, class III consists of 40.0%, and 13.3% was unclassified. Based on the Cramer classification and number of detected compounds, it appears that the PCR sample has a higher probability of toxicity than the virgin sample. However, it is difficult to accurately determine the overall risk severity in the PCR or virgin samples without more information about the specific compounds' total exposure measurements (concentration and diffusion coefficients) and potential toxicological effects. Recycled plastics, including recycled polypropylene, originating from mixed plastic waste streams often exhibit a wider range of contaminants than virgin materials, specifically recycled PP sourced from non-food-grade materials mixed with food and non-food-grade plastics, which may not comply with stringent regulations and can introduce a wide range of unknown contamination [37].

Phthalates are synthetic chemicals widely used in plastic packaging and industrial products to improve the mechanical properties, as well as providing flexibility and softness [59]. Bisphenols have wide applications in the production of polycarbonate plastics, epoxy resins, adhesives, and several additives [60]. However, they are endocrine disruptors/modulators and hazardous to health at very low concentrations (nanogram level) [61]. Analyzing phthalates and bisphenols in non-food-grade recycled PP is important to understand their compliance with phthalate and bisphenol regulations, as well as emerging corporate restrictive substance lists (RSLs). Information regarding the prevalence of endocrine modulators in PCR is also rare in the literature.

In our samples, the virgin PP sample contained diethylhexyl phthalate (DEHP) at a concentration of 0.83 μg/g and in PCR-PP with a concentration of DEHP (1.33 μg/g). While DCHP (2.335 μg/g), DINP (13.617 μg/g), BPA (0.059 μg/g) were only detected in PCR-PP sample (Figure 13). The additional phthalate detected in the PCR samples was as follows: DEP (0.138 μg/g), DIBP (0.181 μg/g), DHEXP (0.116 μg/g), and DBP (2.712 μg/g). In the PCR sample, the concentration of DIBP and DBP (0.181 μg/g and 2.712 μg/g, respectively) was slightly lower compared to the virgin PP sample. Perestrelo et al. investigated various plastic packages from the European market and determined that DIBP was the most prominent phthalate in packaging, with concentrations ranging between 3.61 and 10.7 μg/L [59]. The concentration of DEHP in the PCR sample is higher than virgin at 1.33 μg/g. The higher concentration of DBP in the virgin PP sample suggests its intentional addition during the pellet manufacturing, as no additives were included during bottle manufacturing. However, polymers sourced after recycling can face diversified processing and contamination, which can influence the analyte concentration compared to the virgin sample; this was also prominent in our result. On the other hand, the inherent heterogeneity of recovered post-consumer polymers can contribute to a wide range of standard deviation, which was observed for DBP concentrations. The use of DEHP, DBP, BBP, DIDP, and DINP has a threshold of 0.1% in the final product [62], and Toxics in Packaging Clearinghouse (TPCH) set the limit for incidental ortho-phthalates to no more than 100 μg/g in packaging [63]. Overall, the phthalates detected in the virgin and PCR samples were below the TPCH limit of 100 μg/g and complied with the regulations. Although polypropylene is often considered to be "BPA free" and safe for use as packaging [64], BPA was detected in the PCR sample at a concentration of 0.059 μg/g.

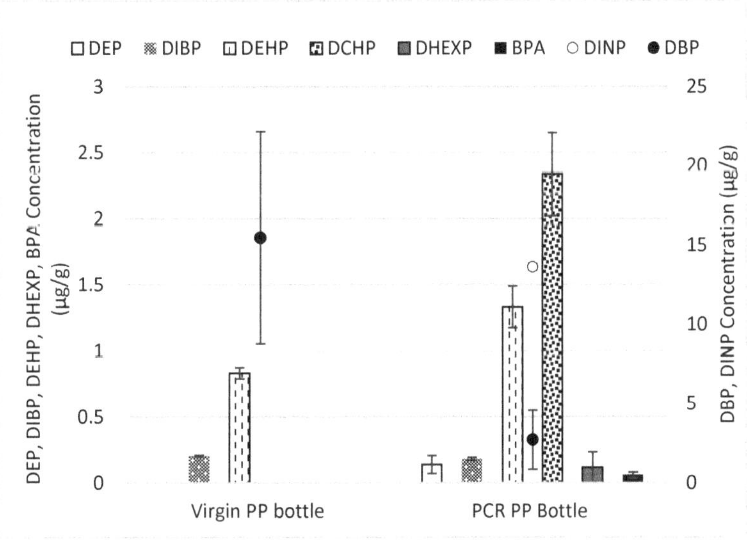

Figure 13. Phthalate and bisphenols in virgin and PCR bottle samples.

Our findings in this study fill the knowledge gaps related to extractables, NIAS/IAS identification, and targeted extractable compounds from material-recovery-facility-sourced PCR-PP materials. The contamination levels in PCR-PP were higher than the virgin sample, according to Cramer classification and quantification results for certain phthalates and BPA. However, they did not exceed the limits set by the Toxics in Packaging Clearinghouse (TPCH). With increasing worldwide and domestic regulations encouraging and requiring the use of recycled plastic with virgin material, our scientific data can provide valuable information to the scientific community and manufacturers regarding potential uses for MRF-recovered PP. Additionally, the study provides information on common additives, antioxidants, and stabilizers that can be present as non-intentionally added substances (NIAS) in both virgin and recycled PP.

4. Conclusions

PP bottles are commonly used to store food products, chemicals, and pharmaceuticals because of their good chemical resistance, superior strength, and cost advantages in blow molding compared to PET [64]. This study investigated the compliance and physical performance of extrusion blow-molded MRF-recovered PCR-PP bottles. Material properties loosely followed the law of mixtures, although some remained statistically the same. This general trend was observed in multiple studies that blended virgin and post-consumer recycled polymers. Therefore, it is important to note that the increase or decrease in properties depends upon both post-consumer and virgin polymers. Key results include:

- Increased crystallization temperature when PCR is present in the blend;
- No practical difference in crystallinity as a function of PCR concentration as-molded (first heat curve);
- No other polymers are present in the thermograms of MRF-recovered PP materials, indicating high polymer purity (melting peaks);
- Oxygen and water-vapor barrier properties remained relatively constant unless the composition was 100% MRF material;
- MRF-recovered PCR-PP significantly (p-value < 0.05) improved the stiffness of virgin PP bottles. On the other hand, the measured yield stress for all treatments was significantly similar;
- A wider range of N/IAS was detected in PCR material compared to the virgin material, which can be attributed to plastic additives, food additives, and degradation byproducts;
- Regulatory compliance (limits of TPCH: 100 μg/g) and maintaining properties up to 75% MRF PCR demonstrates the increased value of MRF materials. Targeted phthalates did not exceed the limits of TPCH, and trace levels of BPA were detected in the MRF PCR-PP.

The results of this study provide critical information for stakeholders making decisions to use MRF recovered in food packaging applications. Moreover, this study demonstrates the viability of a significant source of PP and its notable long-term impacts, increasing profits by using PCR materials. This approach will produce environmentally responsible food plastic packaging in compliance with legislation in the circular economy. However, it must be noted that the material recovery facility post-consumer polypropylene used in this work is only representative of a single, sorted post-consumer bale from one material recovery facility. The reproducibility and variance in the post-consumer polypropylene properties from material recovery facilities is progressing, specifically for bale sources from urban, suburban, and rural areas.

Future work should continuously monitor new PCR sources from different MRFs to understand consistency regarding the US. One should determine levels of PFAS and phthalates of PCR-PP and other recycled plastics to address the increasing concerns regarding toxicity. An additional analysis using GC tandem mass spectrometry to reduce the limits of detection/quantification is in progress. Lastly, a study on the efficiency of different decontamination processes would be vital to emphasize the impacts of using MRF PCR plastics and their viability for direct food-contact applications.

Supplementary Materials: The following supporting information can be downloaded at: https://www.mdpi.com/article/10.3390/polym15163471/s1, Figure S1: DSC OIT thermograms of (a) virgin PP copolymer (b) MRF PCR-PP 25% (c) MRF PCR-PP 50% (d) MRF PCR-PP 75% (e) MRF PCR-PP 100%; Table S1: optimized EBM parameters for post-consumer PP bottle processing; Table S2: parameters of gas chromatographic analysis.

Author Contributions: All authors contributed to the study conception, design, material preparation, and data analysis. The first draft of the manuscript was written by M.C.C.D.I., K.N.T. and M.M., G.W.C. and K.L.V. commented on, revised, and edited previous versions of the manuscript. All authors have read and agreed to the published version of the manuscript.

Funding: This work was supported in part by the Polymer and Food Protection Consortium (PFPC) at Iowa State University, Agriculture and Home Economics Experiment Station HATCH Project 04202, and the Institute for the Advancement of Food and Nutrition Sciences (IAFNS) through a grant from the Food Packaging Safety and Sustainability Committee. IAFNS is a nonprofit science organization that pools funding from industry collaborators and advances science through the in-kind and financial contributions from public and private sector participants.

Institutional Review Board Statement: Not applicable.

Data Availability Statement: Data are contained within the article or Supplementary Material.

Conflicts of Interest: The authors declare no conflict of interest.

References

1. PlasticsEurope. Plastics-The Facts 2022. 2022. Available online: https://plasticseurope.org/knowledge-hub/plastics-the-facts-2022/ (accessed on 17 May 2023).
2. Bora, R.R.; Wang, R.; You, F. Waste Polypropylene Plastic Recycling toward Climate Change Mitigation and Circular Economy: Energy, Environmental, and Technoeconomic Perspectives. *ACS Sustain. Chem. Eng.* **2020**, *8*, 16350–16363. [CrossRef]
3. Geyer, R.; Jambeck, J.R.; Law, K.L. Plastics. Production, use, and fate of all plastics ever made. *Sci. Adv.* **2017**, *3*, e1700782. [CrossRef] [PubMed]
4. Freudenthaler, P.J.; Fischer, J.; Liu, Y.; Lang, R.W. Polypropylene Post-Consumer Recyclate Compounds for Thermoforming Packaging Applications. *Polymers* **2023**, *15*, 345. [CrossRef]
5. Fortune Business Insights. The Global Polypropylene Market Is Projected to Grow from USD 76.00 Billion in 2021 to USD 108.57 Billion in 2028 at a CAGR of 5.2% in Forecast Period, 2021–2028. Available online: https://www.fortunebusinessinsights.com/industry-reports/polypropylene-pp-market-101583 (accessed on 11 July 2023).
6. EPA. National Recycling Strategy. Part One of a Series on Building a Circular Economy for all. 2021. U.S. EPA Office of Resource Conservation and Recovery 53f0-R-21-003. Available online: https://www.epa.gov/system/files/documents/2021-11/final-national-recycling-strategy.pdf (accessed on 17 May 2023).
7. PlasticsEurope. PlasticsEurope-CircularityReport-2022_2804. 2022. Available online: https://plasticseurope.org/knowledge-hub/the-circular-economy-for-plastics-a-european-overview-2/ (accessed on 17 May 2023).
8. Cecon, V.S.; Da Silva, P.F.; Curtzwiler, G.W.; Vorst, K.L. The challenges in recycling post-consumer polyolefins for food contact applications: A review. *Resour. Conserv. Recycl.* **2021**, *167*, 105422. [CrossRef]
9. Curtzwiler, G.W.; Schweitzer, M.; Li, Y.; Jiang, S.; Vorst, K.L. Mixed post-consumer recycled polyolefins as a property tuning material for virgin polypropylene. *J. Clean. Prod.* **2019**, *239*, 117978. [CrossRef]
10. Greenpeace Reports. Circular Claims Fall Flat: Comprehensive U.S. SURVEY of Plastics Recyclability. 2022. Available online: Greenpeace.org/usa/plasticrecycling (accessed on 17 May 2023).
11. Cecon, V.S.; Curtzwiler, G.W.; Vorst, K.L. A Study on Recycled Polymers Recovered from Multilayer Plastic Packaging Films by Solvent-Targeted Recovery and Precipitation (STRAP). *Macromol. Mater. Eng.* **2022**, *307*, 2200346. [CrossRef]
12. Cecon, V.S.; Curtzwiler, G.W.; Vorst, K.L. Evaluation of Mixed #3-7 Plastic Waste from Material Recovery Facilities (MRFs) in the United States. *SSRN* **2023**. [CrossRef]
13. Möllnitz, S.; Feuchter, M.; Duretek, I.; Schmidt, G.; Pomberger, R.; Sarc, R. Processability of Different Polymer Fractions Recovered from Mixed Wastes and Determination of Material Properties for Recycling. *Polymers* **2021**, *13*, 457. [CrossRef]
14. The Recycling Partnership; The Association of Plastic Recyclers. *The Growing Market for Recycled Polypropylene*; The Recycling Partnership: Washington, DC, USA, 2020.
15. Recycling Today. The potential of Polypropylene. Available online: https://www.recyclingtoday.com/article/the-recycling-potential-of-polypropylene/ (accessed on 8 May 2023).
16. The Recycling Partnership. *Polypropylene Recycling Coalition*; The Recycling Partnership: Washington, DC, USA, 2022.
17. TPCH. *Toxics in Packaging Clearinghouse—Fact Sheet*; Toxic in Packaging Clearinghouse: Boston, MA, USA, 2021.
18. TPCH. *Model Toxics in Packaging Legislation*; Toxic in Packaging Clearinghouse: Boston, MA, USA, 2022.

19. Sangroniz, A.; Zhu, J.-B.; Tang, X.; Etxeberria, A.; Chen, E.Y.-X.; Sardon, H. Packaging materials with desired mechanical and barrier properties and full chemical recyclability. *Nat. Commun.* **2019**, *10*, 3559. [CrossRef]
20. APR. Polyolefin Standard Laboratory Processing Practices. The Association of Plastics Recyclers. 2020. Available online: https://plasticsrecycling.org/images/Design-Guidance-Tests/APR-O-P-00-olefin-practices.pdf (accessed on 22 June 2023).
21. *ASTM D1238-20*; Standard Test Method for Melt Flow Rates of Thermo- plastics by Extrusion Plastometer. ASTM International: West Conshohocken, PA, USA, 2020. [CrossRef]
22. Cecon, V.S.; Da Silva, P.F.; Vorst, K.L.; Curtzwiler, G.W. The effect of post-consumer recycled polyethylene (PCRPE) on the properties of polyethylene blends of different densities. *Polym. Degrad. Stab.* **2021**, *190*, 109627. [CrossRef]
23. Mort, R.; Cecon, V.S.; Mort, P.; McInturff, K.; Jiang, S.; Vorst, K.; Curtzwiler, G. Sustainable Composites Using Landfill Bound Materials. *Front. Mater.* **2022**, *9*, 849955. [CrossRef]
24. *ASTM D3418-15*; Standard Test Method for Transition Temperatures and Enthalpies of Fusion and Crystallization of Polymers by Differential Scanning Calorimetry. ASTM International: West Conshohocken, PA, USA, 2015. [CrossRef]
25. *ASTM D3895-19*; Standard Test Method for Oxidative-Induction Time of Polyolefins by Differential Scanning Calorimetry. ASTM International: West Conshohocken, PA, USA, 2015. [CrossRef]
26. Kabir, A.S.; Yuan, Z.; Kuboki, T.; Xu, C.C. De-polymerization of industrial lignins to improve the thermo-oxidative stability of polyolefins. *Ind. Crop. Prod.* **2018**, *120*, 238–249. [CrossRef]
27. *ASTM D3985-17*; Standard Test Method for Oxygen Gas Transmission Rate Through Plastic Film and Sheeting Using a Coulometric Sensor. ASTM International: West Conshohocken, PA, USA, 2015. [CrossRef]
28. *ASTM F1249-20*; Standard Test Method for Water Vapor Transmission Rate through Plastic Film and Sheeting Using a Modulated Infrared Sensor. ASTM International: West Conshohocken, PA, USA, 2015. [CrossRef]
29. *ASTM D638-14*; Standard Test Method for Tensile Properties of Plastics. ASTM International: West Conshohocken, PA, USA, 2015. [CrossRef]
30. FDA. 177.1520 Olefin Polymers. 2023. Available online: https://www.accessdata.fda.gov/scripts/cdrh/cfdocs/cfcfr/cfrsearch.cfm?fr=177.1520 (accessed on 17 May 2023).
31. NIST. NIST Standard Reference Database 1A. 2014. Available online: https://www.nist.gov/system/files/documents/srd/NIST1aVer22Man.pdf (accessed on 3 March 2023).
32. Cramer, G.; Ford, R.; Hall, R. Estimation of toxic hazard—A decision tree approach. *Food Cosmet. Toxicol.* **1976**, *16*, 255–276. [CrossRef] [PubMed]
33. Roberts, D.W.; Aptula, A.; Schultz, T.W.; Shen, J.; Api, A.M.; Bhatia, S.; Kromidas, L. A practical guidance for Cramer class determination. *Regul. Toxicol. Pharmacol.* **2015**, *73*, 971–984. [CrossRef]
34. Engül, Ü. Comparing determination methods of detection and quantification limits for aflatoxin analysis in hazelnut. *J. Food Drug Anal.* **2016**, *24*, 56–62. [CrossRef] [PubMed]
35. Martins, M.H.; De Paoli, M.-A. Polypropylene compounding with post-consumer material: II. Reprocessing. *Polym. Degrad. Stab.* **2002**, *78*, 491–495. [CrossRef]
36. Martínez-Jothar, L.; Montes-Zavala, I.; Rivera-García, N.; Díaz-Ceja, Y.; Pérez, E.; Waldo-Mendoza, M.A. Thermal degra-dation of polypropylene reprocessed in a co-rotating twin-screw extruder: Kinetic model and relationship between Melt Flow Index and Molecular weight. *Rev. Mex. Ing. Química* **2021**, *20*, 1079–1090. [CrossRef]
37. Eriksen, M.K.; Christiansen, J.D.; Daugaard, A.E.; Astrup, T.F. Closing the loop for PET, PE and PP waste from house-holds: Influence of material properties and product design for plastic recycling. *Waste Manag.* **2019**, *96*, 75–85. [CrossRef]
38. Socrates, G. *Infrared and Raman Characteristic Group Frequencies: Tables and Charts*; John Wiley & Sons: West Sussex, UK, 2001.
39. Mistry, B.D. *Handbook of Spectroscopic Data: Chemistry*; Oxford Book Company: Oxford, UK, 2009.
40. Akhoundi, B.; Nabipour, M.; Hajami, F.; Shakoori, D. An Experimental Study of Nozzle Temperature and Heat Treatment (Annealing) Effects on Mechanical Properties of High-Temperature Polylactic Acid in Fused Deposition Modeling. *Polym. Eng. Sci.* **2020**, *60*, 979–987. [CrossRef]
41. Bellucci, D.; Cannillo, V. A novel bioactive glass containing strontium and magnesium with ultra-high crystallization temperature. *Mater. Lett.* **2018**, *213*, 67–70. [CrossRef]
42. Singh, A.A.; Genovese, M.E.; Mancini, G.; Marini, L.; Athanassiou, A. Green Processing Route for Polylactic Acid–Cellulose Fiber Biocomposites. *ACS Sustain. Chem. Eng.* **2020**, *8*, 4128–4136. [CrossRef]
43. Yeo, J.C.C.; Muiruri, J.K.; Thitsartarn, W.; Li, Z.; He, C. Recent advances in the development of biodegradable PHB-based toughening materials: Approaches, advantages and applications. *Mater. Sci. Eng. C* **2018**, *92*, 1092–1116. [CrossRef] [PubMed]
44. Zander, N.E.; Gillan, M.; Lambeth, R.H. Recycled polyethylene terephthalate as a new FFF feedstock material. *Addit. Manuf.* **2018**, *21*, 174–182. [CrossRef]
45. Wang, B.; Liu, X.-M.; Fu, X.-N.; Li, Y.-L.; Huang, P.-X.; Zhang, Q.; Li, W.-G.; Ma, L.; Lai, F.; Wang, P.-F. Thermal stability and safety of dimethoxymethane oxidation at low temperature. *Fuel* **2018**, *234*, 207–217. [CrossRef]
46. Selke, S.; Cutler, E.M.; Hernandez, R.J. *Plastics Packaging: Properties, Processing, Applications, and Regulations*, 2nd ed.; Hanser Gardner Publications: Cincinnati, OH, USA, 2004.
47. Perkins, W. Effect of molecular weight and annealing temperature on the oxygen barrier properties of oriented PET film. *Polym. Bull.* **1988**, *19*, 397–401. [CrossRef]

48. Vannini, M.; Marchese, P.; Celli, A.; Lorenzetti, C. Fully biobased poly(propylene 2,5-furandicarboxylate) for packaging applications: Excellent barrier properties as a function of crystallinity. *Green Chem.* **2015**, *17*, 4162–4166. [CrossRef]
49. De Santis, F.; Pantani, R. Optical Properties of Polypropylene upon Recycling. *Sci. World J.* **2013**, *2013*, 354093. [CrossRef]
50. Demets, R.; Van Kets, K.; Huysveld, S.; Dewulf, J.; De Meester, S.; Ragaert, K. Addressing the complex challenge of understanding and quantifying substitutability for recycled plastics. *Resour. Conserv. Recycl.* **2021**, *174*, 105826. [CrossRef]
51. Nerín, C.; Bourdoux, S.; Faust, B.; Gude, T.; Lesueur, C.; Simat, T.; Stoermer, A.; Van Hoek, E.; Oldring, P. Guidance in selecting analytical techniques for identification and quantification of non-intentionally added substances (NIAS) in food contact materials (FCMS). *Food Addit. Contam. Part A* **2022**, *39*, 620–643. [CrossRef]
52. Sander, M.M.; Nicolau, A.; Guzatto, R.; Samios, D. Plasticiser effect of oleic acid polyester on polyethylene and polypropylene. *Polym. Test.* **2012**, *31*, 1077–1082. [CrossRef]
53. Patti, A.; Lecocq, H.; Serghei, A.; Acierno, D.; Cassagnau, P. The universal usefulness of stearic acid as surface modifier: Applications to the polymer formulations and composite processing. *J. Ind. Eng. Chem.* **2021**, *96*, 1–33. [CrossRef]
54. Sivakumar, P. Extraction and Estimation of antistatic agent Glycerol Mono Stearate in Polypropylene by Gas Chromatography Coupled with Flame Ionisation Detector. *Indian J. Chem. (IJC)* **2023**, *62*, 153–157. [CrossRef]
55. Han, J. What Is Butylated Hydroxytoluene (BHT, e321) in Food: Uses, Safety, Side Effects. 2020. Available online: https://foodadditives.net/antioxidant/butylated-hydroxytoluene-bht/ (accessed on 3 March 2023).
56. ECHA. Substance identity. European Chemicals Agency. 2023. Available online: https://www.echa.europa.eu/registration-dossier/-/registered-dossier/15975 (accessed on 3 March 2023).
57. Trivedi, M.K.; Branton, A.; Trivedi, D.; Nayak, G.; Singh, R.; Jana, S. Physical, Thermal and Spectroscopic Characterization of m-Toluic Acid: An Impact of Biofield Treatment. *Biochem. Pharmacol. Open Access* **2015**, *4*, 1000178. [CrossRef]
58. Sari, A.; Karaipekil, A.; Akcay, M.; Onal, A.; Kavak, F. Polymer/palmitic acid blends as shape-stabilized phase change ma-terial for latent heat thermal energy storage. *Asian J. Chem.* **2006**, *18*, 439–446.
59. Perestrelo, R.; Silva, C.L.; Algarra, M.; Câmara, J.S. Evaluation of the Occurrence of Phthalates in Plastic Materials Used in Food Packaging. *Appl. Sci.* **2021**, *11*, 2130. [CrossRef]
60. Hernández-Fernández, J.; Cano-Cuadro, H.; Puello-Polo, E. Emission of Bisphenol A and Four New Analogs from Industrial Wastewater Treatment Plants in the Production Processes of Polypropylene and Polyethylene Terephthalate in South America. *Sustainability* **2022**, *14*, 10919. [CrossRef]
61. Tumu, K.; Vorst, K.; Curtzwiler, G. Endocrine modulating chemicals in food packaging: A review of phthalates and bisphenols. *Compr. Rev. Food Sci. Food Saf.* **2023**, *22*, 1337–1359. [CrossRef]
62. Alp, A.C.; Yerlikaya, P. Phthalate ester migration into food: Effect of packaging material and time. *Eur. Food Res. Technol. Food Packag. Forum* **2020**, *246*, 425–435. [CrossRef]
63. Food-Packaging-Forum. Toxics in Packaging Legislation to Restrict PFAS, Ortho-Phthalates. 2020. Available online: https://www.foodpackagingforum.org/news/toxics-in-packaging-legislation-to-restrict-pfasortho-phthalate (accessed on 17 May 2023).
64. WebMD. What to Know About the Toxicity of Polypropylene. 2021. Available online: https://www.webmd.com/a-to-z-guides/what-to-know-about-the-toxicity-of-polypropylene (accessed on 12 March 2023).

polymers

MDPI

Article

Microstructural and Mechanical Properties of Calcium-Treated Cactus Pear Mucilage (*Opuntia* spp.), Pectin and Alginate Single-Biopolymer Films

Brandon Van Rooyen [1,*], Maryna De Wit [1], Gernot Osthoff [2], Johan Van Niekerk [1] and Arno Hugo [3]

[1] Department of Sustainable Food Systems and Development, University of the Free State, Bloemfontein 9300, South Africa
[2] Department of Microbiology and Biochemistry, University of the Free State, Bloemfontein 9300, South Africa
[3] Department of Animal Science, University of the Free State, Bloemfontein 9300, South Africa
* Correspondence: vanrooyenbb@ufs.ac.za

Abstract: Pectin and alginate satisfy multiple functional requirements in the food industry, especially relating to natural packaging formulation. The continuous need for economic and environmental benefits has promoted sourcing and investigating alternative biomaterials, such as cactus pear mucilage from the cladodes of *Opuntia* spp., as natural packaging alternatives. The structural and mechanical properties of mucilage, pectin and alginate films developed at a 5% (*w*/*w*) concentration were modified by treating the films with calcium (Ca) in the calcium chloride ($CaCl_2$) form. Scanning electron microscopy (SEM) showed the 5% (*w*/*w*) 'Algerian' and 'Morado' films to display considerable microstructure variation compared to the 5% (*w*/*w*) pectin and alginate films, with calcium treatment of the films influencing homogeneity and film orientation. Treating the alginate films with a 10% (*w*/*w*) stock $CaCl_2$ solution significantly increased ($p < 0.05$) the alginate films' tensile strength (TS) and puncture force (PF) values. Consequently, the alginate films reported significantly higher ($p < 0.05$) film strength (TS and PF) than the pectin + Ca and mucilage + Ca films. The mucilage film's elasticity was negatively influenced by $CaCl_2$, while the pectin and alginate films' elasticity was positively influenced by calcium treatment. These results suggest that the overall decreased calcium sensitivity and poor mechanical strength displayed by 'the Algerian' and 'Morado' films would not make them viable replacements for the commercial pectin and alginate films unless alternative applications were found.

Keywords: natural packaging; cactus pear mucilage; pectin; alginate; biopolymer films; cross-linked; mechanical properties; *Opuntia ficus-indica*

check for updates

Citation: Van Rooyen, B.; De Wit, M.; Osthoff, G.; Van Niekerk, J.; Hugo, A. Microstructural and Mechanical Properties of Calcium-Treated Cactus Pear Mucilage (*Opuntia* spp.), Pectin and Alginate Single-Biopolymer Films. *Polymers* **2023**, *15*, 4295. https://doi.org/10.3390/polym15214295

Academic Editors: César Leyva-Porras and Zenaida Saavedra-Leos

Received: 4 August 2023
Revised: 17 October 2023
Accepted: 21 October 2023
Published: 1 November 2023

1. Introduction

The increased demands for naturally biodegradable packaging solutions have resulted in various biopolymers being investigated for their film-forming potential. Biopolymers displaying functional potential, which can be used in the development of biopolymer films, are of growing interest. Given the nature of biopolymer films' functionality and biocompatibility, they have also been considered for a diverse range of applications other than packaging, specifically relating to medical applications such as intervertebral disc replacement and, consequently, bone regeneration [1,2]. Due to their diverse and desirable functional properties, pectin and alginate have specifically found favourable applications in developing biopolymer films. However, factors that influence the formation of these biopolymer films have been identified, such as the addition of a cross-linker [3,4].

The diverse chemical and physical properties, biodegradability and biocompatibility of pectin have resulted in its use in developing biopolymer films [5]. The chemical composition of different pectins can vary, although about 60–65% of the molecule must be composed of galacturonic acid (GalA), which can display varying degrees of methyl esterification [6,7].

The presence of these charged groups associated with the pectin polymer is predominantly responsible for altering the polymers' functional properties in the presence of acidic and basic pH environments and by the introduction of charged ions, such as calcium. Although calcium ions have the ability to alter the rheology of a pectin solution, the main consequence of pectin cross-linking is the formation of hydrogels, specifically harnessed in the development of biopolymer films [3,7,8]. Due to pectins' ability to successfully form biopolymer films, displaying adequate mechanical and barrier properties, multiple authors have investigated the development of pectin biopolymer films for various applications, with and without the addition of a cross-linker [3,9].

Alginate is another polysaccharide used in the development of biopolymer films. The presence of functional groups, consisting of uronic acid, associated with the alginate polymer is of specific importance, as these charged groups directly determine alginate functionality. These functional groups can be modified to alter alginates' rheological, biochemical and film mechanical properties [9–11]. Typically, the functional properties of alginate can be influenced by the presence of cross-linkers, such as calcium and magnesium. Alginate cross-linking is often described by the 'egg-box' model, characteristic of the formation of a three-dimensional (3D) network. This model is similar to that used to describe pectin–calcium cross-linking, although differences can be expected between the two mechanisms [6–8]. It was reported that alginate–calcium films displayed overall increased mechanical properties regarding tensile strength than pectin–calcium films, indicating that alginate displays increased cross-linking ability due to its structural conformation and chemical composition, ultimately influencing biopolymer films' physical properties [12]. Bierhalz et al. [13] reported on differences between pectin and alginate biopolymer film micrographs, with alginate films showing a more homogenous and regular morphology than pectin films. Consequently, alginate films were, in some instances, associated with superior mechanical properties [13].

One of the most, if not the most, important properties of biopolymer films is their mechanical properties (physical strength and elasticity). These properties are essential to ensure the protection and maintenance of the structural integrity of food during transportation, storage and handling [3]. Biopolymer films developed from polysaccharides, such as pectin and alginate, require a drying step to imitate pre-formed plastic packaging. However, the drying of these films is generally always associated with films displaying brittle and even fragile properties. Glycerol has been well established as a plasticizer to reduce brittleness and improve the ease of handling of pectin and alginate biopolymer films [14,15]. Kang et al. [16] investigated the TS and %E of pectin biopolymer films. The authors formed films by immersion of pectin films into 5 and 10% $CaCl_2$ solutions, acting as a cross-linker. Films were prepared with no addition of calcium, considered the control films. The results confirmed that the films formed using 5% $CaCl_2$ showed an increased TS, 198 MPa higher than that of the control films. Furthermore, the 5% $CaCl_2$-cross-linked films showed the lowest elongation at a break potential of 2.6% [16]. Badita et al. [17] investigated the influence of calcium, used as a cross-linker, on 'dry' alginate biopolymer films' properties. The authors found that the alginate–calcium films' properties were considerably influenced by both the cross-linker as well as the concentration of the cross-linker used. It was further confirmed that hydroxyl and carboxylic groups, associated with the alginate polymer chemical structure, were responsible for the hydrogel formation with the addition of calcium, highlighting the benefits of calcium as a cross-linker in biopolymer films formation [17].

Although pectin and alginate have shown success in the development of biopolymer films, consequential high input costs, in addition to the variability and limitations regarding their availability and functionality, have led to the development and investigation of alternative film-forming polymers. The desirable functional properties displayed by the cactus pear mucilage from *Opuntia ficus-indica* have resulted in its investigation as an alternative biopolymeric material in the development of biopolymer films to address certain shortcomings associated with the current polysaccharides films [14,18,19]. Considering the mucilage precipitate from the cactus pear as a functional polymer can prove beneficial because it is often considered an unwanted by-product from cactus pear processing, resulting in favourable cost implications and general ease of availability.

Although the native mucilage precipitate has been well investigated, variations in its chemical structure and composition have been reported. In general, mucilage is considered a highly flexible heteropolysaccharide with a high molecular weight, which has the potential to carry a negative charge due to the presence of galacturonic acid associated with its chemical structure [20–22]. In addition to charged sugars, various amounts of neutral sugars have also been associated with the mucilage precipitate, which includes L-arabinose, D-galactose, L-rhamnose and D-xylose in varying quantities [20–22]. Structurally, the mucilage precipitate has been reported to be composed of a charged linear core with many natural sugar side chains. These two main fractions of the mucilage precipitate have been referred to as a gelling, pectin-like fraction and a more neutral, pure mucilage fraction. However, great variability in sugar composition for both fractions has been reported by authors [23–25].

In a recent study by van Rooyen et al. [26], the authors showed mucilage, when added at 0.25% and 1.0% to pectin-based composite (blended) films, to enhance certain mechanical properties of pectin films. The authors further suggested these enhanced film mechanical properties could directly be linked to the addition of mucilage to the pectin films [26]. Although this before-mentioned study evaluated mucilages' compatibility in combination with commercially available polymers, limited knowledge is available on mucilages' comparative ability to produce single-biopolymer (homopolymeric) films displaying suitable mechanical properties with sights on using them as biodegradable packaging. Logically, comparing single-polymer mucilage films to well-established pectin and alginate films will further advance the understanding of mucilage's potential to be considered as a homopolymeric film-forming polymer, consequently providing insight into the polymer's functional potential. The factors that have been shown to alter well-established polymer films' properties also remain greatly unexplored for single-polymer mucilage-based biopolymer films. Therefore, this study aimed to gain a better understanding of single-polymer (homopolymeric) mucilage films relative to commercially produced single-polymer pectin and alginate films, specifically, to understand the influence a gelling cation (calcium, in the form of calcium chloride) would have on mucilage films' physical properties so to possibly consider them as alternative biopolymers to pectin and alginate in developing biodegradable packaging. Consequently, exploring the physical properties of these single-polymer mucilage films will contribute to the understanding of the native mucilage precipitate, creating a clearer image of its structural functionality and single-polymer film-forming potential.

2. Materials and Methods

2.1. Materials

2.1.1. Commercial Polymers

Pectin powder with $\leq$10% moisture content, sourced from apple and sodium alginate powder with both glucuronic and mannuronic acid content, both supplied by Sigma-Aldrich, Cape Town, South Africa, was used.

2.1.2. Mucilage Precipitate and Freeze-Dried Mucilage Powders

The 'Algerian' and 'Morado' cultivars of *Opuntia ficus-indica* were used to prepare freeze-dried mucilage powders from a liquid mucilage precipitate with $\leq$10% moisture content. Following a well-established extraction procedure, described by Du Toit and De Wit [27], which is cost-effective and easily replicated, native mucilage precipitates were prepared. The resultant mucilage precipitates were well characterized in previous studies [28,29]. We used ~24-month-old cactus pear plant cladodes from the 'Algerian' and 'Morado' cultivars sourced from an orchard (University of the Free State, 29.1076° S, 26.1925° E) with a cactus pear density of 666 cactus pear/h, without irrigation. Firstly, to promote liquid mucilage precipitation, whole cladodes were cubed and microwaved at 900 W for 4 min and then macerated. Secondly, the macerated cladodes were centrifuged using a Beckman® Centrifuge (2315, Brea, CA, USA) at 8000 rpm for 15 min at a constant temperature of 4 °C, which effectively separated the mucilage precipitate from the solids. The liquid mucilage precipitate was then used to prepare the freeze-dried mucilage powders used in this current work. Freeze-drying involved moisture removal from the mucilage precipitate at a low temperature (−30 °C to −40 °C) and constant vacuum until a 95% sample weight loss was obtained. The freeze-dried samples were then milled finely, using a standard mortar and pestle, obtaining a consistent powder with a moisture content of ≤10%.

2.1.3. Cross-Linker and Plasticizer

Granular calcium chloride, anhydrous, purchased from Sigma-Aldrich, Cape Town, South Africa, with excellent water solubility, and glycerol (Merck, Johannesburg, South Africa) as a plasticizer, with a purity > 99% were used.

2.2. Film Preparation and Development

2.2.1. Film-Forming Solutions

Using film-forming methods, as described by van Rooyen et al. [26], all film-forming solutions were prepared. The basic method involved dispersing the desired amounts of polymer powder into distilled water containing glycerol. The pectin and alginate films required a minimum of 60% glycerol (*w*/*w*, based on the polymer weight used in the films). The mucilage films required 20% glycerol addition.

Making use of a magnetic stirrer (Freed Electric-Model MH-4, Rehovot, Israel), all film-forming solutions were allowed to homogenize for 30 min, promoting polymer rehydration at ambient temperature (~25 °C). The rehydrated samples were then further mechanically homogenized by means of a Stick Blender (Mellerwave-Model 85200, Cape Town, South Africa) for 10 s. To reduce the potential volume and structural differences caused by excess entrapped air, the homogenized samples were degassed under vacuum (Genesis Vacuum Sealer, Verimark (Pty) Ltd., Pretoria, South Africa) [26].

2.2.2. Development of Single-Polymer Films

The films used in this study were developed using a standardized batch film casting method with some modifications [5,12,26,30,31]. According to the casting method, the film-forming solution was evenly spread onto a non-stick surface. Upon drying, measurable films formed [5]. Using 140 mm diameter Petri dishes, 70 g of film-forming solution was evenly spread onto the Petri dishes. The dried films were formed by placing the Petri dishes, containing 70 g of film-forming solution, into a ventilated oven (EcoTherm—Model 920, 1000 W, Labotec) at 40 °C for 24 h [26]. Before film evaluation could be completed, all dried films were equilibrated in a closed container at room temperature (~25 °C) for 24 h to achieve ~52% RH.

2.2.3. Calcium Treatment of Single-Polymer Films

The calcium treatment of the pectin and alginate films required a calcium chloride ($CaCl_2$) solution to be directly poured into a polymer solution contained in Petri dishes. After a 5-minute reaction time, the excessive $CaCl_2$ solution was removed from the Petri dishes, with the excess $CaCl_2$ gently being dabbed off the resultant films with a paper towel [26]. The stock $CaCl_2$ solution was prepared at 10% (*w*/*w*) with distilled water; this was considered a more than adequate amount of calcium to react with pectin and alginate [32]. The mucilage films, however, required $CaCl_2$ to be directly mixed into the prepared, fully rehydrated, film-forming solution before casting and drying the films. The $CaCl_2$ addition was calculated at 30% (*w*/*w*, based on the polymer weight used in the films). Espino-Díaz et al. [18] described similar procedures and calcium concentrations for the calcium treatment of mucilage films. Trachtenberg and Mayer [33] also suggested

30% $CaCl_2$ additions to be more than adequate to interact with mucilage. Once the various polymers had been treated with calcium, the films were dried.

2.3. Film Characterization

2.3.1. Scanning Electron Microscopy

All biopolymer films were subjected to scanning electron microscopy (SEM) imaging. The various samples were first mounted on carbon tape and splutter-coated with iridium. Imaging was then carried out on the various samples at 3.0 KV by making use of a JEOL (JSM-7800F Field Emission, Tokyo, Japan) scanning electron microscope.

2.3.2. Film Microstructure Evaluation

The images used in this current work were evaluated and selected by a team of five researchers with experience in microscopy, so to represent the overall microstructures of the different images taken of the various films at 10,000× magnification. Additionally, differences observed by the five researchers regarding the differently treated films' surface roughness/smoothness were also considered by examining visual differences in the surface morphology of the resultant film SEM images.

2.3.3. Mechanical Properties

Tensile and puncture tests were completed to determine the various films' mechanical properties using a Texture Analyzer CT3™ (Brookfield AMETEK®, Westville, South Africa) with similar methods as those described by van Rooyen et al. [26]. All the film mechanical tests were approached and completed by thoroughly consulting the ASTM international standard methods (ASTM-D882; 2010), as also described by Harper [34]. The various films were cut into 25 × 80 mm rectangular strips, which could then be used to determine their mechanical properties. Twelve films were tested per treatment. The different film thicknesses were measured on all the conditioned films before completing the mechanical tests using a digital micrometre (Grip, South Africa).

Film Tensile Test

The Roller Cam Accessory grips (TA-RCA), set at a spacing of 50 mm and a test speed of 0.80 $mm{\cdot}s^{-1}$, measured the films' tensile strength and elongation at break values (Brookfield AMETEK®, Westville, South Africa). Specifically, a film's tensile strength is representative of the maximum stress (force/area) any given film can withstand when a force is applied to it [12]. The film tensile strength is specifically calculated by dividing the maximum load (N) by the initial cross-sectional area of a film and is expressed in MPa [14,18,35]. A second tensile test, that is, the elongation at break percentage of a film, specifically measures a film's maximum capacity to extend before reaching its breaking point [12], calculated by dividing the difference in length of the film at rupture by the initial film length [36].

Puncture Test

The various films' puncture force and distance to puncture were determined by making use of the texture analyzer, accompanied by a Probe TA44 (4 mm diameter probe). A film's puncture force is defined as the maximum force (N) required to tear the film [37]. All puncture tests were completed by selecting a test speed of 0.80 $mm{\cdot}s^{-1}$ on the texture analyzer.

2.4. Experimental Design

The impact $CaCl_2$ had on mucilage, pectin and alginate films' microstructures and mechanical properties was considered. Firstly, using SEM, the various films' surface microstructures were also investigated. Non-calcium-treated and calcium-treated micrographs were investigated for mucilage, pectin and alginate at 10,000× magnifications. Secondly, the influence $CaCl_2$ had on the mucilage, pectin and alginate films' mechanical properties was evaluated. Due to the calcium sensitivity displayed by the pectin and alginate polymers, $CaCl_2$ solutions were prepared and used to cross-link the liquid film-forming pectin and alginate solutions once they had been cast into the Petri dishes. However, Espino-Díaz et al. [18] suggested mucilage's lower sensitivity to calcium requires $CaCl_2$ to be directly mixed into the prepared mucilage film-forming solutions, whereafter film casting, drying, and evaluation could be completed.

2.5. Statistical Analysis

The results of the various trials were captured using Microsoft Excel (2016), and the data were subjected to statistical analysis (ANOVA) using one-way analysis of variance (NCSS Statistical Software package, version 11.0.20). Using the Tukey–Kramer multiple comparison test ($\alpha = 0.05$), significant differences between the treatment means (NCSS Statistical Software package, version 11.0.20) were identified accordingly.

3. Results

3.1. Film Microstructure Characterization

A biopolymer film's microstructure is often directly associated with its resultant mechanical behaviour [38,39]. Investigating scanning electron microscopy (SEM) micrographs were considered to provide a more holistic insight into the homogeneity and microstructures of non-calcium-treated and calcium-treated mucilage, pectin and alginate films, as calcium is specifically known for its ability to physically alter a film network. Firstly, the surface morphology of 'Algerian' and 'Morado' mucilage films was considered, as displayed in Figure 1.

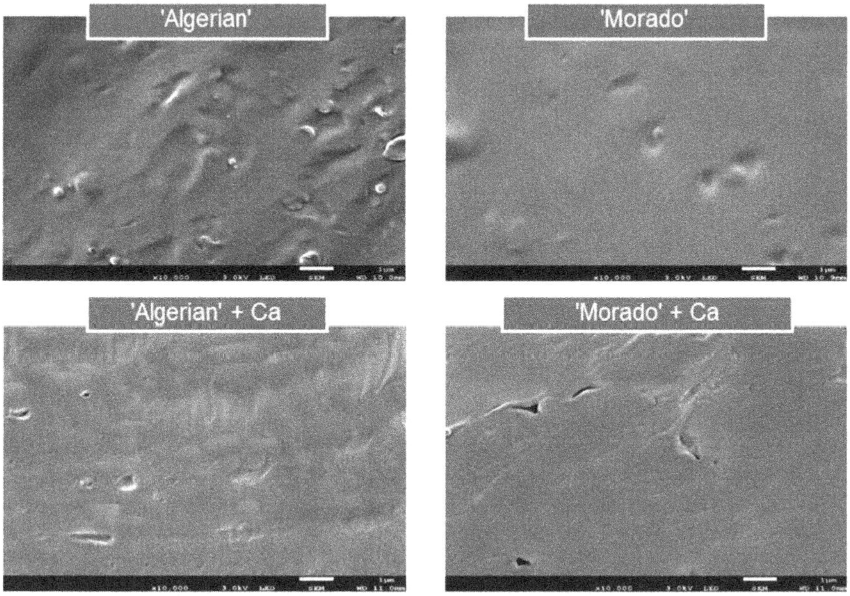

Figure 1. Surface scanning electron microscopy (SEM) images of 'Algerian' and 'Morado' mucilage 'dry' films, with and without calcium (Ca) treatment at 10,000× magnification.

The non-calcium-treated 'Algerian' films showed a lack of homogeneity, with a surface roughness, compared to the calcium-treated 'Algerian' films. The non-calcium-treated 'Morado' films showed a smoother, more homogeneous surface morphology than the non-calcium-treated 'Algerian' films. However, some cracks and breaks in film homogeneity were observed when treating the 'Morado' films with calcium. Guadarrama-Lezama et al. [40] suggested that the appearance of cracks and holes observed in a film microstructure could indicate a denser film formation. The breaks in structure observed in the calcium-treated 'Morado' films could thus be related to films of increased density compared to calcium-treated 'Algerian' films. Furthermore, treating the mucilage films with calcium showed indications of the development of a more organized mucilage film network (Figure 1). Authors suggested that the presence of a pectin-like fraction that displays sensitivity to calcium could be the reason for the development of these more organized film networks when the mucilage films were treated with calcium [19,23].

The SEM imaging of the surface morphology of the pectin and alginate films, with and without calcium treatment, is presented in Figure 2.

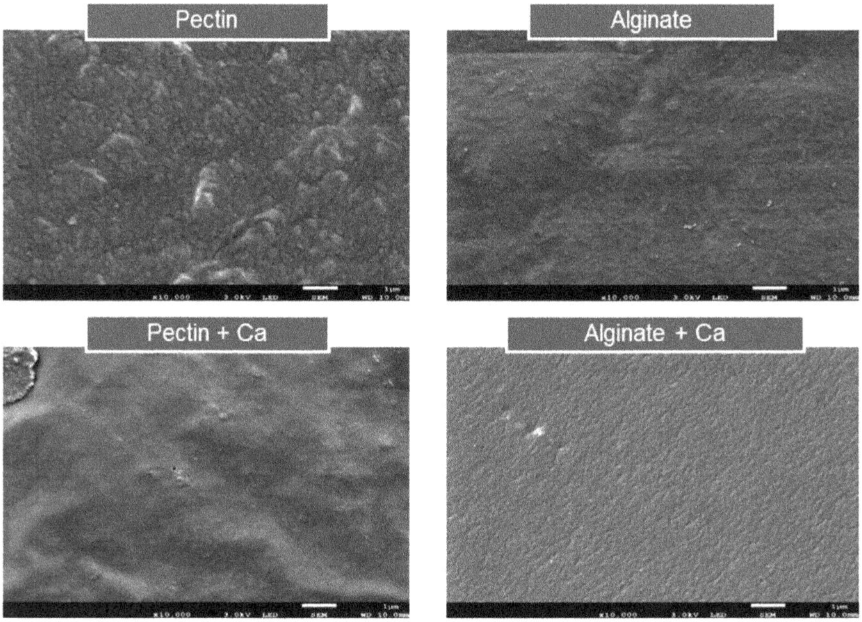

Figure 2. SEM imaging of the surface microstructures of 'dry' pectin and alginate films, with and without calcium (Ca) treatments at 10,000× magnification.

Less homogeneous and more rough surfaces were observed in both the non-calcium-treated pectin and the alginate films' microstructures compared to their calcium-treated counterparts (Figure 2). Specifically, the calcium treatment was shown to change the surface morphology of the pectin and alginate films noticeably. However, the calcium-treated alginate films showed homogeneity, represented by a more uniform microstructure than the calcium-treated pectin films. This observation indicates the different internal arrangements of the different polymers during film formation, with alginates displaying a more organized network than pectin films [41].

Paşcalau et al. [42] also reported that alginate films treated with calcium resulted in the formation of a more homogeneous film network, typically expected for films that underwent cross-linking (which was not observed for the uncross-linked alginate films), similar to that observed in the current research for uncross-linked and calcium cross-linked alginate films (Figure 2). As observed in the current research, the pectin and alginate films' microstructures were, therefore, similar to those observed in the literature, with alginate films forming a more organized, homogeneous film network when treated with calcium than pectin films (Figure 2).

The results proved that treating the mucilage, pectin and alginate films with calcium led to noticeable morphological changes in the films' structures. Microstructure differences were also clearly observed between the mucilage, pectin and alginate films. More homogeneous and organized film networks were found in both pectin and alginate films' microstructures. In contrast, the mucilage films appeared to be more non-homogeneous and irregular and sometimes contained aggregates and pores.

3.2. *Film Mechanical Properties*

3.2.1. The Influence of Calcium on Film Thickness

The influence calcium treatments had on mucilage, pectin and alginate films' thickness is presented in Figure 3. Most films displayed only minimal differences in film thickness. Similar trends in film thickness were reported for films treated with calcium developed at a 5% (w/w) polymer concentration [26]. However, the 'Algerian' + Ca films were significantly thicker ($p < 0.05$) than the 'Morado' + Ca films. A decrease in film thickness with the addition of calcium could be related to a higher degree of cross-linking observed in the films due to a possible decrease in inter-chain polymer spacing [9,13].

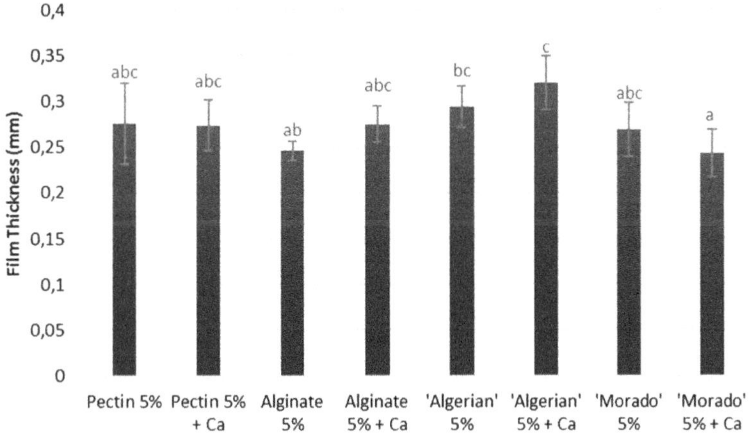

Figure 3. Thickness of the pectin, alginate and mucilage films with and without calcium (Ca) treatment. The mean values of 8 treatments are displayed together with their standard deviations. Error bars represented by differet superscripts differ significantly ($p < 0.05$).

3.2.2. Tensile Test

Further differences were observed between the mechanical properties of the mucilage, pectin and alginate films treated with calcium, as indicated in Table 1. When evaluating the influence the calcium treatment had on the various films' tensile strength (TS), it was seen that only the alginate film's TS values were significantly increased ($p < 0.05$) (Table 1). Bierhalz et al. [13] also found alginate films to display superior TS due to the alginate's linear polymer chains, allowing for a more efficient cross-linking with calcium compared to the cross-linking of the pectin polymer chains. Additionally, alginate also produced films displaying significantly higher ($p < 0.05$) TS values in comparison to the various other films investigated. These findings are also supported by the literature, as authors found alginate to produce films of superior strength due to a more efficient calcium cross-linking of the polymer [12,43,44]. Although calcium-treating the mucilage films showed to have no significant ($p > 0.05$) influence on the film TS, the 'Morado' + Ca films did show trends of increased strength, as the TS value increased from 0.31 (TS of the non-calcium-treated 'Morado' film) to 1.01 MPa. These findings indicated a slight potential of the 'Morado' mucilage to cross-link with calcium.

Table 1. Tensile test mechanical properties of non-calcium-treated and calcium-treated mucilage, pectin and alginate films.

Treatments/Films	TensileStrength (MPa)	Elongation at Break %
Pectin 5%	6.41 ± 0.50 ^{b}	14.31 ± 1.88 ^{cd}
Pectin 5% + Ca	7.01 ± 0.61 ^{b}	20.46 ± 2.76 ^{ef}
Alginate 5%	17.57 ± 0.90 ^{c}	7.79 ± 1.03 ^{ab}
Alginate 5% + Ca	20.10 ± 1.07 ^{d}	15.68 ± 0.60 ^{de}
'Algerian' 5%	0.26 ± 0.05 ^{a}	33.10 ± 6.10 ^{g}
'Algerian' 5% + Ca	0.37 ± 0.10 ^{a}	4.98 ± 0.67 ^{a}
'Morado' 5%	0.31 ± 0.10 ^{a}	21.58 ± 1.76 ^{f}
'Morado' 5% + Ca	1.01 ± 0.10 ^{a}	10.41 ± 0.70 ^{bc}
Significance level	$p < 0.005$	$p < 0.005$

The mean values of 8 treatments, together with their standard deviations (±), are presented. The mean values with different superscripts in the same column differ significantly ($p < 0.05$).

The film %E showed the calcium treatments to have a considerable and varying influence on film elasticity. The calcium treatments significantly increased ($p < 0.05$) both pectin and alginate films' %E values whilst significantly decreasing ($p < 0.05$) the 'Algerian' and 'Morado' mucilage films' elasticity. It has come to be expected that the addition of calcium would aid in increasing the cohesion forces between the polymer chains of alginate and pectin, increasing film strength and elasticity [13,43]. As regards the decreases in the mucilage film %E values observed in Table 1, research has suggested that, due to the low occurrence of carboxyl groups in the native mucilage structure, the addition of calcium could induce polymer chain contraction or the ionic condensation of the mucilage polymers rather than cross-link the polymer chains, as adequate binding sites for calcium would not be available, in turn reducing the film's cohesion forces and thus its elasticity [18,23]. The 'Morado' + Ca film showed significantly higher ($p < 0.05$) %E values when compared to the 'Algerian' + Ca film.

3.2.3. Puncture Test

Similar observations were reported when investigating the various films' puncture force (PF) values compared to those observed for the films' TS (Table 2). The calcium treatment of the alginate films significantly increased ($p < 0.05$) the films' PF values, with the Alginate + Ca films showing the overall highest strength compared to the various other films investigated (Table 2). Again, both 'Algerian' and 'Morado' mucilage films showed to be only minimally influenced by calcium treatment, producing films with the lowest PF. Interestingly, the puncture tests indicated that the Alginate + Ca films had the highest DTP values compared to the other films. The mucilage films showed the lowest DTP values, with calcium treatment of the films further significantly lowering ($p < 0.05$) them. These trends were similar to those reported for the tensile test %*E* evaluation.

Table 2. Puncture test mechanical properties of non-calcium-treated and calcium-treated mucilage, pectin and alginate films.

Treatments/Films	Puncture Force (N)	Distance to Puncture (mm)
Pectin 5%	31.75 ± 2.38 ^{b}	4.04 ± 0.38 ^{b}
Pectin 5% + Ca	35.94 ± 4.32 ^{b}	4.60 ± 0.38 ^{b}
Alginate 5%	72.17 ± 4.68 ^{c}	5.61 ± 0.37 ^{cd}
Alginate 5% + Ca	83.30 ± 5.81 ^{d}	6.26 ± 0.54 ^{d}
'Algerian' 5%	2.43 ± 0.26 ^{a}	4.24 ± 0.63 ^{b}
'Algerian' 5% + Ca	2.38 ± 0.39 ^{a}	1.89 ± 0.33 ^{a}
'Morado' 5%	1.82 ± 0.21 ^{a}	4.71 ± 0.89 ^{bc}
'Morado' 5% + Ca	5.43 ± 0.58 ^{a}	2.60 ± 0.47 ^{a}
Significance level	$p < 0.005$	$p < 0.005$

The mean values of 8 treatments, together with their standard deviations (±), are presented. The mean values with different superscripts in the same column differ significantly ($p < 0.05$).

4. Discussion

When considering the scanning electron microscopy (SEM) imaging micrographs (Figures 1 and 2) and the mechanical properties (Tables 1 and 2) of the various films investigated, treating the alginate films with calcium resulted in producing films displaying superior strength compared to the pectin and mucilage films. Guadarrama-Lezama et al. [40] suggested that films with smoother surfaces, lacking pores in the film network, would display superior mechanical properties. Specifically, the calcium-treated alginate films displayed the greatest homogeneity among all the various films investigated. The Alginate + Ca films' microstructures showed a highly 'organized' film network, compared to the calcium-treated pectin films, which were characterized by a less homogeneous surface morphology with visible rough surfaces. Bierhalz et al. [13] also reported that 'dry' alginate films' microstructures presented a more homogenous and regular network than those of 'dry' pectin films treated with calcium.

Furthermore, the authors also suggested that alginate's more linear structure would allow for a more efficient calcium cross-linking than that present in pectin polymers, which consequently would also influence films' mechanical properties [13]. Therefore, the Pectin + Ca films reported significantly lower ($p < 0.05$) tensile strength (TS) and puncture force (PF) values in comparison to the Alginate + Ca films. Differences in the chemical structures between pectin and alginate have been strongly suggested to be the main reason for the morphological and mechanical differences observed between these two commercially available polymers. As the alginate polymer's linear chains allow for a more efficient and organized cross-linking, the pectin polymer generally results in a more random orientation with calcium due to its branched chemical structure. Additionally, varying degrees of esterification of the pectin polymer will influence its ability to react with calcium [33–35].

When comparing the influence calcium had on the mucilage films, it was seen that both 'Algerian' and 'Morado' films showed the poorest mechanical properties when compared to the commercially available pectin and alginate films. Authors also reported on calcium-treated mucilage films with low TS and %E values [18]. The lack of homogeneity and the irregular nature of the 'Algerian' and 'Morado' films must be strongly considered, as these morphological features have been thought to account for films displaying reduced mechanical properties. The expected highly branched chemical structure of mucilage has also been suggested to be strongly associated with the lack of homogeneity and the irregular appearance of mucilage films [19,40]. Furthermore, variations in the surface morphology and mechanical properties were also observed between the two mucilage cultivars. The 'Morado' films showed the possibility of increased calcium sensitivity, resulting in films displaying trends of increased strength and elasticity when compared to the 'Algerian' films treated with calcium. These findings were supported by examining the microstructural variations between the 'Algerian' and the 'Morado' films, with less rough film surfaces observed for the 'Morado' films.

Lastly, the 'Algerian' + Ca films presented a greater film thickness compared to the 'Morado' + Ca films. These findings strongly suggest that the 'Algerian' films, being rougher, denser and thicker, would display decreased strength and elasticity compared to the 'Morado' mucilage films. Research has suggested that increased film thickness and roughness and decreased film homogeneity could strongly indicate reduced polymer chain interaction in a film matrix, resulting in films displaying reduced strength and elasticity [9,13].

5. Conclusions

With global efforts aimed at reducing single-use plastics, the need to develop biopolymer packaging from sustainable sources is being strongly considered. Mucilage, pectin and alginate films displayed varying degrees of mechanical alteration by adding calcium. Although the alginate films showed superior strength, the mucilage films had exceptional elasticity. Cross-linking the various films with calcium showed to enhance further only the alginate films' tensile strength and puncture strength whilst having only a minimal influence on the pectin and mucilage films' strength. The superior homogeneity observed in the scanning electron microscopy (SEM) images of the alginate films was clearly identifiable compared to the less organized pectin and mucilage films' microstructures. The pectin and alginate films' elasticity was further enhanced by treating the films with calcium, with the mucilage film elongation at break percentage and distance to puncture values negatively influenced by calcium. The various polymers' diverse interactions with calcium accounted for possible structural variations between the polymers. However, the 'Algerian' mucilage polymer showed the likeliness to carry less of a charge than the 'Morado' mucilage, showing an overall reduced functionality in this work. Regardless of the mucilage cultivar, the results indicated that the mucilage films were much weaker overall than the pectin and alginate films, although they had, in most instances, superior elasticity. The native mucilage films display commercial viability. Therefore, evaluating the water and gas barrier, crystallization, and anti-fogging properties of these films would be a logical next step in a phased approach to establish a holistic view of cactus pear mucilage to be used as an alternative packaging solution. Regardless thereof, the mechanical properties of these films suggest their success as an alternative, application-specific packaging, likely not finding similar applications as alginate and pectin films.

Author Contributions: Conceptualization, B.V.R. and M.D.W.; methodology, B.V.R., M.D.W. and G.O.; formal analysis, A.H., M.D.W. and G.O.; investigation, B.V.R.; resources, M.D.W. and J.V.N.; data curation, B.V.R., A.H., G.O. and M.D.W.; writing—original draft preparation, B.V.R.; writing—review and editing, B.V.R. and J.V.N.; supervision, M.D.W., G.O., A.H. and J.V.N.; project administration, B.V.R., M.D.W. and J.V.N. All authors have read and agreed to the published version of the manuscript.

Funding: This research received no external funding.

Institutional Review Board Statement: Not applicable.

Informed Consent Statement: Not applicable.

Data Availability Statement: Not applicable.

Acknowledgments: The authors wish to thank Herman Fouché for his assistance with the harvesting of the raw cactus pear cladode material.

Conflicts of Interest: The authors declare no conflict of interest.

References

1. Zhang, Y.; Gao, H.; Luo, H.; Chen, D.; Zhou, Z.; Cao, X. High Strength HA-PEG/NAGA-Gelma Double Network Hydrogel for Annulus Fibrosus Rupture Repair. *Smart Mater. Med.* **2022**, *3*, 128–138. [CrossRef]
2. Huang, C.; Ye, Q.; Dong, J.; Li, L.; Wang, M.; Zhang, Y.; Zhang, Y.; Wang, X.; Wang, P.; Jiang, Q. Biofabrication of Natural Au/Bacterial Cellulose Hydrogel for Bone Tissue Regeneration via in-Situ Fermentation. *Smart Mater. Med.* **2023**, *4*, 1–14. [CrossRef]
3. Otoni, C.G.; Avena-Bustillos, R.J.; Azeredo, H.M.C.; Lorevice, M.V.; Moura, M.R.; Mattoso, L.H.C.; McHugh, T.H. Recent Advances on Edible Films Based on Fruits and Vegetables—A Review. *Compr. Rev. Food Sci. Food Saf.* **2017**, *16*, 1151–1169. [CrossRef] [PubMed]
4. Gawkowska, D.; Ciesla, J.; Zdunek, A.; Cybulska, J. The Effect of Concentration on the Cross-Linking and Gelling of Sodium Carbonate-Soluble Apple Pectins. *Molecules* **2019**, *24*, 1635. [CrossRef] [PubMed]
5. Espitia, P.J.P.; Du, W.-X.; de Avena-Bustillos, R.J.; Soares, N.; de Fátima Ferreira Soares, N.; McHugh, T.H. Edible Films from Pectin: Physical-Mechanical and Antimicrobial Properties—A Review. *Food Hydrocoll.* **2014**, *35*, 287–296. [CrossRef]
6. Ridley, B.L.; O'Neill, M.A.; Mohnen, D. Pectins: Structure, Biosynthesis, and Oligogalacturonide-Related Signaling. *Phytochemistry* **2001**, *57*, 929–967. [CrossRef]
7. Willats, W.G.T.; Knox, J.P.; Mikkelsen, J.D. Pectin: New Insights into an Old Polymer Are Starting to Gel. *Trends Food Sci. Technol.* **2006**, *17*, 97–104. [CrossRef]
8. Gawkowska, D.; Cybulska, J.; Zdunek, A. Structure-Related Gelling of Pectins and Linking with Other Natural Compounds: A Review. *Polymers* **2018**, *10*, 762. [CrossRef]
9. Yang, J.S.; Xie, Y.J.; He, W. Research Progress on Chemical Modification of Alginate: A Review. *Carbohydr. Polym.* **2011**, *84*, 33–39. [CrossRef]
10. Bouhadir, K.H.; Lee, K.Y.; Alsberg, E.; Damm, K.L.; Anderson, K.W.; Mooney, D.J. Degradation of Partially Oxidized Alginate and Its Potential Application for Tissue Engineering. *Biotechnol. Prog.* **2001**, *17*, 945–950. [CrossRef]

11. Larsen, B.; Salem, D.M.S.A.; Sallam, M.A.E.; Mishrikey, M.M.; Beltagy, A.I. Characterization of the Alginates from Algae Harvested at the Egyptian Red Sea Coast. *Carbohydr. Res.* **2003**, *338*, 2325–2336. [CrossRef] [PubMed]
12. Da Silva, M.A.; Bierhalz, A.C.K.; Kieckbusch, T.G. Alginate and Pectin Composite Films Crosslinked with Ca2+ Ions: Effect of the Plasticizer Concentration. *Carbohydr. Polym.* **2009**, *77*, 736–742. [CrossRef]
13. Bierhalz, A.C.K.; Silva, M.A.D.; Kieckbusch, T.G. Natamycin Release from Alginate/Pectin Films for Food Packaging Applications. *J. Food Eng.* **2012**, *110*, 18–25. [CrossRef]
14. Gheribi, R.; Puchot, L.; Verge, P.; Jaoued-Grayaa, N.; Mezni, M.; Habibi, Y.; Khwaldia, K. Development of Plasticized Edible Films from Opuntia Ficus-Indica Mucilage: A Comparative Study of Various Polyol Plasticizers. *Carbohydr. Polym.* **2018**, *190*, 204–211. [CrossRef]
15. Zibaei, R.; Hasanvand, S.; Hashami, Z.; Roshandel, Z.; Rouhi, M.; de Guimarães, J.T.; Mortazavian, A.M.; Sarlak, Z.; Mohammadi, R. Applications of Emerging Botanical Hydrocolloids for Edible Films: A Review. *Carbohydr. Polym.* **2021**, *256*, 117554. [CrossRef]
16. Kang, H.J.; Jo, C.; Lee, N.Y.; Kwon, J.H.; Byun, M.W. A Combination of Gamma Irradiation and CaCl2 Immersion for a Pectin-Based Biodegradable Film. *Carbohydr. Polym.* **2005**, *60*, 547–551. [CrossRef]
17. Badita, C.R.; Aranghel, D.; Burducea, C.; Mereuta, P.; Engineering, N. Characterization of Sodium Alginate based Films. *Rom. J. Phys.* **2019**, *602*, 1–8.
18. Espino-Díaz, M.; Ornelas-Paz, J.D.J.; Martínez-Téllez, M.A.; Santillán, C.; Barbosa-Cánovas, G.V.; Zamudio-Flores, P.B.; Olivas, G.I. Development and Characterization of Edible Films Based on Mucilage of *Opuntia ficus-indica* (L.). *J. Food Sci.* **2010**, *75*, 347–352. [CrossRef]
19. Lira-Vargas, A.A.; Corrales-Garcia, J.J.E.; Valle-Guadarrama, S.; Peña-Valdivia, C.B.; Trejo-Marquez, M.A. Biopolymeric Films Based on Cactus (Opuntia Ficus-Indica) Mucilage Incorporated with Gelatin and Beeswax. *J. Prof. Assoc. Cactus Dev.* **2014**, *16*, 51–70.
20. Majdoub, H.; Roudesli, S.; Picton, L.; Cerf, D.L.; Muller, G.; Grisel, M. Prickly Pear Nopals Pectin from Opuntia Ficus-Indica Physico-Chemical Study in Dilute and Semi-Dilute Solutions. *Carbohydr. Polym.* **2001**, *46*, 69–79. [CrossRef]
21. Sáenz, C.; Sepúlveda, E.; Matsuhiro, B. *Opuntia* Spp. Mucilage's: A Functional Component with Industrial Perspectives. *J. Arid Environ.* **2004**, *57*, 275–290. [CrossRef]
22. Sepúlveda, E.; Sáenz, C.; Aliaga, E.; Aceituno, C. Extraction and Characterization of Mucilage in *Opuntia* Spp. *J. Arid Environ.* **2007**, *68*, 534–545. [CrossRef]
23. Goycoolea, F.M.; Cárdenas, A. Pectins from *Opuntia Spp.*: A Short Review. *J. Prof. Assoc. Cactus Dev.* **2003**, *5*, 17–29.
24. Cárdenas, A.; Goycoolea, F.M.; Rinaudo, M. On the Gelling Behaviour of "nopal" (Opuntia Ficus Indica) Low Methoxyl Pectin. *Carbohydr. Polym.* **2008**, *73*, 212–222. [CrossRef]
25. Rodríguez-González, S.; Martínez-Flores, H.E.; Chávez-Moreno, C.K.; Macías-Rodríguez, L.I.; Zavala-Mendoza, E.; Garnica-Romo, M.G.; Chacõn-García, L. Extraction and Characterization of Mucilage from Wild Species of Opuntia. *J. Food Process Eng.* **2014**, *37*, 285–292. [CrossRef]
26. van Rooyen, B.; de Wit, M.; Osthoff, G.; Van Niekerk, J.; Hugo, A. Effect of Native Mucilage on the Mechanical Properties of Pectin-Based and Alginate-Based Polymeric Films. *Coatings* **2023**, *13*, 1611. [CrossRef]
27. Du Toit, A.; De Wit, M. Patent PA153178P A Process for Extracting Mucilage from Opuntia Ficus-Indica, Aloe Barbadensis and Agave Americana. Ph.D. Thesis, University of the Free State, Bloemfontein, South Africa, 2021. [CrossRef]
28. Du Toit, A. Selection, Extraction, Characterization and Application of Mucilage from Cactus Pear (Opuntia Ficus-Indica and Opuntia Robusta) Cladodes. Ph.D. Thesis, University of the Free State, Bloemfontein, South Africa, August 2016; pp. 1–13.
29. Du Toit, A.; De Wit, M.; Fouché, H.J.; Taljaard, M.; Venter, S.L.; Hugo, A. Mucilage Powder from Cactus Pears as Functional Ingredient: Influence of Cultivar and Harvest Month on the Physicochemical and Technological Properties. *J. Food Sci. Technol.* **2019**, *56*, 2404–2416. [CrossRef]
30. Campos, C.; Gerschenson, L.; Flores, S. Development of Edible Films and Coatings with Antimicrobial Activity. *Food Bioprocess Technol.* **2010**, *4*, 849–875. [CrossRef]
31. Du, W.X.; Avena-Bustillos, R.J.; Sheng, S.; Hua, T.; McHugh, T.H. Antimicrobial Volatile Essential Oils in Edible Films for Food Safety. *Sci. Against Microb. Pathog. Commun. Curr. Res. Technol. Adv.* **2011**, *2*, 1124–1134.
32. Fazilah, A.; Maizura, M.; Karim, A.A.; Bhupinder, K.; Rajeev, B.; Uthumporn, U.; Chew, S.H. Physical and Mechanical Properties of Sago Starch—Alginate Films Incorporated with Calcium Chloride. *Int. Food Res. J.* **2011**, *18*, 1027–1033.
33. Trachtenberg, S.; Mayer, A.M. Mucilage Cells, Calcium Oxalate Crystals and Soluble Calcium in Opuntia Ficus-Indica. *Ann. Bot.* **1982**, *50*, 549–557. [CrossRef]
34. Harper, B.A. *Understanding Interactions in Wet Alginate Film Formation Used for In-Line Food Processes*; Doctor of Philosophy in Food Science, The University of Guelph: Guelph, ON, Canada, 2013.
35. Xu, Y.X.; Kim, K.M.; Hanna, M.A.; Nag, D. Chitosan-Starch Composite Film: Preparation and Characterization. *Ind. Crops Prod.* **2005**, *21*, 185–192. [CrossRef]
36. Rhim, J.W. Physical and Mechanical Properties of Water Resistant Sodium Alginate Films. *LWT—Food Sci. Technol.* **2004**, *37*, 323–330. [CrossRef]
37. Wang, L.Z.; Liu, L.; Holmes, J.; Kerry, J.F.; Kerry, J.P. Assessment of Film-Forming Potential and Properties of Protein and Polysaccharide-Based Biopolymer Films. *Int. J. Food Sci. Technol.* **2007**, *42*, 1128–1138. [CrossRef]

38. Gheribi, R.; Gharbi, M.A.; Ouni, M.E.; Khwaldia, K. Enhancement of the Physical, Mechanical and Thermal Properties of Cactus Mucilage Films by Blending with Polyvinyl Alcohol. *Food Packag. Shelf Life* **2019**, *22*, 100386. [CrossRef]
39. Gheribi, R.; Habibi, Y.; Khwaldia, K. Prickly Pear Peels as a Valuable Resource of Added-Value Polysaccharide: Study of Structural, Functional and Film Forming Properties. *Int. J. Biol. Macromol.* **2019**, *126*, 238–245. [CrossRef]
40. Guadarrama-Lezama, A.Y.; Castaño, J.; Velázquez, G.; Carrillo-Navas, H.; Alvarez-Ramírez, J. Effect of Nopal Mucilage Addition on Physical, Barrier and Mechanical Properties of Citric Pectin-Based Films. *J. Food Sci. Technol.* **2018**, *55*, 3739–3748. [CrossRef] [PubMed]
41. Galus, S.; Lenart, A. Development and Characterization of Composite Edible Films Based on Sodium Alginate and Pectin. *J. Food Eng.* **2013**, *115*, 459–465. [CrossRef]
42. Paşcalau, V.; Popescu, V.; Popescu, G.L.; Dudescu, M.C.; Borodi, G.; Dinescu, A.; Perhaiţa, I.; Paul, M. The Alginate/k-Carrageenan Ratio's Influence on the Properties of the Cross-Linked Composite Films. *J. Alloys Compd.* **2012**, *536* (Suppl. 1), 418–423. [CrossRef]
43. Fang, Y.; Al-Assaf, S.; Phillips, G.O.; Nishinari, K.; Funami, T.; Williams, P.A. Binding Behavior of Calcium to Polyuronates: Comparison of Pectin with Alginate. *Carbohydr. Polym.* **2008**, *72*, 334–341. [CrossRef]
44. Sriamornsak, P.; Kennedy, R.A. Swelling and Diffusion Studies of Calcium Polysaccharide Gels Intended for Film Coating. *Int. J. Pharm.* **2008**, *358*, 205–213. [CrossRef] [PubMed]

polymers

MDPI

Article

Effect of pH on the Mechanical Properties of Single-Biopolymer Mucilage (*Opuntia ficus-indica*), Pectin and Alginate Films: Development and Mechanical Characterisation

Brandon Van Rooyen [1,*], Maryna De Wit [1], Gernot Osthoff [2], Johan Van Niekerk [1] and Arno Hugo [3]

1 Department of Sustainable Food Systems and Development, University of the Free State, Bloemfontein 9301, South Africa
2 Department of Microbiology and Biochemistry, University of the Free State, Bloemfontein 9301, South Africa
3 Department of Animal Science, University of the Free State, Bloemfontein 9301, South Africa
* Correspondence: vanrooyenbb@ufs.ac.za

Abstract: Pectin and alginate are well-established biopolymers used in natural film development. Single-polymer mucilage films were developed from freeze-dried native mucilage powder of two cultivars, 'Algerian' and 'Morado', and the films' mechanical properties were compared to single-polymer pectin and alginate films developed from commercially available pectin and alginate powders. The casting method prepared films forming solutions at 2.5%, 5%, and 7.5% (*w*/*w*) for each polymer. Considerable variations were observed in the films' strength and elasticity between the various films at different polymer concentrations. Although mucilage films could be produced at 5% (*w*/*w*), both cultivars could not produce films with a tensile strength (TS) greater than 1 MPa. Mucilage films, however, displayed > 20% elongation at break (%*E*) values, being noticeably more elastic than the pectin and alginate films. The mechanical properties of the various films were further modified by varying the pH of the film-forming solution. The various films showed increased TS and puncture force (PF) values, although these increases were more noticeable for pectin and alginate than mucilage films. Although single-polymer mucilage films exhibit the potential to be used in developing natural packaging, pectin and alginate films possess more suitable mechanical attributes.

Keywords: packaging; cactus pear mucilage; *Opuntia ficus-indica*; pectin; alginate; biopolymer films; cross-linked; mechanical properties; pH; polymer concentration

Citation: Van Rooyen, B.; De Wit, M.; Osthoff, G.; Van Niekerk, J.; Hugo, A. Effect of pH on the Mechanical Properties of Single-Biopolymer Mucilage (*Opuntia ficus-indica*), Pectin and Alginate Films: Development and Mechanical Characterisation. *Polymers* **2023**, *15*, 4640. https://doi.org/10.3390/polym15244640

Academic Editors: Zenaida Saavedra-Leos and César Leyva-Porras

Received: 4 August 2023
Revised: 28 November 2023
Accepted: 5 December 2023
Published: 7 December 2023

1. Introduction

The development of biopolymer films, making use of naturally biodegradable materials, has been gaining attention in the food industry. The ever-increasing environmental and human health concerns around the use of petroleum-based plastic packaging have been a driving force surrounding recent developments in this field [1]. Specifically, functional properties associated with the pectin and alginate polysaccharides have been explored in the formulation of natural packaging [2,3]. Pectin and alginate are considered anionic polymers, displaying varying degrees of charge. The chemical structures of pectin and alginate are well-documented and accepted, with their structures directly influencing their functionality. Structural differences between pectin and alginate films have resulted in the different polymer films displaying variations between their mechanical properties, specifically relating to their film strength and elasticity [4–8]. In order to provide adequate protection for food, evaluating the mechanical properties of films is considered essential to ensure the structural integrity of potential food packaging. The mechanical properties of biopolymer films have further been shown to be influenced by several factors, ultimately determining their applications [4,9].

Polymer concentration, changes in pH, and the addition of cross-linking agents have specifically been shown to influence biopolymer films' physical and mechanical properties.

Additionally, film preparation methods and the addition of plasticizers also play a role in determining the specific physical properties displayed by these films. Adding glycerol or sorbitol as a plasticizer can be considered essential to biopolymer film development to ensure films presented by uniform microstructures exhibit desirable physical properties [10–13].

Although pectin and alginate polymers are well-known for their film-forming abilities, which meet multiple food packaging requirements, the continuous need for natural packaging possessing favorable economic and environmental benefits has promoted the sourcing and investigation of alternative biomaterials. Of specific interest are biopolymers produced in large surplus quantities, which are easily accessible and have relatively low-cost implications, such as cactus pear mucilage precipitate from the cladodes of *Opuntia ficus-indica* [11,14,15]. The functional potential of native mucilage precipitate has allowed for its investigation in developing biopolymer films, although limited research is available thereon [16–18].

While native mucilage precipitate has been well-investigated from a molecular level, diversity in its composition has been reported. Generally, cactus pear mucilage is considered a highly flexible heteropolysaccharide with a high molecular weight. Multiple researchers have studied the chemical composition of *Opuntia* spp. mucilage precipitates, often reporting considerable variations thereof [19–21]. Although variations in sugar composition and consequential molecular weight have been reported, mucilage molecules are considered to be quite large. This long-chain polymer is represented by varying proportions of D-galacturonic acid and neutral sugars of L-arabinose, D-galactose, L-rhamnose, and D-xylose. The primary mucilage structure is said to be comprised of a linear core chain of repeating (1→4) D-galacturonic acid and (1→2) L-rhamnose [20,21].

Additionally, the mucilage polymer is associated with many side chains of D-galactose attached to the L-rhamnose residues. These complex peripheral chains have shown the potential to be composed of various types of sugars. The galactose side chains can further branch into either arabinose or xylose. Native mucilage chemical composition can vary considerably, depending on the efficacy and type of extraction method used [20–24].

Different fractions present within the native mucilage precipitate have been further investigated. Generally, a charged, pectin-like fraction and a neutral fraction, considered the purer mucilage fraction, represent native mucilage precipitate. Multiple similarities are observed regarding the chemical composition between these two fractions [21]. In general, these two fractions of native mucilage extract display similar sugar composition. However, there are differences concerning the percentages of the sugars between the pectin-like and pure mucilage fractions associated with whole cladodes precipitate of *Opuntia* spp. [21].

The presence of charged sugars in native mucilage is responsible for its structuring capabilities, granted that specific cross-linking parameters are met, behaving similarly to that of pectin and alginate [25]. The presence of charged sugars also allows for the mucilage molecule to be influenced by changes in solution pH. Consequently, changes in solution pH have been shown to alter the functional properties of a native mucilage solution, ultimately influencing its structuring capabilities [16,26].

As with pectin and alginate biopolymer films, a plasticizer is strongly recommended to develop mucilage biopolymer films. 'Dried' mucilage films generally displayed brittle and fragile properties if no plasticizer was used in their development [12,27,28].

Recent work published by Van Rooyen et al. [13,29] showed mucilage precipitates' potential to be used in the development of single-polymer and blended biopolymer films. However, factors such as polymer concentration and the influence that pH would have on single-polymer mucilage films still remain relatively unexplored compared to other well-established biopolymer films, such as biopolymer pectin and alginate films. Van Rooyen et al. [13] specifically reported mucilage precipitate, in some instances, to enhance certain important pectin films' mechanical properties when added at concentrations lower than that of the pectin-base polymer used. For native mucilage precipitate to be considered a possible viable, novel, and natural packaging alternative to pectin and alginate, it is essential to better understand its homopolymeric (single-polymer) film-forming potential and influencing factors, such as pH. Therefore, this study aimed to determine the mechanical properties of different *Opuntia ficus-indica* cultivars' native mucilage single-biopolymer (homopolymeric) films to similarly developed pectin and alginate single-biopolymer films while considering the influence of polymer concentration and pH.

2. Materials and Methods

2.1. Materials

2.1.1. Commercial Film-Forming Components

Pectin powder from apples (Sigma-Aldrich, Cape Town, South Africa) represented by a 50–75% degree of esterification and a ≤10% moisture content, sodium alginate powder (Sigma-Aldrich, Cape Town, South Africa) with both glucuronic and mannuronic acid content, glycerol, and >99% purity (Merck, Johannesburg, South Africa) was used.

2.1.2. pH Regulators

The pH was adjusted with either acetic acid glacial (100%) or sodium hydroxide (98%) (Merck, Johannesburg, South Africa) were used.

2.1.3. Mucilage Precipitation and Freeze-Drying of Precipitate into Powder

Following a patented method by Du Toit and De Wit [30], as described by Van Rooyen et al., 2023, native mucilage precipitations were prepared from the mature cladodes (~24 months old) of 'Algerian' and 'Morado' cultivars of *Opuntia ficus-indica*. Cladodes were sourced from a well-established cactus pear orchard at the University of the Free State, South Africa, with a cactus pear plant density of 666 cactus pear/ha, unirrigated. This specific, patented method is well-suited for this work due to its cost-effectiveness and repeatability. Furthermore, the resultant precipitate has been well-characterized in previous studies conducted [31,32]. A simplified version of this procedure describes the entire, unpeeled cladode being sliced into approximately 2–3 cm cubes and microwaved on high (900 w) for a time of 4 min at 900 W, followed by maceration of the cubes in a pulp. The pulp was then more easily centrifuged (Beckman® Centrifuge 2315, Brea, CA, USA) to separate the mucilage precipitate effectively from the solids. The pulp was specifically centrifuged at 8000 rpm for 15 min at 8000 rpm at a constant temperature of 4 °C. The freeze-drying process of the mucilage precipitate to form the mucilage powder used in this current work required moisture removal from the liquid precipitate until a 95% weight loss was achieved. Using a mortar and pestle, samples were then milled into a consistent, fine powder. All freeze-drying required that samples were kept under constant vacuum and low temperatures of −30 °C to −40 °C.

2.2. Preparation of Various Film Forming Solutions

With some modifications, standardized approaches were used to form all film-forming solutions prepared, as described by Van Rooyen et al. [29]. Film-forming solutions were prepared by manually dispersing the required amounts of polymer into distilled water. Distilled water used to prepare all film-forming solutions required the addition of glycerol. Glycerol additions were added at 20% (*w*/*w*, based on the weight of the polymer used to develop the films) to the mucilage film-forming solutions and 60% for pectin and alginate film-forming solutions. A minimum of 30 min was allowed for all dispersed polymer solution rehydration at room temperature (~25 °C) under constant stirring (Freed Electric-Model MH-4, Rehovot, Israel). Ensuring homogeneity, all solutions were mechanically mixed for 10 s (Mellerwave Stick Blender-Model 85200, Cape Town, South Africa), and potential entrapped air was removed from the solutions using a Genesis vacuum sealer (Verimark (Pty) Ltd., Pretoria, South Africa) [13,29].

2.3. Single-Polymer Film Development

A standardized film casting method was used to develop the films used in this study, with some modifications, as described by Van Rooyen et al. [29]. The casting method required 70 g of prepared film-forming solution to be evenly spread into 140 mm Petri dishes, forming a film that once dried could be removed for further evaluation [29]. In this current work, films were dried at 40 °C for 24 h using an EcoTherm ventilated oven (Model 920, 1000 W, Labotec, Johannesburg, South Africa). Films were equilibrated for 24 h at room temperature (~25 °C) in a closed container before evaluation.

2.4. pH Adjustment

The process of pH adjustment involved altering the pH of the homogenous film-forming solution by adding either 1 M sodium hydroxide or 1 M acetic acid until the desired pH ranges of pH 9–10 and pH 3–3.5 were achieved.

2.5. Evaluation of Film Mechanical Properties

Making use of a Brookfield AMETEK® CT3™ Texture Analyzer (Westville, South Africa), both tensile and puncture tests were completed on a total of 12 conditioned films cut into 25 × 50 mm strips per treatment. The ASTM International standard methods (ASTM-D882:2010) [33] were followed, similar to that described by Van Rooyen et al. [29], for all mechanical tests completed. The maximum stress (force/area) a film can withstand before it breaks when an external force is applied to the film is considered its tensile strength, with film thickness determined using a digital micrometer (Grip, Johannesburg, South Africa) [4,29]. Elongation at break measures a film's potential to extend until the point where it ruptures [4]. The tensile strength and elongation at break percentages of films are represented by the maximum load divided by the cross-sectional areas of the initial film and the change in the initial length of the films at rupture, respectively [29]. Film tensile tests were approached using the texture analyzer fitted with the Brookfield AMETEK® TA-RCA Roller Cam Accessory grips. A test speed of 0.80 $mm \cdot s^{-1}$, grip spacing of 50 mm, and test distance of 35 mm were selected.

Puncture Test Evaluation

Using a 4 mm Brookfield AMETEK® TA44 probe and accompanying texture analyzer, puncture tests (puncture force and distance to puncture) were completed for the various films. A test speed of 0.80 $mm \cdot s^{-1}$ was selected for this work, similar to that selected in recent studies by Van Rooyen et al. [13,29]. The puncture force and distance to puncture values were determined by a software-generated force-distance graph using the texture analyzers' accompanying software.

2.6. Experimental Design

In the first part of this study, mucilage, pectin, and alginate films were developed at 2.5%, 5%, and 7.5% concentrations at their respective native (unchanged) pH with the addition of glycerol as a plasticizer to determine the effect polymer concentration had on the different film mechanical properties.

Secondly, mucilage, pectin, and alginate film-forming solutions, prepared at 5% (*w*/*w*) together with the prescribed glycerol inclusions, were all subjected to pH alterations at both pH 3–3.5 and pH 9–10 using either acetic acid or sodium hydroxide. Thereafter, the different pH-adjusted film-forming solutions were cast, dried, and tested.

2.7. Statistical Analysis

Upon completion of the various trials, the data were captured using Microsoft Excel (2016) and subjected to analysis of varience statistical analysis (ANOVA). Specifically, a one-way analysis of variance (NCSS Statistical Software package, version 11.0.20) was completed on the data. Using the Tukey–Kramer multiple comparison test ($\alpha = 0.05$), significant differences were identified between the treatment means (NCSS Statistical Software package —v11.0.20, Salt Lake City, UT, USA).

3. Results

3.1. Film Concentration

3.1.1. Film Thickness

Not all tested concentrations allowed for the successful formation of mucilage films at their native (unchanged) pH, as both the 'Algerian' and 'Morado' mucilage could not successfully form films at a concentration of 2.5%. Increasing the mucilage concentration to 5% and 7.5% allowed the successful formation of both 'Algerian' and 'Morado' films. Pectin and alginate formed films at 2.5%, 5%, and 7.5% polymer concentrations. Pectin and alginate film thickness significantly increased by increasing the concentration from 2.5% to 5% ($p < 0.05$). Concentration did not significantly influence the film thickness when increasing the concentration from 5% to 7.5% for the mucilage, pectin, and alginate films ($p > 0.05$) (Figure 1).

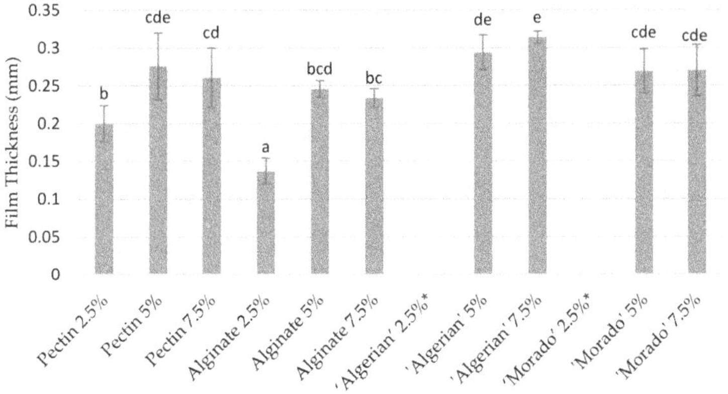

Figure 1. Film thickness of pectin and alginate, as well as, 'Algerian' and 'Morado' mucilage films at concentrations of 2.5%, 5%, and 7.5%. The mean values of the different treatments, together with their standard deviation error bars, are displayed. Error bars with different superscripts differ significantly ($p < 0.05$). * Below minimum measurable film thickness.

The influence of varying the polymer concentrations on the tensile and puncture tests of mucilage, pectin, and alginate films was considered, as presented in Tables 1 and 2.

Table 1. The tensile tests of mucilage, pectin, and alginate films at increasing polymer concentrations.

Treatments	Tensile Strength (MPa)	Elongation at Break %
Pectin 2.5%	3.31 ± 0.43^{b}	5.59 ± 1.51^{a}
Pectin 5%	6.41 ± 0.50^{c}	14.32 ± 1.88^{bc}
Pectin 7.5%	6.27 ± 1.11^{c}	18.98 ± 5.66^{cd}
Alginate 2.5%	10.20 ± 1.38^{d}	4.91 ± 0.75^{a}
Alginate 5%	17.57 ± 0.90^{e}	7.79 ± 1.03^{a}
Alginate 7.5%	16.88 ± 0.69^{e}	21.99 ± 0.69^{d}
'Algerian' 2.5% *		
'Algerian' 5%	0.26 ± 0.05^{a}	33.10 ± 6.10^{e}
'Algerian' 7.5%	1.17 ± 0.08^{a}	49.82 ± 1.53^{f}
'Morado' 2.5% *		
'Morado' 5%	0.32 ± 0.10^{a}	21.58 ± 1.80^{d}
'Morado' 7.5%	1.36 ± 0.15^{a}	47.79 ± 5.84^{f}
Significance level	$p < 0.005$	$p < 0.005$

* Below minimum measurable film tensile mechanical properties. Where applicable, the mean values of 12 different treatments are displayed together with their standard deviations (±). The mean values represented by different superscripts in the same column differed significantly ($p < 0.05$).

Table 2. Puncture tests of mucilage, pectin, and alginate films at increasing polymer concentrations.

Treatments	Puncture Force (N)	Distance to Puncture (mm)
Pectin 2.5%	25.67 ± 5.51 [b]	3.51 ± 0.78 [ab]
Pectin 5%	31.75 ± 2.38 [bc]	4.04 ± 0.38 [abc]
Pectin 7.5%	47.85 ± 4.17 [d]	5.98 ± 0.62 [e]
Alginate 2.5%	34.47 ± 2.72 [c]	4.02 ± 0.23 [abc]
Alginate 5%	72.17 ± 4.68 [e]	5.61 ± 0.37 [de]
Alginate 7.5%	72.22 ± 8.23 [e]	8.63 ± 0.75 [f]
'Algerian' 2.5% *		
'Algerian' 5%	2.43 ± 0.26 [a]	4.24 ± 0.63 [abc]
'Algerian' 7.5%	5.67 ± 0.64 [a]	3.09 ± 0.93 [a]
'Morado' 2.5% *		
'Morado' 5%	1.82 ± 0.21 [a]	4.71 ± 0.89 [bcd]
'Morado' 7.5%	8.30 ± 0.70 [a]	4.99 ± 0.45 [cde]
Significance level	$p < 0.005$	$p < 0.005$

* Below minimum measurable film puncture mechanical properties. Where applicable, the mean values of 12 different treatments are displayed together with their standard deviations (±). The mean values represented by different superscripts in the same column differed significantly ($p < 0.05$).

3.1.2. Tensile Tests

Differences in the mechanical properties between the various polymer films at increasing concentrations were observed.

Increasing the polymer concentration from 2.5% to 5% significantly increased film tensile strength ($p < 0.05$) (Table 1). However, increasing the polymer concentration from 5% to 7.5% did not significantly influence the tensile strength (TS) of mucilage, pectin, and alginate films ($p > 0.05$). Alginate films showed significantly higher TS values than their pectin film counterparts ($p < 0.05$). Galus and Lenart [7] also reported alginate films to have a greater TS than pectin films. These differences were accounted for by polymer variation, with pectin displaying a less organized polymer network than alginate [7]. 'Algerian' and 'Morado' films showed significantly poorer TS than pectin and alginate films, irrespective of the concentration ($p < 0.05$). Gheribi et al. [17] also showed that 4% (w/w) films produced from purified *Opuntia ficus-indica* mucilage, with the addition of glycerol as a plasticizer, were represented by low TS values of ±1 MPa. Espino-Díaz et al. [16] also reported mucilage films to have TS values similar to those reported in the current research, with low TS values accounting for the mucilage films' high molecular weight distribution. Differences observed between the TS of different polymer films have been attributed to differences in polymer molecular weight and inter- and intramolecular association between polymer chains [16,34].

In addition to the film TS, film elongation at break percentage (%E) is also considered essential when evaluating a film's mechanical properties, as an adequate film needs to display both resistance and flexibility [7,17]. Increasing the polymer concentration showed trends in increasing the %E values for the various polymer films. At a 7.5% polymer concentration, all mucilage films showed significantly better ($p < 0.05$) %E values when compared to pectin and alginate polymer films. Interestingly, mucilage films showed higher %E values but relatively low TS compared to pectin and alginate films. Espino-Díaz et al. [16] also investigated mucilage films with low TS values, which showed excellent elasticity.

3.1.3. Puncture Tests

The puncture tests were used to evaluate the mechanical properties of the various polymer films (mucilage, pectin, and alginate) at increasing polymer concentrations (Table 2).

Considering the puncture force (PF) values, increasing the polymer concentration of the various films tested resulted in an increase in the force required to puncture the films (Table 2). Pectin and alginate films showed significantly greater PF values than 'Algerian' and 'Morado' mucilage films at all concentrations tested ($p < 0.05$). Similar findings were reported for the mucilage films when evaluating their tensile strength, confirming mucilage films to have inferior mechanical strength compared to the pectin and alginate films, regardless of the different polymer concentrations tested in this current work. Sandoval et al. [14] suggested that low puncture force resistance values for mucilage films would be influenced by concentration, with low mucilage concentrations being associated with decreased film strength. Unlike the tensile data, the puncture test data suggested that the mucilage films possessed no clear superior elastic properties compared to the pectin and alginate films. In fact, at 7.5% concentration, alginate films showed significantly greater distance to puncture (DTP) values than any of the other films investigated ($p < 0.05$) (Table 2).

3.2. *Film pH*

3.2.1. Film Thickness and the Influence of pH

Changes in pH showed little influence on the pectin and alginate films' thickness compared to their native pH films ($p > 0.05$). These trends were not observed for mucilage films, as both 'Algerian' and 'Morado' film thickness significantly decreased ($p < 0.05$) at a lower pH, with only minimal changes in film thickness reported at pH 9–10 in comparison to their native pH counterpart films (Figure 2).

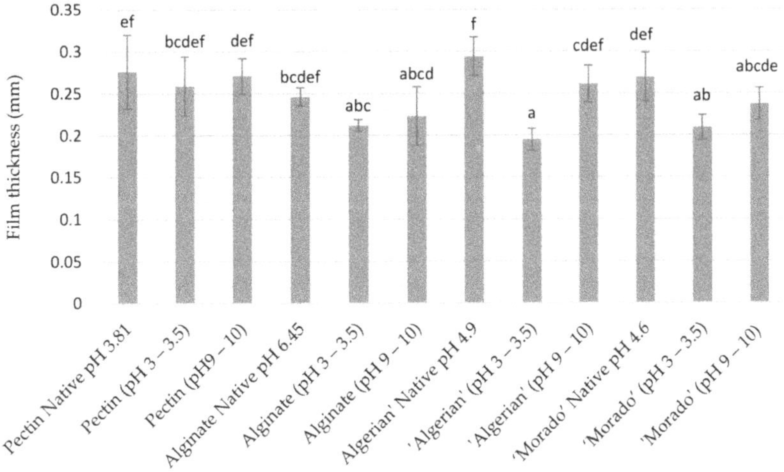

Figure 2. Film thickness of pectin and alginate films, as well as, 'Algerian' and 'Morado' mucilage films at varying pH. The mean values of 12 different treatments are displayed together with their standard deviation error bars. Error bars represented by different superscripts differed significantly ($p < 0.05$).

3.2.2. pH and Film Tensile Measurements

Pectin and alginate films were shown to be significantly influenced by a decrease in pH ($p < 0.05$), resulting in an increase in the film's TS values, which was not observed for mucilage films (Table 3). Only pectin films showed a significant increase in TS values ($p < 0.05$) compared to their native pH counterpart films when the films' pH was increased.

The films' %*E* values were also considered. It was observed that both 'Algerian' and 'Morado' mucilage produced films displaying the highest %*E* values at all pH ranges tested. Specifically, decreasing the film pH resulted in significant increases ($p < 0.05$) in film %*E* values compared to their counterpart films at native pH. Espino-Díaz et al. [16] also reported trends of increases in a mucilage film %*E* when the pH of the film was decreased.

Table 3. The mechanical properties of mucilage, pectin, and alginate biopolymer films at their native pH and at pH 3–3.5 and pH 9–10.

Treatments/Films	Tensile Strength (MPa)	Elongation at Break %
Pectin Native pH 3.81	6.41 ± 0.50 [b]	14.31 ± 1.88 [b]
Pectin (pH 3–3.5)	11.52 ± 1.06 [d]	2.39 ± 0.34 [a]
Pectin (pH9–10)	8.51 ± 1.30 [c]	4.68 ± 1.04 [a]
Alginate Native pH 6.45	17.57 ± 0.90 [e]	7.79 ± 1.03 [a]
Alginate (pH 3–3.5)	20.71 ± 0.85 [f]	5.38 ± 0.23 [a]
Alginate (pH 9–10)	17.58 ± 0.73 [e]	5.17 ± 0.48 [a]
'Algerian' Native pH 4.9	0.262 ± 0.05 [a]	33.10 ± 6.10 [d]
'Algerian' (pH 3–3.5)	1.01 ± 0.15 [a]	55.01 ± 6.32 [e]
'Algerian' (pH 9–10)	0.27 ± 0.09 [a]	14.19 ± 0.61 [b]
'Morado' Native pH 4.6	0.31 ± 0.10 [a]	21.58 ± 1.76 [c]
'Morado' (pH 3–3.5)	0.71 ± 0.10 [a]	36.53 ± 3.69 [d]
'Morado' (pH 9–10)	0.48 ± 0.03 [a]	35.42 ± 5.13 [d]
Significance level	$p < 0.005$	$p < 0.005$

The mean values of 12 different treatments are displayed together with their standard deviations ($\pm$). The mean values represented by different superscripts in the same column differ significantly ($p < 0.05$).

3.2.3. pH and Film Puncture Tests

Considering the various films' PF values, decreasing the pH of mucilage, pectin, and alginate films showed trends of increased PF values compared to their counterpart films at native pH (Table 4). However, only the pectin and alginate films' PF values were significantly increased ($p < 0.05$) when decreasing the film pH (pH 3–3.5). Lastly, the DTP values of only 'Algerian' mucilage films at pH 9–10 were significantly increased ($p < 0.05$) compared to 'Algerian' films at native pH. 'Morado' mucilage films' DTP values were not significantly influenced ($p > 0.05$) by changes in pH (Table 4).

Table 4. Puncture tests of pH-altered mucilage, pectin, and alginate films at their native pH, pH 3–3.5, and pH 9–10.

Treatments/Films	Puncture Force (N)	Distance to Puncture (mm)
Pectin Native pH 3.81	31.75 ± 2.38 [b]	4.04 ± 0.38 [bc]
Pectin (pH 3–3.5)	44.19 ± 4.32 [c]	2.68 ± 0.48 [a]
Pectin (pH9–10)	41.95 ± 8.77 [c]	3.77 ± 0.83 [b]
Alginate Native pH 6.45	72.17 ± 4.68 [d]	5.61 ± 0.37 [de]
Alginate (pH 3–3.5)	80.29 ± 2.12 [e]	5.00 ± 0.41 [cde]
Alginate (pH 9–10)	73.78 ± 4.54 [de]	5.89 ± 0.49 [e]
'Algerian' Native pH 4.9	2.43 ± 0.26 [a]	4.24 ± 0.63 [bc]
'Algerian' (pH 3–3.5)	2.65 ± 0.56 [a]	4.17 ± 0.48 [bc]
'Algerian' (pH 9–10)	2.10 ± 0.67 [a]	5.99 ± 0.45 [e]
'Morado' Native pH 4.6	1.82 ± 0.21 [a]	4.71 ± 0.88 [bcd]
'Morado' (pH 3–3.5)	3.43 ± 0.34 [a]	3.78 ± 0.16 [b]
'Morado' (pH 9–10)	3.15 ± 0.51 [a]	4.22 ± 0.25 [bc]
Significance level	$p < 0.005$	$p < 0.005$

The mean values of 12 different treatments are displayed together with their standard deviations ($\pm$). The mean values represented by different superscripts in the same column differ significantly ($p < 0.05$).

4. Discussion

The mechanical properties of biopolymer films are fundamental as these parameters directly evaluate the physical integrity expected from potential food packaging. Biopolymer films envisaged to be used for packaging need to provide strength and elasticity [17,34]. The tensile tests (tensile strength and elongation at break percentage) and puncture tests (puncture force and distance to puncture) were used to evaluate the film strength and elasticity. Overall, increasing the polymer concentration in mucilage, alginate, and pectin films directly influenced film strength and elasticity. Furthermore, the mechanical properties of the various polymer films investigated were influenced differently by increasing the polymer concentration in the films. Increasing the mucilage concentration for both 'Algerian' and 'Morado' films enhanced both film strength and elasticity, potentially due to the increased likeness of polymer association and entanglement, which can be expected when increasing a polymer concentration in film development. Although mucilage films showed lower film strength when compared to commercially available pectin and alginate films, mucilage films still displayed adequate mechanical properties, specifically displaying excellent elasticity.

At all pH ranges tested, pectin and alginate films showed superior film strength, as they reported higher tensile strength (TS) and puncture force (PF) values when compared to the mucilage films. Furthermore, pectin and alginate films showed that pH considerably influenced the resultant film strength. Regardless of the pH range tested, only minimal changes were reported for the various mucilage film strengths (Tables 3 and 4). At a pH of 3–3.5, pectin and alginate film TS and PF values were significantly increased ($p < 0.05$) compared to their counterparts at their native pH. The literature has suggested that certain charged polymers have the ability to undergo a type of acid gelation, forming structured hydrogels by increasing polymer chain association [35]. Although minimal, trends of increased film strength were reported for both 'Algerian' and 'Morado' films at pH 3–3.5, compared to their counterparts at native pH. It is thus postulated that the mucilage used in developing these biopolymer films has the potential to undergo a type of acid gelation due to the increased film strength observed at a decreased pH. These findings were further supported by a decrease in mucilage film thicknesses at pH 3–3.5, as films that display decreased thickness might be linked to an increased polymer organization within the film matrix. Consequently, increased polymer organization would result in less intermolecular spacing between the polymer chains, which is strongly associated with the gelation of charged polymers [5].

Similarly, due to a type of acid gelation, pectin films displayed increased film strength at a decreased pH as there was an expected reduction in electrostatic repulsion between their polymer chains, resulting in a consequent association of their polymer chains [36]. Furthermore, acid gelation is responsible for developing a more organized polymer network, as in the case of alginate films, as a more organized polymeric structure would allow for increased cohesion forces between the polymer chains, ultimately responsible for enhancing film mechanical properties [4,5]. Although both pectin and alginate polymers are able to undergo acid gelation, differences in their mechanical properties, specifically regarding elasticity, highlight differences between the two polymer film-forming abilities. Mucilage films generally showed superior $\%E$ values. However, when evaluating their DTP values, little elastic advantage was observed for the mucilage films compared to the pectin and alginate films.

5. Conclusions

Cactus pear mucilage was successfully used to develop biopolymer films at >5% concentrations. Decreasing the pH of the different biopolymer films influenced their mechanical properties differently. Pectin and alginate films reported increased tensile strength and puncture force values at a decreased pH, while mucilage film strength was only minimally influenced by changes in pH.

It is postulated that the acid gelation of the pectin and alginate films was responsible for these increases in film strength observed in this current work. Although no significant differences ($p > 0.05$) were reported for mucilage films at a decreased pH, in comparison to the mucilage films at native pH, trends of increased film strength were observed. As the pectin and alginate films' mechanical properties showed a greater dependency on pH than mucilage films, the native mucilage precipitate was therefore believed to be presented by a less charged polymer than the pectin and alginate polymers used in this current work.

Overall, mucilage was successfully used in developing biopolymer films, highlighting its potential to be used in developing natural packaging. Mucilage films were, however, much weaker overall than pectin and alginate films, although they displayed excellent elasticity. Mucilage shows some commercial viability when comparing its mechanical properties to those of pectin and alginate. It is, however, recommended that future research should investigate additional film properties, specifically the water and gas barrier properties of these films. However, mucilage films are not likely to find similar packaging applications as expected from alginate and pectin films.

Author Contributions: Conceptualization, B.V.R. and M.D.W.; methodology, B.V.R., M.D.W. and G.O.; formal analysis, A.H., M.D.W. and G.O.; investigation, B.V.R.; resources, M.D.W. and J.V.N.; data curation, B.V.R., A.H., G.O. and M.D.W.; writing—original draft preparation, B.V.R.; writing—review and editing, B.V.R. and J.V.N.; supervision, M.D.W., G.O., A.H. and J.V.N.; project administration, B.V.R., M.D.W. and J.V.N. All authors have read and agreed to the published version of the manuscript.

Funding: This research received no external funding.

Institutional Review Board Statement: Not applicable.

Informed Consent Statement: Not applicable.

Data Availability Statement: Data are contained within the article.

Acknowledgments: The authors wish to thank Herman Fouché for his help with harvesting the raw cactus pear cladode material.

Conflicts of Interest: The authors declare no conflict of interest.

References

1. Van Rooyen, B.; De Wit, M.; Osthoff, G.; Van Niekerk, J. Cactus Pear (*Opuntia* spp.) Crop Applications and Emerging Biopolymer Innovations. In *Acta Horticulturae*; International Society for Horticultural Science (ISHS): Leuven, Belgium, 2023; pp. 129–134. [CrossRef]
2. Mohamed, S.A.A.; El-Sakhawy, M.; El-Sakhawy, M.A.M. Polysaccharides, Protein and Lipid -Based Natural Edible Films in Food Packaging: A Review. *Carbohydr. Polym.* **2020**, *238*, 116178. [CrossRef] [PubMed]
3. Asgher, M.; Qamar, S.A.; Bilal, M.; Iqbal, H.M.N. Bio-Based Active Food Packaging Materials: Sustainable Alternative to Conventional Petrochemical-Based Packaging Materials. *Food Res. Int.* **2020**, *137*, 109625. [CrossRef] [PubMed]
4. Da Silva, M.A.; Bierhalz, A.C.K.; Kieckbusch, T.G. Alginate and Pectin Composite Films Crosslinked with Ca2+ Ions: Effect of the Plasticizer Concentration. *Carbohydr. Polym.* **2009**, *77*, 736–742. [CrossRef]
5. Bierhalz, A.C.K.; Da Silva, M.A.; Kieckbusch, T.G. Natamycin Release from Alginate/Pectin Films for Food Packaging Applications. *J. Food Eng.* **2012**, *110*, 18–25. [CrossRef]
6. Lee, K.Y.; Mooney, D.J. Alginate: Properties and Biomedical Applications. *Prog. Polym. Sci.* **2012**, *37*, 106–126. [CrossRef] [PubMed]
7. Galus, S.; Lenart, A. Development and Characterization of Composite Edible Films Based on Sodium Alginate and Pectin. *J. Food Eng.* **2013**, *115*, 459–465. [CrossRef]
8. Gao, C.; Pollet, E.; Avérous, L. Innovative Plasticized Alginate Obtained by Thermo-Mechanical Mixing: Effect of Different Biobased Polyols Systems. *Carbohydr. Polym.* **2017**, *157*, 669–676. [CrossRef] [PubMed]
9. Otoni, C.G.; Avena-Bustillos, R.J.; Azeredo, H.M.C.; Lorevice, M.V.; Moura, M.R.; Mattoso, L.H.C.; McHugh, T.H. Recent Advances on Edible Films Based on Fruits and Vegetables—A Review. *Compr. Rev. Food Sci. Food Saf.* **2017**, *16*, 1151–1169. [CrossRef]
10. Tosif, M.M.; Najda, A.; Bains, A.; Zawiślak, G.; Maj, G.; Chawla, P. Starch–Mucilage Composite Films: An Inclusive on Physicochemical and Biological Perspective. *Polymers* **2021**, *13*, 2588. [CrossRef]
11. Gheribi, R.; Khwaldia, K. Cactus Mucilage for Food Packaging Applications. *Coatings* **2019**, *9*, 655. [CrossRef]
12. Damas, M.S.P.; Pereira Junior, V.A.; Nishihora, R.K.; Quadri, M.G.N. Edible Films from Mucilage of Cereus Hildmannianus Fruits: Development and Characterization. *J. Appl. Polym. Sci.* **2017**, *134*, 45223. [CrossRef]
13. Van Rooyen, B.; De Wit, M.; Osthoff, G.; Van Niekerk, J.; Hugo, A. Effect of Native Mucilage on the Mechanical Properties of Pectin-Based and Alginate-Based Polymeric Films. *Coatings* **2023**, *13*, 1611. [CrossRef]
14. Sandoval, D.C.G.; Sosa, B.L.; Martínez-Ávila, G.C.G.; Fuentes, H.R.; Abarca, V.H.A.; Rojas, R. Formulation and Characterization of Edible Films Based on Organic Mucilage from Mexican *Opuntia ficus-indica*. *Coatings* **2019**, *9*, 506. [CrossRef]
15. Kurek, M.; Galus, S.; Debeaufort, F. Surface, Mechanical and Barrier Properties of Bio-Based Composite Films Based on Chitosan and Whey Protein. *Food Packag. Shelf Life* **2014**, *1*, 56–67. [CrossRef]

16. Espino-Díaz, M.; De Jesús Ornelas-Paz, J.; Martínez-Téllez, M.A.; Santillán, C.; Barbosa-Cánovas, G.V.; Zamudio-Flores, P.B.; Olivas, G.I. Development and Characterization of Edible Films Based on Mucilage of *Opuntia ficus-indica* (L.). *J. Food Sci.* **2010**, *75*, 347–352. [CrossRef]
17. Gheribi, R.; Puchot, L.; Verge, P.; Jaoued-Grayaa, N.; Mezni, M.; Habibi, Y.; Khwaldia, K. Development of Plasticized Edible Films from *Opuntia ficus-indica* Mucilage: A Comparative Study of Various Polyol Plasticizers. *Carbohydr. Polym.* **2018**, *190*, 204–211. [CrossRef] [PubMed]
18. Lira-Vargas, A.A.; Lira-Vargas, A.A.; Corrales-Garcia, J.J.E.; Valle-Guadarrama, S.; Peña-Valdivia, C.B.; Trejo-Marquez, M.A. Biopolymeric Films Based on Cactus (*Opuntia ficus-indica*) Mucilage Incorporated with Gelatin and Beeswax. *J. Prof. Assoc. Cactus Dev.* **2014**, *16*, 51–70.
19. Madera-Santana, T.J.; Vargas-Rodríguez, L.; Núñez-Colín, C.A.; González-García, G.; Peña-Caballero, V.; Núñez-Gastélum, J.A.; Gallegos-Vázquez, C.; Rodríguez-Núñez, J.R. Mucilage from Cladodes of Opuntia Spinulifera Salm-Dyck: Chemical, Morphological, Structural and Thermal Characterization. *CYTA J. Food* **2018**, *16*, 650–657. [CrossRef]
20. Rodríguez-González, F.; Pérez-González, J.; Muñoz-López, C.N.; Vargas-Solano, S.V.; Marín-Santibáñez, B.M. Influence of Age on Molecular Characteristics and Rheological Behavior of Nopal Mucilage. *Food Sci. Nutr.* **2021**, *9*, 6776–6785. [CrossRef]
21. Garfias Silva, V.; Cordova Aguilar, M.S.; Ascanio, G.; Aguayo, J.P.; Pérez-Salas, K.Y.; Susunaga Notario, A.D.C. Acid Hydrolysis of Pectin and Mucilage from Cactus (*Opuntia ficus*) for Identification and Quantification of Monosaccharides. *Molecules* **2022**, *27*, 5830. [CrossRef]
22. Matsuhiro, B.; Lillo, L.E.; Sáenz, C.; Urzúa, C.C.; Zárate, O. Chemical Characterization of the Mucilage from Fruits of *Opuntia ficus indica*. *Carbohydr. Polym.* **2006**, *63*, 263–267. [CrossRef]
23. Rodríguez-González, S.; Martínez-Flores, H.E.; Chávez-Moreno, C.K.; Macías-Rodríguez, L.I.; Zavala-Mendoza, E.; Garnica-Romo, M.G.; Chacõn-García, L. Extraction and Characterization of Mucilage from Wild Species of Opuntia. *J. Food Process Eng.* **2014**, *37*, 285–292. [CrossRef]
24. Monrroy, M.; García, E.; Ríos, K.; García, J.R. Extraction and Physicochemical Characterization of Mucilage from *Opuntia cochenillifera* (L.) Miller. *J. Chem.* **2017**, *2017*, 4301901. [CrossRef]
25. Soukoulis, C.; Gaiani, C.; Hoffmann, L. Plant Seed Mucilage as Emerging Biopolymer in Food Industry Applications. *Curr. Opin. Food Sci.* **2018**, *22*, 28–42. [CrossRef]
26. Du Toit, A.; de Wit, M.; Seroto, K.D.; Fouche, H.; Hugo, A.; Venter, S. Rheological Characterization of Cactus Pear Mucilage for Application in Nutraceutical Food Products. *Acta Hortic.* **2019**, *1247*, 63–72. [CrossRef]
27. Allegra, A.; Inglese, P.; Sortino, G.; Settanni, L.; Todaro, A.; Liguori, G. The Influence of *Opuntia ficus-indica* Mucilage Edible Coating on the Quality of "Hayward" Kiwifruit Slices. *Postharvest Biol. Technol.* **2016**, *120*, 45–51. [CrossRef]
28. Zibaei, R.; Hasanvand, S.; Hashami, Z.; Roshandel, Z.; Rouhi, M.; Guimarães, J.D.T.; Mortazavian, A.M.; Sarlak, Z.; Mohammadi, R. Applications of Emerging Botanical Hydrocolloids for Edible Films: A Review. *Carbohydr. Polym.* **2021**, *256*, 117554. [CrossRef]
29. Van Rooyen, B.; De Wit, M.; Osthoff, G.; Van Niekerk, J.; Hugo, A. Microstructural and Mechanical Properties of Calcium-Treated Cactus Pear Mucilage (*Opuntia* spp.), Pectin and Alginate Single-Biopolymer Films. *Polymers* **2023**, *15*, 4295. [CrossRef]
30. Du Toit, A.; De Wit, M. Patent PA153178P A Process for Extracting Mucilage from *Opuntia ficus-indica, Aloe barbadensis* and *Agave americana*. Ph.D. Thesis, University of the Free State, Bloemfontein, South Africa, 2021. [CrossRef]
31. Du Toit, A. Selection, Extraction, Characterization and Application of Mucilage from Cactus Pear (*Opuntia ficus-indica* and *Opuntia robusta*) Cladodes. Ph.D. Thesis, University of the Free State, Bloemfontein, South Africa, 2016. Volume 2. pp. 1–13.
32. Du Toit, A.; De Wit, M.; Fouché, H.J.; Taljaard, M.; Venter, S.L.; Hugo, A. Mucilage Powder from Cactus Pears as Functional Ingredient: Influence of Cultivar and Harvest Month on the Physicochemical and Technological Properties. *J. Food Sci. Technol.* **2019**, *56*, 2404–2416. [CrossRef]
33. *ASTM-D882:2010; Standard Test Method for Tensile Properties of Thin Plastic Sheeting. ASTM: West Conshohocken, PA, USA, 2010.*
34. Guadarrama-Lezama, A.Y.; Castaño, J.; Velázquez, G.; Carrillo-Navas, H.; Alvarez-Ramírez, J. Effect of Nopal Mucilage Addition on Physical, Barrier and Mechanical Properties of Citric Pectin-Based Films. *J. Food Sci. Technol.* **2018**, *55*, 3739–3748. [CrossRef]
35. Gawkowska, D.; Cybulska, J.; Zdunek, A. Structure-Related Gelling of Pectins and Linking with Other Natural Compounds: A Review. *Polymers* **2018**, *10*, 762. [CrossRef] [PubMed]
36. Espitia, P.J.; Du, W.X.; de Jesús Avena-Bustillos, R.; Soares, N.D.; McHugh, T.H. Edible Films from Pectin: Physical-Mechanical and Antimicrobial Properties—A Review. *Food Hydrocoll.* **2014**, *35*, 287–296. [CrossRef]

polymers MDPI

Article

Colorimetric Indicator Based on Gold Nanoparticles and Sodium Alginate for Monitoring Fish Spoilage

Lissage Pierre [1,2,*], Julio Elías Bruna Bugueño [1,2,*], Patricio Alejandro Leyton Bongiorno [3], Alejandra Torres Mediano [1,2] and Francisco Javier Rodríguez-Mercado [1,2]

[1] Packaging Innovation Center (LABEN), Food Science and Technology Department (DECYTAL), Technological Faculty, University of Santiago de Chile (USACH), Ave. Víctor Jara 3769, Santiago 9170124, Chile; alejandra.torresm@usach.cl (A.T.M.); francisco.rodriguez.m@usach.cl (F.J.R.-M.)
[2] Center for the Development of Nanoscience and Nanotechnology (CEDENNA), University of Santiago de Chile (USACH), Santiago 9170022, Chile
[3] Instituto de Química, Pontificia Universidad Católica de Valparaíso, Valparaíso 2340025, Chile; patricio.leyton@pucv.cl
* Correspondence: lissage.pierre@usach.cl (L.P.); julio.bruna@usach.cl (J.E.B.B.); Tel.: +56-937564320 (L.P.)

Abstract: In this work, a colorimetric indicator based on gold nanoparticles (AuNP) and a biodegradable and eco-friendly polymer (sodium alginate, Alg.), was developed for the real-time detection of fish spoilage products. The AuNPs and the colorimetric indicator were characterized using UV-VIS, FTIR spectroscopies, TGA, DSC, XRD, TEM, and colorimetry. The UV-VIS spectrum and TEM showed the successful synthesis, the spherical shape, and the size of AuNPs. The results indicated color changes of the indicator in packaged fish on day 9 of storage at a refrigerated temperature (5 °C. These results showed the successful application of the colorimetric indicator in the detection of TVB-N in packaged fish.

Keywords: gold nanoparticles; polymer; colorimetric indicator; sodium alginate; biogenic amines; ethylenediamine

check for updates

Citation: Pierre, L.; Bruna Bugueño, J.E.; Leyton Bongiorno, P.A.; Torres Mediano, A.; Rodríguez-Mercado, F.J. Colorimetric Indicator Based on Gold Nanoparticles and Sodium Alginate for Monitoring Fish Spoilage. *Polymers* **2024**, *16*, 829. https://doi.org/10.3390/polym16060829

Academic Editors: César Leyva-Porras and Zenaida Saavedra-Leos

Received: 15 December 2023
Revised: 7 February 2024
Accepted: 12 February 2024
Published: 17 March 2024

1. Introduction

Fish is an excellent source of nutrients and is easily digestible in the human diet [1] because it contains healthy fats, omega-3 (EPA, DHA), high-quality protein, iron, phosphorus, zinc, selenium, vitamins, and minerals. However, it is highly perishable, vulnerable, and very prone to deterioration due to enzymatic reactions or microbial contamination during handling, distribution, and/or storage, leading to the formation of different metabolic products, such as alcohols, ketones, some aldehydes, organic acids, and sulfides, which usually occurs after the death of the fish. Storage in inappropriate places and/or poor hygiene conditions will lead to deterioration and the formation of biogenic amines (BA) [2], among which the most abundant are putrescine, cadaverine, tyramine, and histamine, some of which are dangerous to the human organism [3]. In the same way, the other factors that play an important role in the deterioration of fish through the consequent production of BA are bacteria, mainly Gram-positive and Gram-negative bacteria. In general, they are located in different parts of the fish body, particularly in the skin, gills, or gastrointestinal tract [4–6], and can also spread to the muscle mass during evisceration through the rupture or loss of gastric contents. The species most frequently found in this decomposition process are *Enterobacteriaceae*, including mesophilic and psychrotolerant bacteria, such as *Morganella*, *Enterobacter*, and Gram-negative bacteria of the *Hafniaceae* family: *Hafnia*, *Proteus*, and *Photobacteria* [7]. Furthermore, the genus *Pseudomonas* and lactic acid bacteria belonging to the genera *Lactobacillus* and *Enterococcus* can cause BA formation [8].

In this sense, in recent times, various techniques have been used to detect the quality of fish, but they are mainly based on the analysis of its structure (tenderness, color, texture, etc.), using methods of counting bacteria, determining the total volatile basic nitrogen (TVB-N), and measuring pH values. Other techniques, such as optical spectroscopy, Nuclear Magnetic Resonance (NMR), Fourier-transform infrared spectroscopy (FTIR), and gas chromatography–mass spectrometry (GC/MS), are also used for evaluation. However, there are disadvantages associated with the use of these techniques because, in some cases, the sample is destroyed; a complex, slow, expensive sample preparation is required; and, in general, an expert is required to execute the methods [9]. Therefore, the intelligent packaging system is elaborated by the incorporation of a device that can interact internally or externally, monitoring the changes that may occur in the packaged product. This can be separated or listed into three groups: (1) indicators are established to provide information about the quality of the products to consumers; (2) sensors are prepared to detect specific analytes in packaged foods; and (3) data carriers (including barcodes and radio frequency identification labels) are designed to carry out traceability and/or monitoring in the food supply chain [10].

In this context, Alg. is an anionic polymer produced by brown algae and bacteria. It is a biocompatible, biodegradable, non-toxic, low-cost, and readily available polymer. It consists of α-L-guluronic acid (G) and β-D-mannuronic acid (M) residues linked linearly by 1,4-glycosidic bonds [11]. It is widely used in various fields and in the food industry because it is capable of producing strong gels in the presence of metal cations. Thus, it is considered convenient to develop an indicator based on the metallic nanoparticles inserted in Alg. In this sense, colorimetric indicators are valuable and interesting for application due to their low cost, their ease of use, their simplicity, and, above all, they offer high legibility with the naked eye [12]. Therefore, in previous studies, researchers have developed a colorimetric label based on bacterial cellulose with the incorporation of grape anthocyanin that allows the monitoring of the freshness of stored minced meat [13]. Wang et al. [14] reported the development of an elementary colorimetric sensor that can be used as a cheaper indicator to detect the freshness of fish using PANI films, which can be regenerated using acid solution. Furthermore, a pH-sensitive sensor based on cellulose-modified polyvinyl alcohol (PVA) was developed. Its evaluation in deteriorated fish produced color changes [15].

On the other hand, other groups of researchers have developed an indicator matrix with 16 diverse detection components to monitor fish spoilage [16]. Although the colorimetric indicator based on pH-sensitive dyes is a simple way to control the quality of food, which can be observed with the naked eye, it is essential to develop a new detection system based on metal nanoparticles inserted in a biodegradable and eco-friendly polymer for the real-time monitoring of fish deterioration. In this research work, a colorimetric indicator based on gold nanoparticles and a natural polymer was developed, capable of indicating, through a color change, when the fish was unsuitable for consumption.

2. Materials and Methods

2.1. Materials

To obtain gold nanoparticles (AuNP), sodium tetrachloroaurate (III) dihydrate ($NaAuCl_4{\cdot}2H_2O$) 99% CAS No. 13874-02-07 and trisodium citrate dihydrate (NaCit) ($C_6H_5Na_3O_7{\cdot}2H_2O$) CAS No. 6132-04-3 were used as reagents; they were purchased from Sigma-Aldrich and used without further purification and distilled water.

2.2. Synthesis of Gold Nanoparticles (AuNP)

Gold nanoparticles (AuNPs) were synthesized using the method described by El-Nour et al. [17] with some modifications. Briefly, 40 mL of distilled water was poured into a 100 mL flask and heated until boiling, then 1 mL of 1% trisodium citrate was gradually added with continuous stirring. Then, 100 µL of 1% sodium tetrachloroaurate (III) dihydrate solution was added and the solution was stirred and heated until the color of the solution changed from yellow to deep red at approximately 5 min, indicating the formation of AuNP. Gold nanoparticles were gradually formed as the citrate reduced Au_{3+} to Au_0 as indicated by the red color. The AuNP solution was cooled down at room temperature and stored at 4 °C until further use.

2.3. Colorimetric Indicator Preparation

Colorimetric indicators were prepared according to the method described by Dudnik et al. [18]. Briefly, an aqueous solution was added to Alg. (2 wt%) and stirred at 1000 rpm for 3 h or until complete dissolution. Then, 3 mL of dispersion of gold nanoparticles was added to 1 mL of Alg. with continuous stirring. Then, the colloidal suspension was cast into a Petri dish containing 0.10 mol/L solution of calcium chloride to obtain an Alg. gel and allowed to dry in ambient conditions. The prepared colorimetric indicators were then packed in a 20 mL screw-cap vial and stored at room temperature until further use.

2.4. Characterization

The synthesized AuNP was characterized using the UV-VIS spectrophotometer Spectroquant® pharo 300M, from Merck KGaA, Darmstadt, Germany in the range of 300–800 nm. The samples were placed in the cuvette, and the UV-VIS spectrum was measured at different time intervals.

An FTIR analysis was performed to characterize the chemical structure of the materials (Alg. and AuNP). For this, a colloidal solution of nanoparticles was concentrated under a vacuum (vacuum system) in a rotary evaporator (Heidolph, Schwabach, Germany) and dried using lyophilization. For the analysis, pellets were obtained by mixing the dry gold nanoparticles with KBr 1.0 wt% of the sample. The FTIR equipment model, Alpha II Bruker, Waltham, MA, USA was used in the following conditions: a range of 4000–500 1/cm and a resolution of 4 1/cm, which was obtained after cumulating 64 scans to determine the FTIR of the AuNP and the indicators, respectively. To determine the thermal properties of gold nanoparticles (AuNP) and the colorimetric indicator, a thermogravimetric analysis was performed using a thermogravimetric analyzer (METTLER TOLEDO Gas Controller GC20 Stare system TGA/DSC) at a rate of 10 °C min^{-1} under nitrogen gas, flowing at 20 mL min^{-1} in the range of 30–800 °C. The DSC analyses were carried out using (DSC1 equipment, model STAR System 822, MA, USA) operating at the following conditions: a heating rate of 10 °C min^{-1}, a nitrogen flow rate of 20 mL min^{-1}, and a temperature range from 0 to 400 °C and 0–200 °C for the AuNP and indicators, respectively.

An X-ray diffraction (XRD) analysis was carried out using an X-ray diffractometer (D8 Advance, Bruker, Germany) in the range of 2–80° at an angle of 2θ with CuKα radiation. The tube current and voltage were 30 mA and 40 kV, respectively, and the scan rate was 1°/min. The morphology of the AuNP was studied using Transmission Electron Microscopy (TEM) images (Talos F200X (Thermo Fisher Scientific)), Waltham, MA, USA. The microscope was operated at an accelerating voltage of 200 kV. The sample (10 μL) was mounted on a copper grid covered by a carbon–formvar film and allowed to dry at room temperature for 24 h before the TEM analysis.

2.5. Detection Capacity and Colorimetric Analysis

The detection capacity of gold nanoparticles was evaluated by preparing a stock solution of ethylenediamine (ETD) with a concentration of 2000 ppm (mg/L), from which the following concentrations were obtained: 10–400 ppm. A colorimetric analysis of the AuNP and the colorimetric indicator was performed by obtaining photographs with a digital camera. The photographs obtained were evaluated using the ImageJ software (2.9.0/1.53t/Java 1. 8. 0_322 (64-bit)), determining the color parameters (L^*, a^*, and b^*) based on the CIELab color system, and the total color difference according to the following formula:

$$\Delta E = \sqrt{(\Delta L*)^2 + (\Delta a*)^2 + (\Delta b*)^2}$$

L^* = lightness (0 = black, 100 = white), $+b^*$ (yellow), $-b$ (blue), $+a^*$ (red), $-a^*$ (green).

2.6. Evaluation of Indicators under Simulated Conditions (In Vitro)

The in vitro evaluation of the colorimetric indicator was performed using ETD, according the method described by Zhai et al. [19] with slight modifications. For this, the aforementioned indicator was taken and placed in a clean and smooth plastic Petri dish with a 90 mm diameter and with the ETD solution at room temperature, observing the possible color changes of the indicator. Then, the indicators were peeled from the Petri dishes and observed in an optical microscope, taking photos with a Samsung smartphone (Galaxy A21S). Then, the total color differences were analyzed using ImageJ software in triplicate and the data were processed using the CIELAB system to determine the color parameters (L^*, a^*, and b^*) [20].

2.7. Evaluation of the Indicator in Real Packaging Conditions

For the evaluation of the indicator in real packaging conditions, Atlantic salmon was used as a model food, which was purchased in the local trade. For this, the fish were aseptically filleted with a knife to obtain more-or-less uniform rectangular samples of equal weight (approximately 50 g). All the fish samples were introduced into sterile vacuum stomacher bags, in which the colorimetric indicator to be evaluated was inserted, using a polyethylene/polyamine support so that the indicator was kept in direct contact with the muscle food. The stomacher bags were stored at a temperature of 5 °C for 9 days to monitor the spoilage and the color change of the indicator. At specified intervals during the storage period, the samples were analyzed for any chemical or microbial changes. Reference stomacher bags containing fish samples without indicators were also maintained for the same experimental period as those containing indicators.

2.8. Characterization of Fish Samples during Storage

In all the fish samples, the pH and the total volatile basic nitrogen (TVB-N) were determined in duplicate ($n = 2$). The pH measurements were collected using a Meat pH Tester (Hanna Instruments, Smithfield, RI, USA). The calibration was previously carried out using buffers 4 and 7, then the pH was measured by inserting the tip, in the form of a glass cone, inside the fish and obtaining the corresponding values.

2.9. Determination of Total Volatile Basic Nitrogen (TVB-N)

The total volatile basic nitrogen (TVB-N) values of salmon were determined in accordance with the Chilean standard norm NCh 2668:2018 for hydrobiological products. Briefly, 10 g of the fish sample, 2 g of magnesium oxide, and 100 mL of water were added to the distillation flask, followed by the distillation of the mixture. The distillate was taken up in 25 mL of a 3% m/v aqueous solution of boric acid and 5–7 drops of Tashiro's indicator (mixed indicator composed of a solution of methylene blue (0.1%) and methyl red (0.03%) in ethanol or methanol). Then, the boric acid solution was titrated to the endpoint with 0.1 N sulfuric acid. The TVB-N content (mg of N/100 g) of the fish sample was determined as follows:

$$\text{TVB} - \text{N} = \frac{\text{mL } \text{H}_2\text{SO}_4 \text{ 0.1N} \times 1.4 \times 100}{\text{Sample weight(g)}}$$

2.10. Microbiological Analysis

As the fish spoilage was being monitored using colorimetric indicators, samples were subjected to a microbiological analysis at regular intervals. For the total microbial count determination, 25 g of the salmon sample was added to 225 mL of 0.1% m/v peptone water. The mixture was homogenized in a homogenizer for 1 min and the 0.1 dilution was obtained. From this dilution, further decimal dilutions were prepared and plated on Petri dishes in the appropriate media. The enumeration of the total viable aerobic bacteria counts was performed according to the pour plate method, using plate count agar (PCA) purchased from Merck. The inoculated Petri dishes were incubated at 35 ± (1 °C) for 48 h to determine the mesophilic counts. The colony-forming unit (CFU) counts were expressed as CFU/g. All the microbiological experiments were performed in duplicate.

2.11. Statistical Analysis

The analyses and experiments were performed in triplicate and the data were evaluated using analysis of variance (ANOVA) (comparison of several samples) using the PROGRAM-STATGRAPHICS Centurion XIX.v.64, followed by Fisher's Least Significant Difference (LSD) procedure. A probability level of $p < 0.05$ was considered statistically significant.

3. Results

3.1. Characterization of Gold Nanoparticles (AuNPs)

Gold nanoparticles were characterized using UV-VIS, transmission electron microscopy (TEM), colorimetric analysis, infrared spectroscopy (FTIR), thermogravimetric analysis (TGA), differential scanning calorimetry (DSC), and X-ray diffraction (XRD).

In Figure 1, a color change of the solution was observed, it changed from yellow to a red dispersion with the addition of trisodium citrate (NaCit), indicating the formation of gold nanoparticles [21]. In addition, it was possible to observe through the UV-VIS spectrum an absorption band at a wavelength of 520 nm in the red color of the spectrum that gives its hue; this phenomenon is related to the surface plasmon resonance (SPR) because the AuNP conduction electrons interact with incident photons to produce a resonance effect, manifested as SPR. This interaction depends on the size, the shape, and the composition of the metallic nanoparticles, as well as the type and content of the dispersion medium [22]. Similar results were obtained by [23].

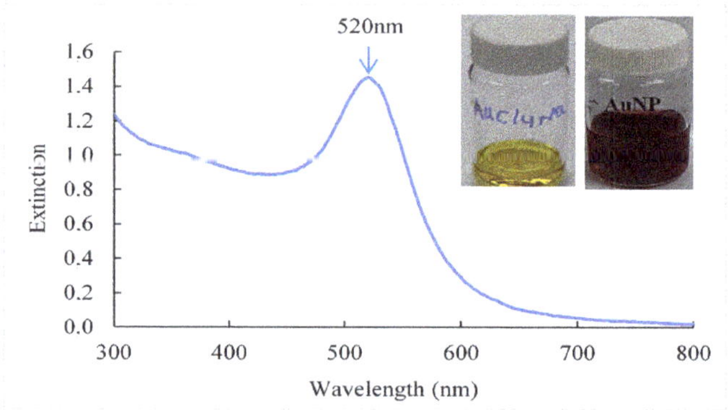

Figure 1. UV-VIS spectrum and photographs of gold nanoparticles (AuNP) obtained. Sodium tetrachloroaurate ($NaAuCl_{4}.2H_2O$) was used as a precursor of AuNP and NaCit ($C_6H_5Na_3O_7{\cdot}2H_2O$), as a reducer and a stabilizer.

Figure 2 shows the results of the detection capacity of the AuNPs by UV-VIS spectrum, observing two absorption peaks. The first peak was between 520 nm and 531 nm, which may be associated with the aggregation of the AuNPs caused by their contact with the ETD solution. The second absorption peak was located at a wavelength between 660 nm and 702 nm, which is associated with a change in the morphology of the AuNPs because its union with the ETD produces a strong interaction between particles, varying their shape and size, as well as their color [24]. Similar results were reported by Sun et al. [25]. It should be noted that the bathochromic shift of AuNP results in a large displacement towards the red region of the Localized Surface Plasmon Resonance (LSPR) peak [26] Furthermore, Mahatnirunkul et al. [27]. affirmed that when the analyte of interest binds to the surface of AuNPs, the Localized Surface Plasmon Resonance (LSPR) spectrum will shift to a longer wavelength. Similar results were reported by [28]. Additionally, the aggregation of gold nanoparticles, after their contact with ethylenediamine, occurred through the interaction of attractive van der Waals (VA) and repulsive Coulomb (VR) forces. Certainly, the gold nanoparticles in suspension remain stable when VR > VA. On the other hand, when VR < VA, the nanoparticles clump together [29]. It has been detected that a wide variety of factors intervene in this process, such as particle size, surface tension, and the electrical double layer, which have a high participation in reducing the stability of the NPs and their possible aggregation. However, in this case, the citrate ions adsorb on the surface of the prepared AuNPs, creating a negative surface charge that stabilizes the particles; the energy barrier was powerful enough to prevent a strong interaction between the particles and, with that, it prevented them from aggregating. However, the addition of ETD to NaCit-stabilized AuNPs disturbed the stability of the nanoparticles, leading to their aggregation. Additionally, we used ETD as a test molecule for the in vitro evaluation of the detection ability of AuNPs, because it is an analogous molecule to TVB-N, such as NH_3, dimethylamine (DMA), and trimethylamine (TMA); they have in common the amino group, which contains nitrogen. Since biogenic amines are one of the main spoilage products of fish that we need to detect with the colorimetric indicator.

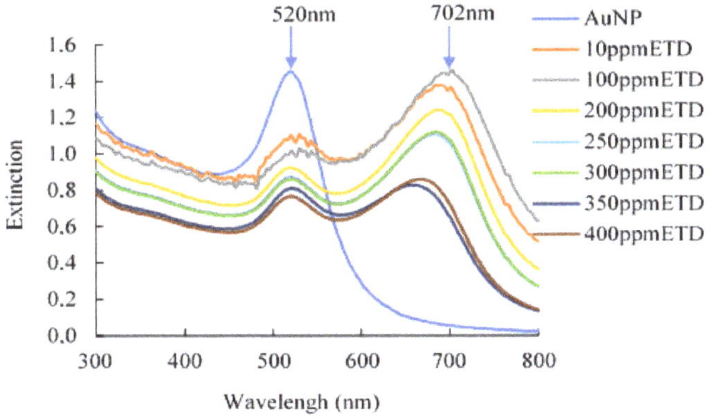

Figure 2. UV-VIS evaluation of the synthesized gold nanoparticles, based on different ETD concentrations (10–400 ppm).

Furthermore, when a higher AuNP extinction is observed, this is an indication of a higher concentration of nanoparticles. It should be noted that, as the nanoparticles clump together, they increase in size, gradually settling to the bottom and resulting in less extinction. Similar results were obtained in a previous study by Ranjan et al. [30].

In order to study the size and morphology of the synthesized AuNPs, TEM measurements were carried out. In Figure 3a, it can be observed that all the particles presented a homogeneous spherical shape, as expected for this type of nanoparticle [31]. Furthermore, it presented a wide size distribution from 10 to 16 and 50 nm, approximately. On the other hand, in Figure 3b, it is shown that by adding ETD, the AuNPs aggregated and formed clusters due to a strong interparticle interaction. In addition, in the UV-VIS spectrum, it was possible to corroborate a displacement of the wavelength towards the red region of the spectrum, indicating larger nanoparticles as demonstrated in the TEM analysis.

The FTIR analysis of the citrate-capped AuNPs is presented in Figure 4, where it is possible to observe the presence of the absorption bands at 1696, 1580, 1387, 1281, and 1079 1/cm. The band at 1696 1/cm is related to the COO^- group, originating due to the adsorption of the citrate anions on gold nanoparticles through the central carboxylate group [29]. The two bands with their peaks at around 1387 and 1580 1/cm, respectively, corresponded to the symmetric ($\nu_s(COO^-)$) and antisymmetric ($\nu_{as}(COO^-)$) stretching bands of the carboxyl groups of NaCit. For the citrate-capped AuNP, the $\nu_{as}(COO^-)$ peak was found to be largely a high-frequency shift from the original peak position. In addition, absorption bands were observed between 1281 and 1079 1/cm, which corresponded to the stretching vibration of the C–O bond, as well as the C–C bond of NaCit [32]. Similar results were reported elsewhere by [33]. The 3344 1/cm and 3446 1/cm absorption bands that appeared between 3600 and 3000 1/cm were related to the stretching vibration of the O–H bond of NaCit. On the other hand, changes in the stretching vibration of the O–H group of the citrate were observed with the addition of AuNP. This was due to the interaction of the nanoparticles surrounded by the NaCit, giving rise to the formation of a new absorption band at wavenumber 3344 1/cm. This indicated the disappearance of the citrate band (3446 cm^{-1}), evidencing the presence of gold nanoparticles [34].

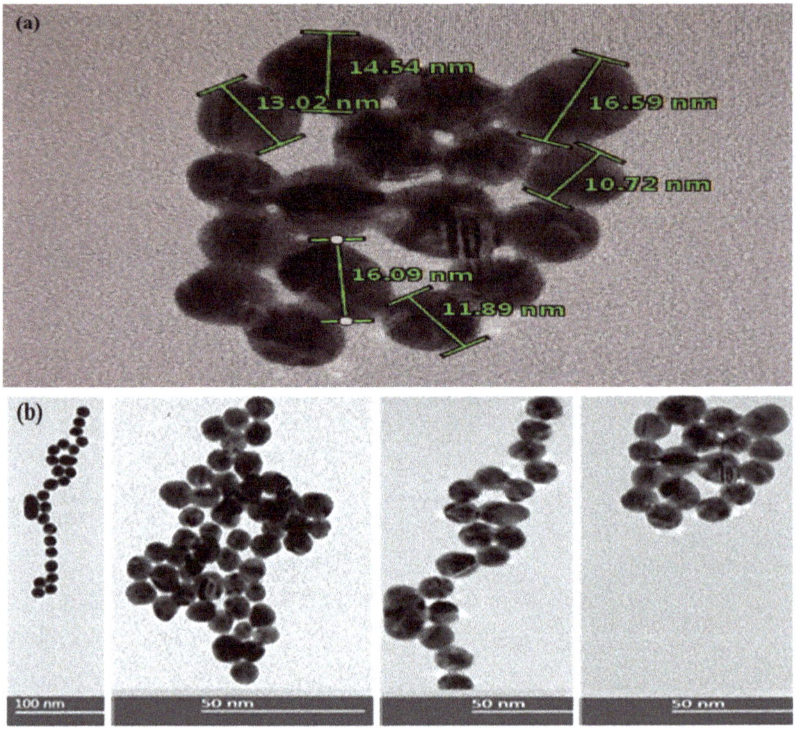

Figure 3. Images of (**a**) AuNPs and (**b**) AuNP+ ETD (aggregated form).

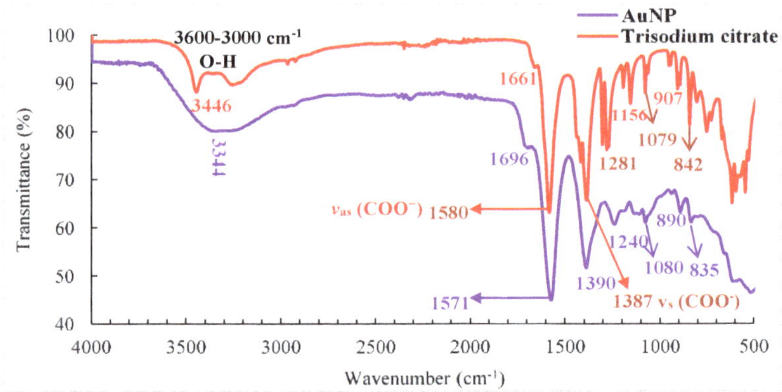

Figure 4. Infrared spectroscopy (FTIR) of AuNP functionalized with trisodium citrate.

In this experiment, the AuNPs covered by a layer of negatively charged citrate ions were used to keep the nanoparticles well dispersed and stable in the colloidal solution. The addition of 10 ppm ETD in 1 mL of AuNP caused an aggregation of AuNPs (Figure 5 and Table 1), mainly affecting the extinction spectrum with color changes of the AuNPs with a total color difference of ΔE = 28.81. This aggregation could be due to an electrostatic interaction between the positively charged amine groups and the negatively charged trisodium citrate ion groups surrounding the AuNPs, or could also be the result of some exchange between the trisodium citrate ions and amines that probably can directly adhere to the AuNPs. These results indicated the ability of the metallic nanoparticles to detect fish spoilage products by properly interacting with the test molecule, causing the formation of larger nanoparticles with a subsequent color change in the dispersion; however, by adding 200 ppm of ETD in 1mL of AuNP, a ΔE = 36.9 was obtained. This indicates that the higher the concentration of ethylenediamine, the greater the color changes observed.

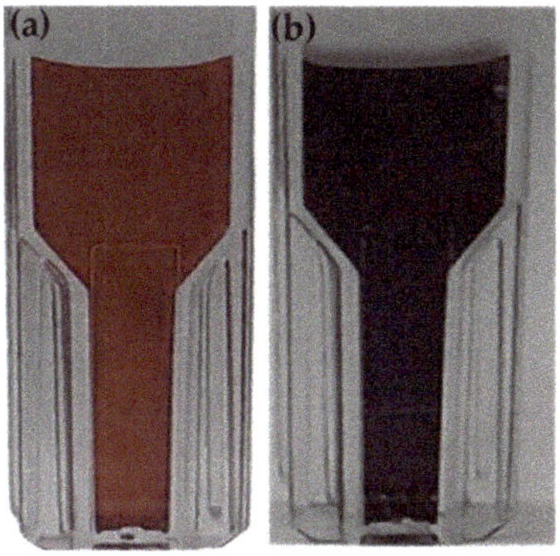

Figure 5. Colloidal solution of nanoparticles (**a**) AuNP (red) and (**b**) AuNP + ETD (black).

Table 1. Color measurement of gold nanoparticles.

Color Parameters	AuNP	AuNP + ETD
L^*	10.56 ± 0.48	9.96 ± 0.40
a^*	24.47 ± 0.29	3.17 ± 0.24
b^*	13.32 ± 0.45	−6.06 ± 0.48
ΔE	–	28.81

L^* = lightness (0 = black, 100 = white), $+b^*$ (yellow), $-b^*$ (blue), $+a^*$ (red), $-a^*$ (green).

Figure 6 shows the TGA/DTGA of citrate-capped AuNP, where five mass losses were observed. The first mass loss corresponds to the release of water from the sample and began at (166 °C) to (250 °C), representing 7.83% of the initial mass. The second mass loss was between (256 °C) and (336.5 °C), with a peak at (288.5 °C) corresponding to a loss of 5.5%. The third endothermic peak was observed in the range of (375.3 °C) to (482.6 °C), with a peak at (437.3 °C) corresponding to a mass loss of 5.1%. The second and third peaks after the dehydration peak probably correspond to the degradation of the sample and/or the organic matter covering the trisodium citrate. The fourth and fifth mass losses were observed in the temperature range of (547 °C) to (695.9 °C), with peaks at (54.8 °C) and (695.8 °C) respectively. It was observed that in these last degradations, the losses are lower and are in the order of 0.18% and 0.27%, respectively, which indicates that they coincide with the final degradation of the trisodium citrate-capped gold nanoparticles. The total mass loss was equivalent to 18.9%. Similar results were obtained in previous studies [35].

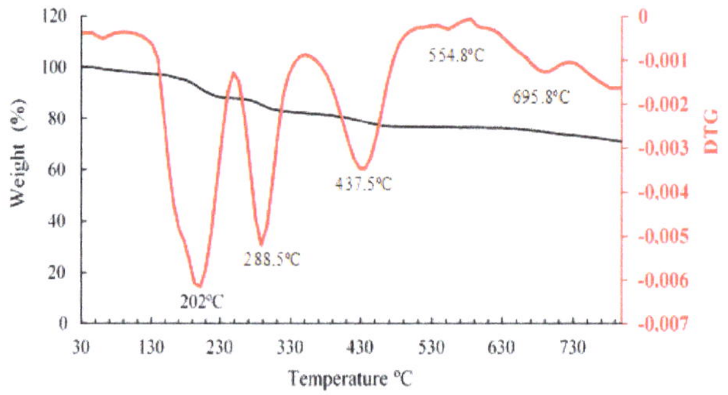

Figure 6. Thermogravimetric analysis TGA (black) and DTG (red) curves of citrate-capped AuNP.

According to the DSC curve (Figure 7), it was observed that the citrate-capped AuNP in the first heating showed endothermic peaks at (159.4°C), (76.4°C), (19.1°C), and (298.2 °C) and an exothermic peak at (323.7 °C) where the latter corresponds to the crystallization of trisodium citrate. The first endothermic peak corresponds to the evaporation of adsorbed water. The second and third peaks could correspond to the beginning of the degradation of NaCit. The endothermic peak at (298.2 °C) is related to the melting point of the NaCit that coats the surface of the gold nanoparticles. Similar results were obtained in previous studies by [36]. The XRD patterns of the AuNPs (Figure 8a) showed diffraction peaks at 2θ = 38.1°, 44.3°, 64.5°, and 77.5°. These values match the corresponding (111), (200), (220), and (311) lattice planes of AuNP and were in accordance with the Joint Committee on Powder Diffraction Standards (JCPDS) database no. 04-0784, confirming the formation of AuNP. According to the results obtained, it is clear that the AuNPs formed were crystalline in nature. Similar results were reported by [37] and [38]. The strong and high diffraction peak found here indicates the high crystallinity of the synthesized gold nanoparticles. To determine the average size of the sample crystals, the Debye–Scherrer formula was applied for the four diffraction peaks: (111) at 38.1°, (200) at 44.3°, (220) at 64.5°, and (311) at 77.5°, obtaining an average size of 15.4 nm. The diffraction patterns obtained demonstrated that the synthesized AuNPs were composed of pure crystalline gold nanoparticles. These results were in perfect agreement with the particle size obtained from the TEM analysis. Related results were reported by [39]. Furthermore, the appearance of several absorption peaks at 2θ = 9.1°, 11.2°, 17.5°, 18.6°, 19.9°, 23.9°, 25.1°, 27.3°, 31.6°, 33.4°, 35.4°, 45.3°, 56.5°, 66.1°, and 75.2°. All these absorption peaks correspond to the NaCit that covers the AuNP surface. Also, it was observed that the diffraction peaks localized at 2θ = 56°, 66°, and 75.2°, belonging to NaCit, showed a higher intensity in AuNP (Figure 8a) than in citrate alone (Figure 8b). This may indicate a weak influence on the AuNP surface to maintain the stability.

As can be observed, the diffraction patterns (XRD) of NaCit used as a reducer and a stabilizer in the synthesis of gold nanoparticles are presented in Figure 8b, where it is observed that it presented several crystalline peaks. The most pronounced peaks were visualized at 2θ = 11.2°, 17.5°, 23.9°, 27.3°, 32.6°, 34.1°, 36.7°, 40.1°, 45°, and 46.3°; it is noted that the main peaks obtained were found in the XRD patterns of ICDD-00-056-0123 and correspond to the crystallinity of NaCit. The corresponding results were achieved in previous studies by [35]. Other lower intensity crystalline peaks were observed at 2θ = 48.5° to 58.5°. Finally, the crystalline peaks of minimum intensity perceived after 2θ = 60° are meaningless.

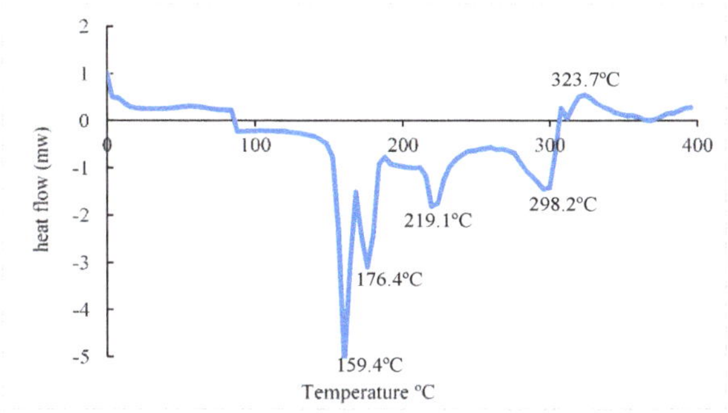

Figure 7. Differential scanning calorimetry analysis of gold nanoparticles (AuNP) capped by trisodium citrate.

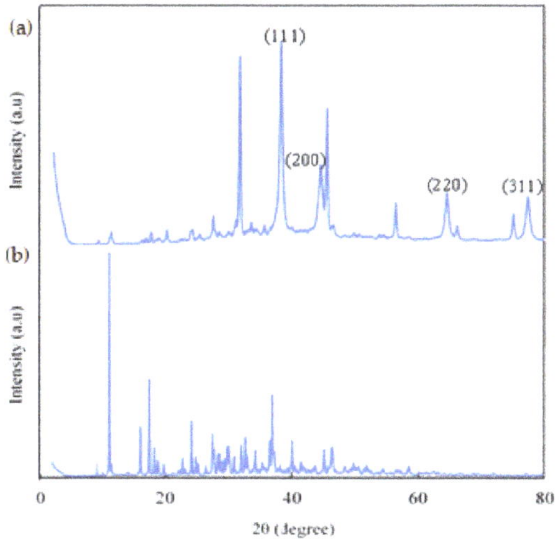

Figure 8. X-ray diffraction of (**a**) AuNP and (**b**) NaCit.

3.2. The Development and Characterization of Alg. Bead-Shaped Indicator Film

Figure 9 shows the results of the development of the indicator film (Alg. bead shape) using the casting technique, observing the complete formation of the film based on the gold nanoparticles and their in vitro evaluation with ethylenediamine, and observing the color changes after the reaction with the test molecule.

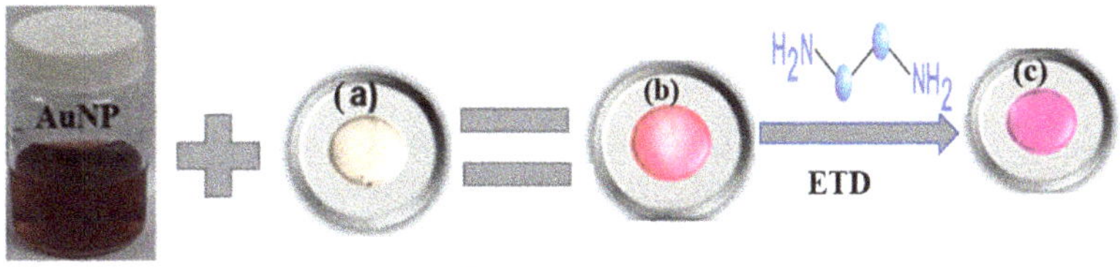

Figure 9. (**a**) Sodium Alginate (Alg.), (**b**) the colorimetric indicator (Alg/AuNP), and (**c**) the colorimetric indicator + ETD. AuNP: gold nanoparticles. ETD: Ethylenediamine.

In the FTIR spectra of the Alg. beads (Figure 10), a large absorption band was observed in the range of 3600 to 3000 1/cm, related to the stretching vibration of the OH group and the C-H vibration bands at 2935 1/cm. The bands observed at 1594 1/cm and 1415 1/cm were attributed to the asymmetric and symmetric stretching vibrations of the COO^- groups of Alg., respectively, and are specific for ionic bonding. The shoulder located at 1080 1/cm, which was related to C–C and C–O stretching, can also be attributed to crossover. The absorption band detected at (1027 ± 4 1/cm) shows a higher intensity in relation to the 1080 cm^{-1} band, suggesting a stronger O-H binding vibration or a stronger binding of Ca^{2+} to the guluronic acids of Alg. In contrast, the stretching vibration bands observed at approximately 939 1/cm and 889 1/cm were specific to the guluronic and mannuronic acids. Also, small displacements of the carboxyl groups were observed, which may be indicative of an ionic union between Ca^{2+} and the Alg. chains [40]. In addition, in the evaluation of the colorimetric indicator with ETD, a crossover was observed in the stretching of the O-H group in the range of 3600–3000 1/cm. This was due to the reaction of the ETD with the carboxyl group of the NaCit on the surface of the gold nanoparticles, causing the nanoparticles to clump together. This led to the appearance of new bands in the indicated range.

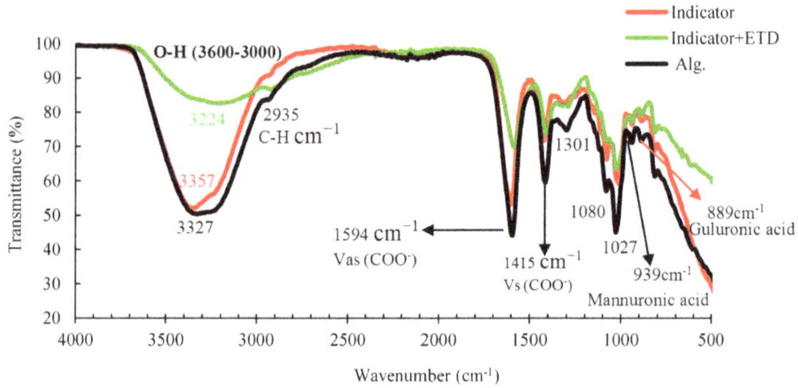

Figure 10. Infrared spectroscopy (FTIR) of the colorimetric indicator (Alg./AuNP), the colorimetric indicator + ETD, and Alg.

On the other hand, it is known that Alg. is a polymer with a strong hydrophilic character [41], so its contact with the ethylenediamine solution and the modifications provided by the gold nanoparticles in its structure could produce a crossover.

In Figure 11 and Table 2, the four or five main stages or successive losses can be seen in the thermograms. In the first stage, the weight loss occurred at the beginning of the heat treatment, in the range of (47.8 °C) to (168.8 °C) which was mainly attributed to the loss of free water absorbed by Alg. and the indicator [42]. The second stage of thermal degradation was sensed at the onset of the temperature ranging from about (172.1 °C) to (259.3 °C) associated with the thermal degradation of the polymer as well as the polymer capping around the nanoparticles. The third and fourth weight losses could be related to the conversion of the remaining polymer into carbon residues [43] in the case of Alg.; however, in the colorimetric indicator, the third and fourth losses could correspond to the degradation of the citrate and the residual compounds present in the sample. It was observed that the colorimetric indicator with a weight loss of 47% presented certain thermal stability in relation to Alg. and its evaluation with ETD. This is because the addition of AuNP could have affected the stability of the Alg. beads due to the increase in the negative charge density in the hydrogel matrix [44]. Moreover, it was observed that there was a greater weight loss (54.7%) in the control Alg. with respect to the colorimetric indicator and its reaction with ETD, respectively, because Alg. is a hydrophilic polyanion and is sensitive to pH [45]. On the other hand, a weight loss in the order of 55.7% was observed in the colorimetric indicator evaluated with ETD. This weight loss could be due to the interaction of the amino group (NH2) of the ETD with the carboxyl group (COO^-) of the oxygen-hungry NaCit on the surface of the AuNPs, which results in destabilization of the citrate and aggregation of the nanoparticles that could, to a certain extent, decrease the temperature resistance of the colorimetric indicator until it degrades above (700 °C). No significant differences were observed between the weights lost ($p \geq 0.05$).

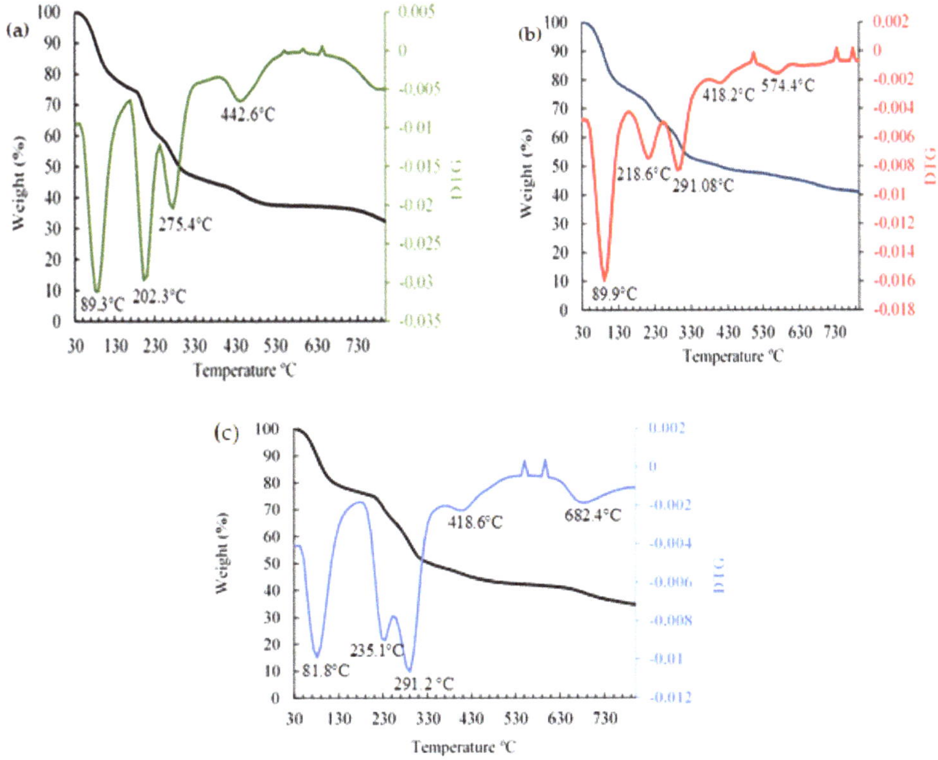

Figure 11. Thermogravimetric analysis of: (**a**) Alg. TGA–black DTG–green, (**b**) the colorimetric indicator (Alg./AuNP) TGA–black, DTG–red, and (**c**) the colorimetric indicator + ETD TGA–black, DTG–blue.

Table 2. TGA analysis results of Alg., the colorimetric indicator, and the colorimetric indicator + ETD.

Samples	Temperature (°C)	T-Onset (°C)	T50 (°C)	T Endset (°C)	Weight Lost (%)
Sodium Alginate (Alg.)	30–800	49.4	89.3	168.8	23.4 ± 1.3 [a]
		172.8	202.3	243.7	15.8 ± 1.6 [a]
		246.5	275.4	316.7	11.1 ± 0.4 [a]
		388.3	442.6	460.3	4.4 ± 0.5[a]
Indicator (Alg. + AuNP)	30–800	47.8	89.9	153.9	22.2 ± 4 [a]
		172.1	218.6	237.4	9.4 ± 0,7 [a]
		239.1	291.08	314.4	12 ± 1.6 [a]
		347.8	418.2	397.2	1.7 ± 0.3 [a]
		527.8	574.4	620.8	1.7 ± 0.08 [a]
Indicator + ETD	30–800	49.5	81.8	141.2	19.9 ± 4 [a]
		207.1	235.1	259.3	19 ± 4.6 [a]
		267.3	291.2	363.1	11.3 ± 0.4 [a]
		375.2	418.6	445.5	3.6 ± 0.6 [a]
		637.6	682.4	729.8	1.9 ± 0.3 [a]

[a]: Same lower-case letters in the same row indicate no statistically significant differences.

The DSC analysis curves reflect the thermal properties of the control Alg., the colorimetric indicator, and the colorimetric indicator evaluation with ETD. The endothermic melting peak of water crystallization appeared at 100.7, 103.6, and 106.9 °C in Alg., the colorimetric indicator, and the colorimetric indicator with ETD, respectively (Figure 12). The results obtained from the current experiment did not present significant differences ($p \geq 0.05$) in any of the analyzed indicators nor in the control, with the exception of a greater broadening of the water absorption peak in the evaluated indicator with ETD, which could be due to the hydrophilic behavior of Alg. [46] and/or the formation of new chemical bonds and the interaction between the components of the in vitro assessed colorimetric indicator. Additionally, the enthalpy of fusion (ΔH) of the control (Alg.) (−536.7 J/g) was higher than that of the colorimetric indicator (−1217.8 J/g) and the colorimetric indicator with ETD (−556.9 J/g). This could indicate that AuNP with the analyte caused a reduction in this parameter. Chen et al., showed analogous results when evaluating the thermal property (DSC) of Alg. with the addition of thymol [47].

The XRD diffractograms of Alg., the colorimetric indicator, and the colorimetric indicator assessment with ETD are presented in Figure 13, where it was observed that Alg. (Figure 13a) presented diffraction peaks of lower intensity at 2θ = 31.5°, 45.2°, 56.4°, 66.2°, and 75.2°, corresponding to its amorphous nature. Similar results were reported in previous works by [48] and [49]. On the other hand, the colorimetric indicator (Figure 13b) showed two peaks of different intensities at 2θ = 31.7°, corresponding to Alg., and 2θ = 45.5°, which could be related to NaCit and/or Alg. because this same peak was observed with minimal variation in both. Furthermore, the peaks observed at 2θ = 22.4° and 38.3° belonged to NaCit and AuNP, respectively. Additionally, small peaks were observed at 2θ = 56.5°, 66.3°, and 75.2° associated with Alg. In other ways, the results obtained from the colorimetric indicator with ETD (Figure 13c) presented two peaks at 2θ = 31.7° and 45.3°. The first was associated with Alg. and the second may be associated with Alg. and/or NaCit. Likewise, two small peaks were visualized at 2θ = 22.5° and 29.4° related to NaCit. The other small peaks observed at 2θ = 56.4° and 75.2° were related to Alg. Additionally, two characteristic peaks of AuNP located at 2θ = 38.6° and 43.2° were observed, indicating the crystallinity of AuNP.

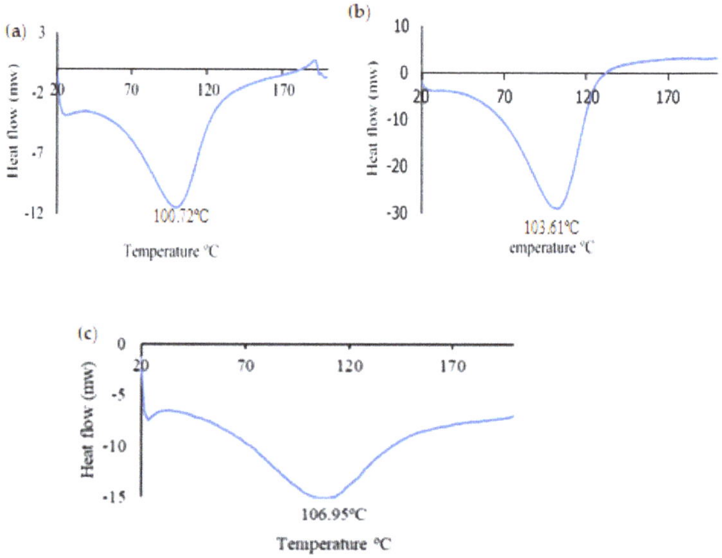

Figure 12. Differential scanning calorimetry DSC. (**a**) Alg., (**b**) the colorimetric indicator (Alg./AuNP), and (**c**) the colorimetric indicator + ETD.

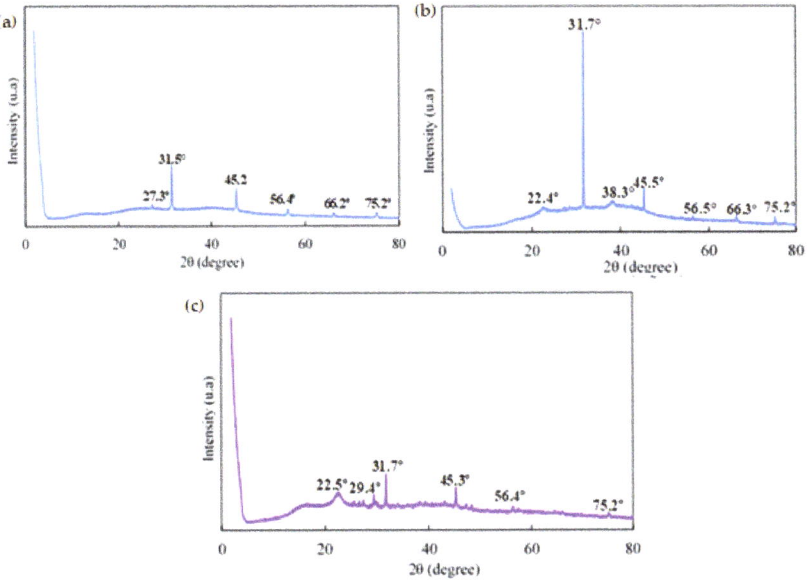

Figure 13. X-ray diffraction of: (**a**) Alg., (**b**) the colorimetric indicator, and (**c**) the colorimetric indicator + ETD.

3.3. Determination of Total Volatile Basic Nitrogen (TVB-N)

Volatile compounds, such as trimethylamine, ammonia, and dimethylamine, produced by the destructive activities of microorganisms, are considered one of the most important parameters to determine the quality and freshness of fish and are known as the total volatile basic nitrogen (TVB-N) [50].According to the results (Figure 14), the initial values of TVB-N were 13.3 and 13.8 for the colorimetric indicator and the control, respectively, and they coincided correctly with the low initial microbial counts. These values are similar to the results obtained in previous research [51]; however, the TVB-N content increased progressively from 13.3 to 45.9 mg/100 g and from 13.8 to 44.8 mg N/100 g in the colorimetric indicator and the control, respectively, after 9 days of storage at 5 °C. This increase may be due to: (1) the activity of the spoilage bacteria that grow in the fish; (2) the enzymatic reaction that can occur in stored muscle food [52]; (3) and autolysis. A comparison of TVB-N values with colorimetric analysis showed that the color change of the colorimetric indicator was proportional to the TVB-N content and the pH change because, immediately after the death of the fish, a series of physical and chemical changes begin to occur in its body that led to its final alteration. Among these changes are mucus production on the body surface, rigor mortis, autolysis, organoleptic changes, and bacterial decompositions. The latter leads to the gradual increase in pH along with the TVB-N value and bacterial growth. All these parameters vary almost at the same time as the stored fish spoilage progresses; therefore, the increase in the amount of TVB-N caused an increase in pH and a color change in the colorimetric indicator.

3.4. Mesophilic Aerobic Count

The changes in aerobic bacteria counting in the salmon samples stored at 5 °C are shown in Figure 15. The initial Aerobic Mesophyll Count (AMC) values for the marine fish prior to cold storage typically range from 2 to 4 $\log_{10}$ CFU/g, and a value of 6 $\log_{10}$ CFU/g is considered the upper limit of acceptability [53]. Therefore, the values obtained in this study are within this range: 2.82 and 2.80 $\log_{10}$ CFU/g for the control and the sample with the colorimetric indicator, respectively. This indicates the freshness of the salmon with a low microorganism count. However, the bacteria grew gradually until reaching a value of 6.8 $\log_{10}$ CFU/g on the seventh day and reached values greater than 9 $\log_{10}$ CFU/g on the ninth day of storage at a temperature of (5 °C). This growth of microorganisms led to a greater production of biogenic amines, changes in the pH value, and the deterioration of the fish. However, there was no significant difference ($p \geq 0.05$) in the growth of the bacteria between the control (fish sample only) and the sample with a colorimetric indicator inserted so that they grew at the same rate. Similar results were obtained when evaluating the spoilage potential of *P. fluorescens* in salmon at different temperatures [54]. According to previous studies, the bacteria that grow more at refrigeration temperatures are the so-called psychotropics. Also, Pseudomonas spp. was suggested as the main specific spoilage organism [55] and it is useful to predict shelf life with a cut-off level of 6.5 $\log_{10}$ CFU/g. Cheng and Sun reported that the main group of bacteria causing the spoilage of refrigerated or modified atmosphere vacuum-packed fish products are lactic acid bacteria, such as *lactobacillus*, *Streptococcus*, *Leuconostoc*, and *Pediococcus* spp. These bacteria are capable of spoiling foods by fermenting sugars and commonly cause undesirable defects, such as off-flavors, discoloration, gas production, slime production, and the lowering of pH [56]. In addition, they are capable of inhibiting the growth of other bacteria due to the formation of lactic acid and bacteriocins, and this facilitates their selective growth during fish spoilage [57].

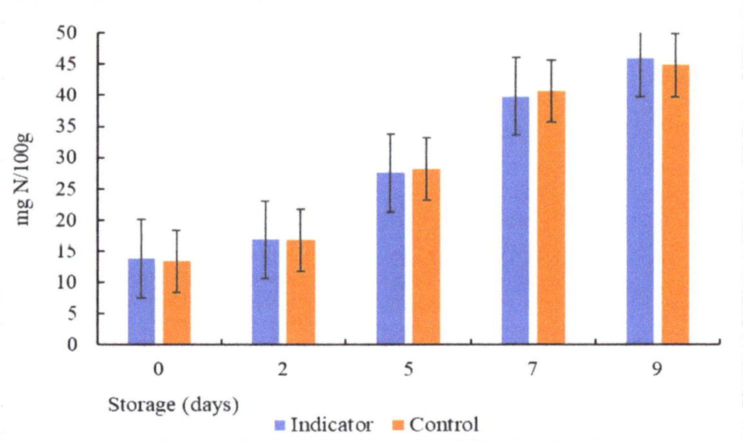

Figure 14. Determination of total volatile basic nitrogen (TVB-N) in fish stored at (5 °C).

3.5. Determination of pH

The changes in the pH values during storage are presented in Figure 16, where a decrease in the pH value was observed on day 2 of storage. This may be related to the production of some acidic substances because both the skin and the digestive system of fish can host a variety of bacteria, among which are lactic acid bacteria, which are facultatively anaerobic and grow very well under microaerophilic conditions [58]. These bacteria produce lactic acid as the main metabolic end product of carbohydrate consumption. When the lactic acid population increased during storage, there was an increase in the lactic acid capable of neutralizing the alkaline amine products, thus reducing the pH [57]. Undoubtedly, lactic acid bacteria predominated in the total natural microflora of the vacuum-packed fish fillets. On the other hand, another factor in the drop in pH could be the by-products of lipid oxidation, caused by the reaction of the amine compounds with the aldehydes. Similar results were obtained in previous investigations [53]; however, on days 7 and 9 of storage, variations in pH values from 7.54 and 7.92 to 9.45 and 9.60 were observed for the colorimetric indicator and the control, respectively. This was due to an increase in the growth of microorganisms in the muscle food, producing an increase in the TVB-N values and pH changes.

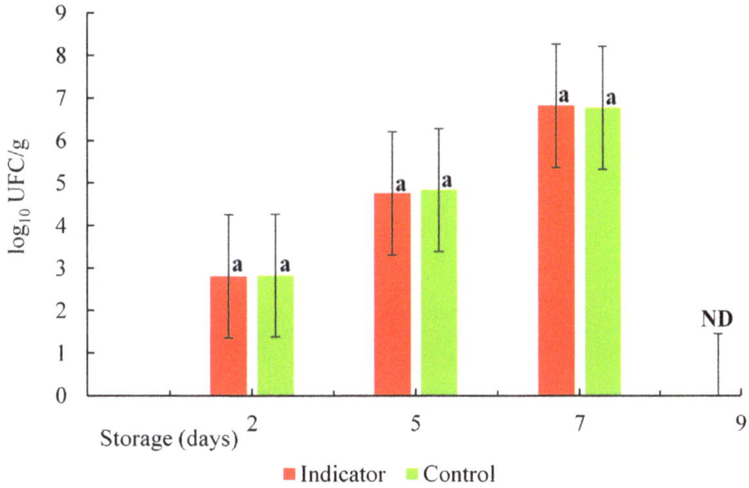

Figure 15. Count of mesophilic aerobes as a function of time. Same lower-case letters in each day indicate no statistically significant differences. ND: not determined.

3.6. Indicator Color Measurement

The color parameters for the colorimetric indicators based on metallic nanoparticles were determined. According to Figure 17 and Table 3, a^* values decreased from 33 (redder) to 13 (greener) after 9 days of storage at refrigeration temperature. This means that the red color had a lower intensity at the end of the storage period. The values recorded for b* decreased from 13 to −24.11, indicating that the blue color appeared more pronounced at the end of the storage period. However, it was observed that the blue color began to appear from days 5 to 9 of storage; therefore, it is obvious that the blue color predominated in the evaluation of the colorimetric indicator. Researchers have reported that a ΔE* value greater than 4 can be easily detected with the naked eye, while values greater than 12 imply a complete color difference that is detectable even by untrained panelists [59]. Therefore, the ΔE* value of the colorimetric indicators obtained in this experiment could be detected by the human eye during fish storage. Generally, the indicators used as a colorimetric sensor exhibit a wide range of color variations depending on the pH, which can be affected by the TVB-N content during the storage period. This has the obvious advantage that color changes can be inspected with the naked eye to detect the rate of spoilage of the packaged fish. On the other hand, a progressive increase in the ΔE parameter was observed until the end of the storage time. This can be attributed to the loss of water that occurs as time passes, which gives rise to greater water deposits on the surface of the fish and inevitably leads to a variation in the value of this parameter.

When comparing the total color differences (ΔE) (Table 3) from the second to the last day of storage, it was observed that there were significant differences in the color of the colorimetric indicators analyzed using $p \leq 0.05$.

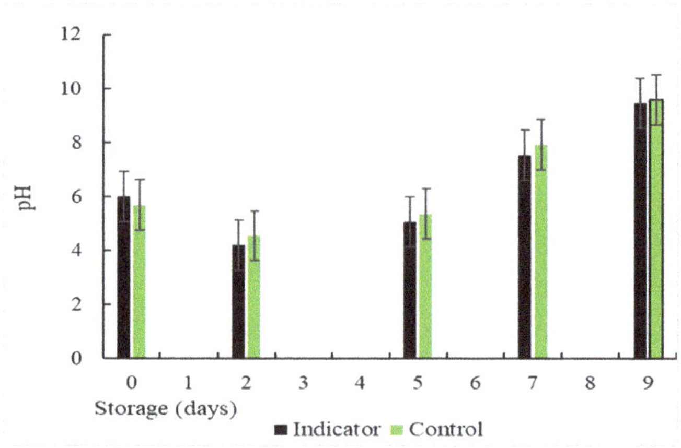

Figure 16. Changes in the pH value during storage at a temperature of (5 °C).

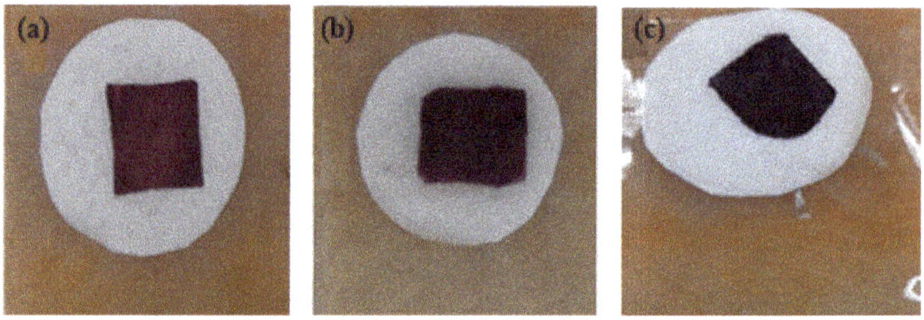

Figure 17. Color changes of the colorimetric indicator in fish stored at (5 °C). (**a**) Day 0, (**b**) day 7, (**c**) day 9.

Table 3. Indicator colorimetric analysis.

Storage Times (Days)	L^*	a^*	b^*	ΔE
0	17.86 ± 3.57	33.47 ± 4.09	13.01±3.94	0
2	9.60 ± 0.32	12.42 ± 2.09	0.8 0± 0.75	25.7
5	24.57 ± 0.87	12.18 ± 1.76	−8.05 ± 0.53	30.6
7	10.26 ± 0.12	13.39 ± 2.32	−19.11 ± 0.72	38.7
9	8.86 ± 0.61	13.73 ± 1.04	−24.11 ± 1.01	43.1

L^*: lightness (0 = black, 100 = white), $+b^*$ (yellow), $-b^*$ (blue), $+a^*$ (red), $-a^*$ (green). ΔE: total color difference.

3.7. Antibacterial Activity of AuNP

The antibacterial properties of AuNP can affect the effectiveness of the colorimetric indicator in the early stages of fish spoilage. However, according to the results obtained from the microbiological analyses (Mesophilic aerobic count), it can be observed that the bacteria grew progressively until the last day of storage, without presenting significant inhibition at the beginning. Nor was any reduction detected in the TVB-N values in the first days of storage, that means the amount of volatile compounds continued to vary on the second day of storage, going from 13.86 ± 0.5 to 16.87 and 16.67 mg of N/100 g on the indicator and the control, respectively. Furthermore, it was noticed that the growth of certain bacteria on the second day of storage reduced the pH values. Hence, it can be concluded that no antibacterial activity of gold nanoparticles was observed at this stage of development of the colorimetric indicator. This may be due to the particles' size because the size and dispersion capacity are extremely necessary for good antibacterial activity of the nanoparticles. According to previous studies, it was confirmed that the size of the nanoparticles has great importance in modifying the cell membrane of microorganisms; therefore, it may affect their antimicrobial activity [60].

On the other hand, other researchers have indicated that citrate-capped gold nanoparticles did not have antibacterial activity when compared with kanamycin [61]. Also, AuNPs have negligible bactericidal effects at high concentrations [62]; therefore, in this study, the AuNP concentration used in the indicator was 0.3% v/v. Based on the results achieved, the synthesized AuNPs did not affect the effectiveness of the indicator in the first days of fish storage.

4. Conclusions

In this study, a new colorimetric indicator based on gold nanoparticles inserted in a biodegradable, non-toxic, and eco-friendly polymer was developed to monitor fish spoilage products. The colorimetric indicator showed the ability to successfully detect total volatile basic nitrogen in the last days of fish storage through a color change; therefore, it can be used to detect TVB-N in spoiled fish.

The ETD (similar to TVB-N) was successfully used as a test molecule in the in vitro evaluation of the detection capacity of AuNP and the colorimetric indicator. This allows us to quickly observe the discovery power of AuNPs and their possible application in smart packaging. The results obtained from the XRD of AuNP synthesized with trisodium citrate dihydrate indicated its crystalline nature and face-centered cubic structure. On the other hand, the semi-crystalline nature of Alg. was not observed in the diffraction pattern.

Given the progressive changes observed in the values of TVB-N, pH, mesophilic aerobic count (RAM), and color, it can be certified that the deterioration of the Atlantic salmon (*salmo salar*) occurred during the last days of storage at refrigeration temperature. Furthermore, significant changes were observed in bacterial growth above the allowable acceptability limit of 6.5 log10 CFU/g, as reported by some researchers. On the other hand, the values obtained in color determination, mainly the total color difference (ΔE), gradually increased over the storage time until a color difference visible to the naked eye was obtained. These results indicated that the colorimetric indicator showed its effectiveness in detecting fish spoilage products through a color change. In conclusion, the gold nanoparticles-based colorimetric indicator system is suitable for monitoring the quality of refrigerated fish at (5 °C).

Author Contributions: Conceptualization, L.P.; methodology, J.E.B.B. and L.P.; validation, J.E.B.B. and P.A.L.B.; formal analysis, L.P.; investigation, L.P.; resources, J.E.B.B. and National Research and Development Agency (ANID).; data curation, L.P.; writing—original draft preparation, L.P.; writing—review and editing, L.P., J.E.B.B., P.A.L.B., A.T.M. and F.J.R.-M.; visualization, J.E.B.B. and P.A.L.B.; supervision, J.E.B.B. and A.G; project administration, L.P.; funding acquisition, J.E.B.B. and L.P. All authors have read and agreed to the published version of the manuscript.

Funding: This research was funded by the National Research and Development Agency (ANID) Folio 21211119 through the National Doctoral Scholarship and the Center of Excellence with Basal Financing (CEDENNA), grant number AFB220001.

Institutional Review Board Statement: Not applicable.

Data Availability Statement: Data are contained within the article.

Acknowledgments: The authors are also grateful to the University of Santiago de Chile for the Foreigner grant and Research Support grant (Lissage Pierre).

Conflicts of Interest: The authors declare no conflict of interest.

References

1. Pataca, J.K.; Porto-Figueira, P.; Pereira, J.A.; Caldeira, H.; Câmara, J.S. Profiling the occurrence of biogenic amines in different types of tuna samples using an improved analytical approach. *LWT* **2020**, *139*, 110804. [CrossRef]

2. Visciano, P.; Schirone, M.; Paparella, A. An overview of histamine and other biogenic amines in fish and fish products. *Foods* **2020**, *9*, 1795. [CrossRef]
3. Zhai, X.; Li, Z.; Shi, J.; Huang, X.; Sun, Z.; Zhang, D.; Zou, X.; Sun, Y.; Zhang, J.; Holmes, M.; et al. A colorimetric hydrogen sulfide sensor based on gellan gum-silver nanoparticles bionanocomposite for monitoring of meat spoilage in intelligent packaging. *Food Chem.* **2019**, *290*, 135–143. [CrossRef]
4. Doeun, D.; Davaatseren, M.; Chung, M.-S. Biogenic amines in foods. *Food Sci. Biotechnol.* **2017**, *26*, 1463–1474. [CrossRef] [PubMed]
5. Barbieri, F.; Montanari, C.; Gardini, F.; Tabanelli, G. Biogenic amine production by lactic acid bacteria: A review. *Foods* **2019**, *8*, 17. [CrossRef] [PubMed]
6. Xu, Y.; Zang, J.; Regenstein, J.M.; Xia, W. Technological roles of microorganisms in fish fermentation: A review. *Crit. Rev. Food Sci. Nutr.* **2021**, *61*, 1000–1012. [CrossRef] [PubMed]
7. Comas-Basté, O.; Latorre-Moratalla, M.L.; Sánchez-Pérez, S.; Veciana-Nogués, M.T.; Vidal-Carou, M.D.C. Histamine and other biogenic amines in food. From scombroid poisoning to histamine intolerance. *Biog. Amines* **2019**, *1*. [CrossRef]
8. Fusek, M.; Michálek, J.; Buňková, L.; Buňka, F. Modelling biogenic amines in fish meat in Central Europe using censored distributions. *Chemosphere* **2020**, *251*, 126390. [CrossRef] [PubMed]
9. Senapati, M.; Sahu, P.P. Onsite fish quality monitoring using ultra-sensitive patch electrode capacitive sensor at room temperature. *Biosens. Bioelectron.* **2020**, *168*, 112570. [CrossRef] [PubMed]
10. Ghaani, M.; Cozzolino, C.A.; Castelli, G.; Farris, S. An overview of the intelligent packaging technologies in the food sector. *Trends Food Sci. Technol.* **2016**, *51*, 1–11. [CrossRef]
11. Majdinasab, M.; Hosseini, S.M.H.; Sepidname, M.; Negahdarifar, M.; Li, P. Development of a novel colorimetric sensor based on alginate beads for monitoring rainbow trout spoilage. *J. Food Sci. Technol.* **2018**, *55*, 1695–1704. [CrossRef]
12. Lin, T.; Wu, Y.; Li, Z.; Song, Z.; Guo, L.; Fu, F. Visual monitoring of food spoilage based on hydrolysis-induced silver metallization of au nanorods. *Anal. Chem.* **2016**, *88*, 11022–11027. [CrossRef]
13. Taherkhani, E.; Moradi, M.; Tajik, H.; Molaei, R.; Ezati, P. Preparation of on-package halochromic freshness/spoilage nanocellulose label for the visual shelf life estimation of meat. *Int. J. Biol. Macromol.* **2020**, *164*, 2632–2640. [CrossRef]
14. Wang, L.; Wu, Z.; Cao, C. Technologies and fabrication of intelligent packaging for perishable products. *Appl. Sci.* **2019**, *9*, 4858. [CrossRef]
15. Ding, L.; Li, X.; Hu, L.; Zhang, Y.; Jiang, Y.; Mao, Z.; Xu, H.; Wang, B.; Feng, X.; Sui, X. A naked-eye detection polyvinyl alcohol/cellulose-based pH sensor for intelligent packaging. *Carbohydr. Polym.* **2020**, *233*, 115859. [CrossRef]
16. Morsy, M.K.; Zór, K.; Kostesha, N.; Alstrøm, T.S.; Heiskanen, A.; El-Tanahi, H.; Sharoba, A.; Papkovsky, D.; Larsen, J.; Khalaf, H.; et al. Development and validation of a colorimetric sensor array for fish spoilage monitoring. *Food Control* **2016**, *60*, 346–352. [CrossRef]
17. El-Nour, K.M.A.; Salam, E.T.A.; Soliman, H.M.; Orabi, A.S. Gold Nanoparticles as a Direct and Rapid Sensor for Sensitive Analytical Detection of Biogenic Amines. *Nanoscale Res. Lett.* **2017**, *12*, 231. [CrossRef] [PubMed]
18. Dudnyk, I.; Janeček, E.-R.; Vaucher-Joset, J.; Stellacci, F. Edible sensors for meat and seafood freshness. *Sens. Actuators B Chem.* **2018**, *259*, 1108–1112. [CrossRef]
19. Zhai, X.D.; Shi, J.Y.; Zou, X.B.; Wang, S.; Jiang, C.P.; Zhang, J.J.; Huang, X.W.; Zhang, W.; Holmes, M. Novel colorimetric films based on starch/polyvinyl alcohol incorporated with roselle anthocyanins for fish freshness monitoring. *Food Hydrocoll.* **2017**, *69*, 308–317. [CrossRef]
20. Chen, H.-Z.; Zhang, M.; Bhandari, B.; Yang, C.-H. Novel pH-sensitive films containing curcumin and anthocyanins to monitor fish freshness. *Food Hydrocoll.* **2020**, *100*, 105438. [CrossRef]
21. Patil, T.; Gambhir, R.; Vibhute, A.; Tiwari, A.P. Gold nanoparticles: Synthesis methods, functionalization and biological applications. *J. Clust. Sci.* **2023**, *34*, 705–725. [CrossRef]
22. Nadaf, S.J.; Jadhav, N.R.; Naikwadi, H.S.; Savekar, P.L.; Sapkal, I.D.; Kambli, M.M.; Desai, I.A. Green synthesis of gold and silver nanoparticles: Updates on research, patents, and future prospects. *OpenNano* **2022**, *8*, 100076. [CrossRef]
23. Herizchi, R.; Abbasi, E.; Milani, M.; Akbarzadeh, A. Current methods for synthesis of gold nanoparticles. *Artif. Cells Nanomed. Biotechnol.* **2016**, *44*, 596–602. [CrossRef]
24. Chang, C.-C.; Chen, C.-P.; Wu, T.-H.; Yang, C.-H.; Lin, C.-W.; Chen, C.-Y. Gold nanoparticle-based colorimetric strategies for chemical and biological sensing applications. *Nanomaterials* **2019**, *9*, 861. [CrossRef]
25. Sun, J.; Lu, Y.; He, L.; Pang, J.; Yang, F.; Liu, Y. Colorimetric sensor array based on gold nanoparticles: Design principles and recent advances. *TrAC Trends Anal. Chem.* **2020**, *122*, 115754. [CrossRef]
26. Kolb, A.N.D.; Johnston, J.H. Colour tuneable anisotropic, water-dispersible gold nanoparticles stabilized by chitosan. *Gold Bull.* **2016**, *49*, 1–7. [CrossRef]
27. Mahatnirunkul, T.; Tomlinson, D.C.; McPherson, M.J.; Millner, P.A. One-step gold nanoparticle size-shift assay using synthetic binding proteins and dynamic light scattering. *Sens. Actuators B Chem.* **2022**, *361*, 131709. [CrossRef]
28. Lin, J.-H.; Tseng, W.-L. Ultrasensitive detection of target analyte-induced aggregation of gold nanoparticles using laser-induced nanoparticle Rayleigh scattering. *Talanta* **2015**, *132*, 44–51. [CrossRef] [PubMed]

29. Rani, M.; Moudgil, L.; Singh, B.; Kaushal, A.; Mittal, A.; Saini, G.S.S.; Tripathi, S.K.; Singh, G.; Kaura, A. Understanding the mechanism of replacement of citrate from the surface of gold nanoparticles by amino acids: A theoretical and experimental investigation and their biological application. *RSC Adv.* **2016**, *6*, 17373–17383. [CrossRef]
30. Ranjan, R.; Kirillova, M.A.; Kratasyuk, V.A. Ethylene diamine functionalized citrate-capped gold nanoparticles for metal-enhanced bioluminescence. *J. Sib. Fed. Univ. Biol.* **2020**, *13*, 322–330. [CrossRef]
31. Contreras-Trigo, B.; Díaz-García, V.; Guzmán-Gutierrez, E.; Sanhueza, I.; Coelho, P.; Godoy, S.E.; Torres, S.; Oyarzún, P. Slight ph fluctuations in the gold nanoparticle synthesis process influence the performance of the citrate reduction method. *Sensors* **2018**, *18*, 2246. [CrossRef] [PubMed]
32. Wulandari, P.; Nagahiro, T.; Fukada, N.; Kimura, Y.; Niwano, M.; Tamada, K. Characterization of citrates on gold and silver nanoparticles. *J. Colloid Interface Sci.* **2015**, *438*, 244–248. [CrossRef] [PubMed]
33. Khalkho, B.R.; Deb, M.K.; Kurrey, R.; Sahu, B.; Saha, A.; Patle, T.K.; Chauhan, R.; Shrivas, K. Citrate functionalized gold nanoparticles assisted micro extraction of L-cysteine in milk and water samples using Fourier transform infrared spectroscopy. *Spectrochim. Acta Part A Mol. Biomol. Spectrosc.* **2022**, *267*, 120523. [CrossRef] [PubMed]
34. Frost, M.S.; Dempsey, M.; Whitehead, D.E. The response of citrate functionalised gold and silver nanoparticles to the addition of heavy metal ions. *Colloids Surf. A Physicochem. Eng. Asp.* **2017**, *518*, 15–24. [CrossRef]
35. Elbeyli, I.Y. Production of crystalline boric acid and sodium citrate from borax decahydrate. *Hydrometallurgy* **2015**, *158*, 19–26. [CrossRef]
36. Gao, J.; Wang, Y.; Hao, H. Investigations on dehydration processes of trisodium citrate hydrates. *Front. Chem. Sci. Eng.* **2012**, *6*, 276–281. [CrossRef]
37. Rajakumar, G.; Gomathi, T.; Rahuman, A.A.; Thiruvengadam, M.; Mydhili, G.; Kim, S.-H.; Lee, T.-J.; Chung, I.-M. Biosynthesis and biomedical applications of gold nanoparticles using eclipta prostrata leaf extract. *Appl. Sci.* **2016**, *6*, 222. [CrossRef]
38. Sadeghi, B.; Mohammadzadeh, M.; Babakhani, B. Green synthesis of gold nanoparticles using Stevia rebaudiana leaf extracts: Characterization and their stability. *J. Photochem. Photobiol. B Biol.* **2015**, *148*, 101–106. [CrossRef]
39. Reddy, G.B.; Ramakrishna, D.; Madhusudhan, A.; Ayodhya, D.; Venkatesham, M.; Veerabhadram, G. Catalytic Reduction of p-Nitrophenol and Hexacyanoferrate (III) by Borohydride Using Green Synthesized Gold Nanoparticles. *J. Chin. Chem. Soc.* **2015**, *62*, 420–428. [CrossRef]
40. Badita, C.R.; Aranghel, D.; Burducea, C.; Mereuta, P. Characterization of sodium alginate based films. *Rom. J. Phys.* **2020**, *65*, 1–8.
41. Liu, S.; Li, Y.; Li, L. Enhanced stability and mechanical strength of sodium alginate composite films. *Carbohydr. Polym.* **2017**, *160*, 62–70. [CrossRef]
42. Islam, N.U.; Amin, R.; Shahid, M.; Amin, M. Gummy gold and silver nanoparticles of apricot (Prunus armeniaca) confer high stability and biological activity. *Arab. J. Chem.* **2019**, *12*, 3977–3992. [CrossRef]
43. Ho, T.C.; Kim, M.H.; Cho, Y.J.; Park, J.S.; Nam, S.Y.; Chun, B.S. Gelatin-sodium alginate-based films with Pseuderanthemum palatiferum (Nees) Radlk. freeze-dried powder obtained by subcritical water extraction. *Food Packag. Shelf Life* **2020**, *24*, 100469. [CrossRef]
44. Martins, A.F.; Facchi, S.P.; Monteiro, J.P.; Nocchi, S.R.; da Silva, C.T.P.; Nakamura, C.V.; Girotto, E.M.; Rubira, A.F.; Muniz, E.C. Preparation and cytotoxicity of *N,N,N*-trimethyl chitosan/alginate beads containing gold nanoparticles. *Int. J. Biol. Macromol.* **2015**, *72*, 466–471. [CrossRef] [PubMed]
45. Ahmad, A.; Mubarak, N.; Jannat, F.T.; Ashfaq, T.; Santulli, C.; Rizwan, M.; Najda, A.; Bin-Jumah, M.; Abdel-Daim, M.M.; Hussain, S.; et al. A critical review on the synthesis of natural sodium alginate based composite materials: An innovative biological polymer for biomedical delivery applications. *Processes* **2021**, *9*, 137. [CrossRef]
46. Ghorbani-Vaghei, R.; Veisi, H.; Aliani, M.H.; Mohammadi, P.; Karmakar, B. Alginate modified magnetic nanoparticles to immobilization of gold nanoparticles as an efficient magnetic nanocatalyst for reduction of 4-nitrophenol in water. *J. Mol. Liq.* **2021**, *327*, 114868. [CrossRef]
47. Chen, J.; Wu, A.; Yang, M.; Ge, Y.; Pristijono, P.; Li, J.; Xu, B.; Mi, H. Characterization of sodium alginate-based films incorporated with thymol for fresh-cut apple packaging. *Food Control* **2021**, *126*, 108063. [CrossRef]
48. Samanta, H.S.; Ray, S.K. Synthesis, characterization, swelling and drug release behavior of semi-interpenetrating network hydrogels of sodium alginate and polyacrylamide. *Carbohydr. Polym.* **2014**, *99*, 666–678. [CrossRef] [PubMed]
49. Brahmi, M.; Essifi, K.; Elbachiri, A.; Tahani, A. Adsorption of sodium alginate onto sodium montmorillonite. *Mater. Today Proc.* **2021**, *45*, 7789–7793. [CrossRef]
50. Moosavi-Nasab, M.; Khoshnoudi-Nia, S.; Azimifar, Z.; Kamyab, S. Evaluation of the total volatile basic nitrogen (TVB-N) content in fish fillets using hyperspectral imaging coupled with deep learning neural network and meta-analysis. *Sci. Rep.* **2021**, *11*, 5094. [CrossRef]
51. Rizo, A.; Mañes, V.; Fuentes, A.; Fernández-Segovia, I.; Barat, J.M. Physicochemical and microbial changes during storage of smoke-flavoured salmon obtained by a new method. *Food Control* **2015**, *56*, 195–201. [CrossRef]
52. Wang, S.; Xiang, W.; Fan, H.; Xie, J.; Qian, Y.-F. Study on the mobility of water and its correlation with the spoilage process of salmon (*Salmo solar*) stored at 0 and 4 °C by low-field nuclear magnetic resonance (LF NMR 1H). *J. Food Sci. Technol.* **2018**, *55*, 173–182. [CrossRef]

53. Jia, Z.; Shi, C.; Wang, Y.; Yang, X.; Zhang, J.; Ji, Z. Nondestructive determination of salmon fillet freshness during storage at different temperatures by electronic nose system combined with radial basis function neural networks. *Int. J. Food Sci. Technol.* **2020**, *55*, 2080–2091. [CrossRef]
54. Xie, J.; Zhang, Z.; Yang, S.-P.; Cheng, Y.; Qian, Y.-F. Study on the spoilage potential of Pseudomonas fluorescens on salmon stored at different temperatures. *J. Food Sci. Technol.* **2018**, *55*, 217–225. [CrossRef]
55. Mikš-Krajnik, M.; Yoon, Y.-J.; Ukuku, D.O.; Yuk, H.-G. Volatile chemical spoilage indexes of raw Atlantic salmon (*Salmo salar*) stored under aerobic condition in relation to microbiological and sensory shelf lives. *Food Microbiol.* **2016**, *53*, 182–191. [CrossRef] [PubMed]
56. Cheng, J.H.; Sun, D.W. Recent Applications of Spectroscopic and Hyperspectral Imaging Techniques with Chemometric Analysis for Rapid Inspection of Microbial Spoilage in Muscle Foods. *Compr. Rev. Food Sci. Food Saf.* **2015**, *14*, 478–490. [CrossRef]
57. Hsiao, H.-I.; Chang, J.-N. Developing a microbial time–temperature indicator to monitor total volatile basic nitrogen change in chilled vacuum-packed grouper fillets. *J. Food Process. Preserv.* **2017**, *41*, e13158. [CrossRef]
58. Wiernasz, N.; Cornet, J.; Cardinal, M.; Pilet, M.-F.; Passerini, D.; Leroi, F. Lactic acid bacteria selection for biopreservation as a part of hurdle technology approach applied on seafood. *Front. Mar. Sci.* **2017**, *4*, 119. [CrossRef]
59. Abel, N.; Rotabakk, B.T.; Rustad, T.; Ahlsen, V.B.; Lerfall, J. Physiochemical and Microbiological Quality of Lightly Processed Salmon (*Salmo salar* L.) Stored Under Modified Atmosphere. *J. Food Sci.* **2019**, *84*, 3364–3372. [CrossRef]
60. Suchak, N.M.; Desai, P.H.; Deshpande, M.P.; Chaki, S.H.; Pandya, S.J.; Kunjadiya, A.; Bhatt, S.V. Study on the concentration of gold nanoparticles for antibacterial activity. *Adv. Nat. Sci. Nanosci. Nanotechnol.* **2021**, *12*, 035003. [CrossRef]
61. Payne, J.N.; Waghwani, H.K.; Connor, M.G.; Hamilton, W.; Tockstein, S.; Moolani, H.; Chavda, F.; Badwaik, V.; Lawrenz, M.B.; Dakshinamurthy, R. Novel synthesis of kanamycin conjugated gold nanoparticles with potent antibacterial activity. *Front. Microbiol.* **2016**, *7*, 607. [CrossRef] [PubMed]
62. Zhang, Y.; Dasari, T.P.S.; Deng, H.; Yu, H. Antimicrobial activity of gold nanoparticles and ionic gold. *J. Environ. Sci. Heath Part C* **2015**, *33*, 286–327. [CrossRef] [PubMed]

polymers

Review

Cactus Pear Mucilage (*Opuntia* spp.) as a Novel Functional Biopolymer: Mucilage Extraction, Rheology and Biofilm Development

Brandon Van Rooyen [1,*], Maryna De Wit [1], Gernot Osthoff [2] and Johan Van Niekerk [1]

[1] Department of Sustainable Food Systems and Development, University of the Free State, Bloemfontein 9301, South Africa
[2] Department of Microbiology and Biochemistry, University of the Free State, Bloemfontein 9301, South Africa
* Correspondence: vanrooyenbb@ufs.ac.za

Abstract: The investigation of novel, natural polymers has gained considerably more exposure for their desirable, often specific, functional properties. Multiple researchers have explored these biopolymers to determine their potential to address many food processing, packaging and environmental concerns. Mucilage from the cactus pear (*Opuntia ficus-indica*) is one such biopolymer that has been identified as possessing a functional potential that can be used in an attempt to enhance food properties and reduce the usage of non-biodegradable, petroleum-based packaging in the food industry. However, variations in the structural composition of mucilage and the different extraction methods that have been reported by researchers have considerably impacted mucilage's functional potential. Although not comparable, these factors have been investigated, with a specific focus on mucilage applications. The natural ability of mucilage to bind water, alter the rheology of a food system and develop biofilms are considered the major applications of mucilage's functional properties. Due to the variations that have been reported in mucilage's chemical composition, specifically concerning the proportions of uronic acids, mucilage's rheological and biofilm properties are influenced differently by changes in pH and a cross-linker. Exploring the factors influencing mucilage's chemical composition, while co-currently discussing mucilage functional applications, will prove valuable when evaluating mucilage's potential to be considered for future commercial applications. This review article, therefore, discusses and highlights the key factors responsible for mucilage's specific functional potential, while exploring important potential food processing and packaging applications.

Keywords: biopolymer; cactus pear mucilage; biofilms; functional properties; rheology; *Opuntia ficus-indica*; mucilage extraction

Citation: Van Rooyen, B.; De Wit, M.; Osthoff, G.; Van Niekerk, J. Cactus Pear Mucilage (*Opuntia* spp.) as a Novel Functional Biopolymer: Mucilage Extraction, Rheology and Biofilm Development. *Polymers* **2024**, *16*, 1993. https://doi.org/10.3390/polym16141993

Academic Editors: Zenaida Saavedra-Leos and César Leyva-Porras

Received: 30 April 2024
Revised: 4 July 2024
Accepted: 9 July 2024
Published: 12 July 2024

1. Introduction

Many researchers and food producers are actively exploring alternative ways to improve food quality and safety while minimising the environmental impact. Identifying and investigating novel biopolymers designated as meeting specific processing and environmental requirements in the food industry has recently gained considerable attention [1,2]. It has become essential to find alternative functional biopolymers that address this current global move towards using more sustainably sourced, natural polymers [1,3–5].

Although some well-known functional biopolymers, such as pectin, alginate and carrageenan, have been identified, biopolymers that are produced in large quantities, easily sourced and associated with relatively low input costs are a growing area of focus in the research [6]. Mucilage, sourced from the cladodes of the cactus pear, is one such biopolymer that is rapidly gaining attention for its desirable functional properties, which can be utilised to meet current industry demands [3,7–10]. However, as with all naturally sourced polymers, specific factors have been shown to influence the functional behaviour of mucilage, such as variations in mucilage's structural composition and the influence of various extraction methods used to obtain the mucilage. These factors should be considered as highly important when evaluating mucilage as a functional biopolymer and ultimately assessing its applications and commercial viability, with a holistic view often lacking in research [11–17]. Furthermore, it is important to note that the investigation of mucilage has further been promoted from a food, health and medicinal perspective, highlighting its human consumption benefits and safety aspects [6].

The cactus pear, specifically *Opuntia ficus-indica*, the source of mucilage, is a succulent plant that can survive under extreme growing conditions, specifically prolonged drought and extreme heat. Additionally, it has adapted to grow in poor soil conditions, requiring little intervention regarding fertilisation and crop management [6,18]. The plant's excellent adaptability and survival in times of drought are largely due to mucilage production. Mucilage is responsible for binding and storing water in the plant leaves (called 'cladodes') [6]. Consequently, the inherent nature of the mucilage polymer is associated with its excellent water-holding potential. It has, therefore, been evaluated by multiple researchers for its rheology-altering properties when introduced into a solution [18–20].

Mucilage, extracted from the cladodes of *Opuntia ficus-indica*, is considered a highly flexible heteropolysaccharide with a high molecular weight [11]. Its primary structure is said to comprise a linear core chain of repeating D-galacturonic acid and L-rhamnose, together with many neutral sugar side chains [15,21,22]. As mucilage has been considered to display polyelectrolyte properties, factors that have been shown to influence other charged biopolymers, such as pectin and alginate, could also be considered to influence the mucilage polymer [19,20,23]. It has generally been well established that mucilage can be associated with two main fractions, differentiated by a charged, pectin-like fraction and a more neutral fraction [15]. The presence of uronic acids associated with the charged fraction of mucilage has specifically been identified, although different authors have reported varying proportions [11,14–16]. Various factors account for the differences in the chemical composition of mucilage. One main factor is the methods used to extract mucilage from the cactus pear [12–14,21,23].

As mucilage powder is not available commercially, like other well-established biopolymers, authors have evaluated and reported on various mucilage extraction methods. It has been seen that the extraction method selection can directly impact the mucilage yield, highlighting the efficiency of the mucilage extraction process [11,13,15,24]. In addition to the extraction efficacy, the chemical composition and nature of the different mucilage fractions can also be influenced by the extraction method employed, having a consequential impact on the functional properties of the mucilage [11,15,16]. Both of these factors have been identified as directly influencing the rheological behaviour of mucilage solutions [11,13–15].

Due to mucilage's structural conformation and inherent water-storing capacity, it has been researched for its viscosity-enhancing potential [19,20,23]. Authors have reported increases in solution viscosity, accounted primarily to the physical entanglement of mucilage's polymer chains in a solution [11]. Several prominent factors have been shown to alter the rheological properties of mucilage solutions further. These include the alteration of the pH of a solution and the incorporation of cross-linkers, such as calcium and magnesium [19,20,23].

More recently, environmental concerns over the use of non-biodegradable plastic packaging in the food industry have further prompted the investigation of functional biopolymers that display a potential to be used in developing biofilms [25–27]. The mucilage from *Opuntia ficus-indica* has specifically been an area of focus in this regard. Research has shown that mucilage has an ever-increasing application in the development of biofilms as a natural packaging [3,8,9,25,28,29].

Mucilage has primarily been explored as a polymer in biofilm formation, either as homopolymeric biofilms or in combination with other biomaterials. These films can further be differentiated by their moisture content, with 'dry' or low-moisture biofilms being the focus area of many authors [3,30]. Changes in pH, polymer concentration and the presence of a cross-linker have further been shown to directly impact the physical- mechanical properties of these biofilms. A biofilm's physical–mechanical properties are often considered key when evaluating its packaging development potential [3]. Although mucilage biofilm development has proved successful, variations in the development procedures and the consequential properties displayed by the biofilms have been reported on by different authors. Therefore, it is important to consider and identify the key aspects of the successful development of mucilage biofilms in order to consider them a viable natural packing solution for future exploration [3,9,10,31].

Overall, many factors have been shown to influence the functional behaviour of mucilage. These factors provide practical insight into a polymer's functional potential and, ultimately, its rheological properties and biofilm formation potential [19,20,32,33]. Considering consumer demands, environmental concerns and the high costs associated with commercially available biopolymers, researchers and food manufacturers have focused on identifying and investigating novel biopolymers, such as cactus pear mucilage [9,25]. Mucilage has been shown to display functional properties that can be utilised to address the food industry's current needs.

However, with the abundant research often showing considerable variations in findings, a comparative overview of the factors influencing mucilage's rheological behaviour and biofilm development potential is still lacking.

Thus, this review article focuses on investigating the specific factors shown to influence the functional behaviour of mucilage sourced from the cactus pear (*Opuntia* spp.), when used as a rheology-altering polymer in a solution and in the application of 'dry' biofilms, to provide an accurate, reliable and insightful holistic view of mucilage as a novel functional biopolymer.

2. Mucilage Chemical Structure

In general, mucilage is considered a highly flexible heteropolysaccharide with a high molecular weight. Multiple researchers have studied the chemical composition of *Opuntia* spp. mucilage, and have often reported considerable variations therein [11,21,34]. Although variations in the sugar composition and consequential molecular weight have been reported, mucilage molecules are considered to be quite large. This long-chain polymer is represented by varying proportions of D-galacturonic acid and neutral sugars of L-arabinose, D-galactose, L-rhamnose and D-xylose [11,21,22]. The primary structure is said to comprise a linear core chain of repeating (1→4) D-galacturonic acid and (1→2) L-rhamnose.

Additionally, the mucilage polymer is associated with many side chains of D-galactose attached to L-rhamnose residues. These complex peripheral chains have the potential to comprise various types of sugars. The galactose side chains can further branch into either arabinose or xylose (Figure 1). Native mucilage's chemical composition can vary considerably, depending on the efficacy and type of extraction method used [12–14,21,23].

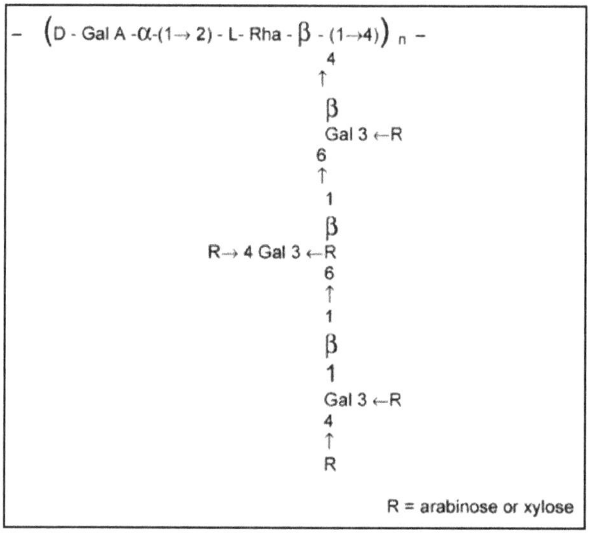

Figure 1. A proposed schematic representation of the mucilage structure of *Opuntia ficus-indica* displaying the charged main linear chain of D-galacturonic acid and L-rhamnose units together with side chains attached to the L-rhamnose residues [21].

2.1. Mucilage Fractions

Investigations of mucilage extracted from cactus pear cladodes have generally shown two prominent, water-soluble fractions present in native mucilage. The first fraction is a pectin-like molecule displaying gelling properties, and the second fraction does not display gel-forming properties [15]. The chemical composition of these two main native mucilage fractions has been shown to differ from one another [23]. The different fractions present within native mucilage extract have been further investigated, and the general conclusions are that native mucilage is represented by both a gelling, pectin-like fraction and a more neutral fraction showing a decreased ability to interact with cross-linkers to form a gel, considered the pure mucilage fraction. Multiple similarities have been observed regarding the chemical composition of these two fractions [15].

In general, the two fractions of native mucilage extract display a similar sugar composition; however, there are differences concerning the percentages of the sugars between the pectin-like and pure mucilage fractions associated with the whole cladodes of *Opuntia* spp. [11]. The differences in sugar percentages are of importance, specifically regarding the content of the charged sugars, such as uronic acid [11,15]. Approximately 20% charged sugars can be present in cladodes, and >50% charged sugars in part of the peel (Table 1). The charged sugars have been identified as galacturonic acid, but the presence of glucuronic acid has also been reported [11,21]. The pectin-like polymer fraction contains a noteworthy greater amount of galacturonic acid compared to the pure mucilage fraction [15]. Uronic acids are important in terms of a polymer's viscosity-enhancement potential, as their carboxyl groups are able to interact with water molecules, charged metal ions and changes in pH [14].

Considering the proposed chemical structure of mucilage (Figure 1), high amounts of L-arabinose could be expected, as observed in Table 1. Mucilage with higher quantities of arabinose is likely to result in higher viscosities when in an aqueous medium, than that of mucilage with lower amounts of arabinose [14]. Majdoub et al. [11] suggested the high molecular weight of mucilage could be a result of branching at the rhamnose sugars, promoting a random coil conformation of the mucilage polymer in a solution. This is evident when considering the higher levels of neutral sugars (Table 1).

Table 1. Sugar composition of the pectin-like and mucilage fractions associated with *Opuntia ficus-indica* cladodes (% in weight).

Sugars Investigated **	Pectin Fraction (A)	Pectin Fraction (B)	Mucilage Fraction (A)	Mucilage Fraction (C)	Mucilage Fraction (D)
Uronic acid *	56.30	85.40	11.00	19.4	13.91
Arabinose	5.60	6.00	17.93	33.10	35.36
Galactose	6.50	7.00	20.99	20.30	27.26
Rhamnose	0.50	0.60	1.75	6.90	1.93
Xylose	0.90	1.00	3.06	18.7	16.32

Research performed by:
(A) Goycoolea and Cárdenas [15];
(B) Cárdenas et al. [16];
(C) Majdoub et al. [11];
(D) Rodriguez-González et al. [14].

* Galacturonic + Glucuronic acid. ** Other component may have been present and reported on by the various authors.

The unique chemical composition and structural orientation of mucilage are, therefore, likely to result in the formation of highly viscous solutions [14]. In addition to the two main fractions associated with native mucilage, proteins have also been associated with native mucilage. These proteins have been suggested to display decreased water solubility and lower molecular weight compared to pure mucilage fractions [11,15].

Lastly, the high molecular weight, pectin-like polymer fraction will likely display a low degree of esterification, as <50% of their carboxyl groups will be methylated. This pectin-like fraction present in native mucilage is represented by a charged chemical structure, with the potential to interact with divalent cations and respond to changes in pH [11,15,16,35].

It is important to recognise the different fractions present in native mucilage as variations in their sugar composition (Table 1). Although the two polymeric fractions associated with native mucilage are not chemically associated, they share a similar composition of neutral sugar residues (Table 1). Specifically, the pectin-like fractions have been linked to higher amounts of uronic acids, with the pure mucilage fractions displaying increased amounts of neutral sugars. The composition of native mucilage will, therefore, greatly determine its observed functional behaviour, consequently influencing its applications [15,16]. Differences in the extraction processes employed by different authors have also been shown to directly affect the sugar composition of the two fractions, as shown in Table 1 [11,12,14,15]. Although similar sugar compositions have been reported for various cultivars of *Opuntia ficus-indica*, differences in the ratios of sugars among different cultivars have been reported, which have a consequential impact on their functional properties [36].

2.2. Mineral Composition of Mucilage

Besides sugars, mucilage contains essential minerals, carbohydrates and dietary fibre [13,14]. The mineral composition includes calcium (Ca), potassium (K), magnesium (Mg), phosphate (P), zinc (Zn), iron (Fe) and sodium (Na), as well as other minerals in lower quantities. Calcium has been reported to be amongst the most abundant minerals, although significant levels of K and Mg have also been reported [13,19,22,37]. Using X-ray diffraction, Madera-Santana et al. [34] identified crystalline structures displaying the typical characteristics of calcium salts in the mucilage extracted from cladodes. These results agree with those of Contreas-Padilla et al. [37], who confirmed the presence of calcium and calcium oxalate in Opuntia ficus-indica using atomic absorption spectroscopy. For 40–135-day-old cladodes, calcium levels of 17.90–30.70 mg/g dry mass were reported, and increasing calcium levels were also expected for older cladodes.

3. Extraction Methods

The different extraction methods used to extract mucilage have been shown to significantly affect the sugar composition of the native extract (NS), the composition of the different fractions and, ultimately, the functional properties displayed by the mucilage [11,15,16]. Understanding the different extraction methods considered by different researchers and identifying their key differences, as displayed in Table 2. are the purpose of Table 2.

Table 2. Summary of key extraction method steps described in various publications by different authors together with their expected mucilage yields. representation of the different extraction methods investigated by various authors and their extraction efficacy.

Publication	Heating/ Microwave Treatment	Maceration/ Blending/ Milling	Filtration	Centrifugation	Ethanol Extraction	Moisture Removal	Mucilage Yield
Monrroy et al. [13]	√	√	√	-	√	Oven-dried	~24–31%
Felkai-Haddache et al. [24]	√	√	√ Cheesecloth	√	√	Freeze-drying	~6.82–25.56%
Du Toit and De Wit PA153178P [17]	√	√	-	√	-	Freeze-drying	~39–62% Native Mucilage ***
Majdoub et al. [11]	-	√	Filtration + Ultrafiltration	√	-	Freeze-drying	Separate Fractions **
A review: Goycoolea and Cárdenas [15]	√	√	Complex Filtration; pH Adjustment	√	√ Multi-step	* DNS	Separate Fractions **

* DNS—Publication did not specify; ** Separate Fractions—extraction procedure described different mucilage fractions present; *** Native Mucilage—mucilage obtained was not separated according to different fractions.

Monrroy et al. [13] evaluated the efficacy of different natural extraction methods to determine mucilage's viability as a functional food ingredient. The authors investigated extraction methods that employ hydration, agitation and a combination thereof for both dried and fresh cladodes. Hydration extraction was tested with and without thermal treatment on both fresh and dried cladodes. The cladode samples were prepared by peeling, cutting into pieces, crushing and homogenising by suspending 100 mg extract in 5 mL water. For the dried samples, the cladode extract was dried at 60 °C for 48 h before homogenising. The samples were then placed into a water bath at 44–86 °C for 54–96 min, followed by filtration. The mucilage was precipitated from the fresh and dried samples using 45 mL and 15 mL ethanol, respectively, and then oven-dried at 60 °C. A similar protocol was followed for non-thermal hydration extraction, except the sample was allowed to rest for 24 h at room temperature and the heat treatment step was eliminated. The solution was then filtered, and mucilage was precipitated by adding 15 mL ethanol and oven-drying at 60 °C. The extraction through agitation followed a similar protocol for sample preparation as described above, and after homogenisation, the sample was stirred using a magnetic stirrer for 30 min, followed by filtration and precipitation with 15 mL ethanol for 30 min and then dried at 60 °C. Lastly, the combination extraction subjected the samples to 24 h hydration, followed by 30 min of stirring, filtration and mucilage precipitation with ethanol. Considering non-thermal extraction methods for O. *cochenillifera* cladodes, Monrroy et al. [13] reported agitation extraction to be less effective than hydration extraction, and little difference with the combination of agitation and hydration extraction methods. Agitation resulted in about a 26% mucilage yield extraction and hydration about 31%. These results were for dried cladode extraction.

Felkai-Haddache et al. [24] described a process that used the cladodes of *Opuntia ficus-indica* where the outer epidermis was removed from the cladodes, macerated with a domestic mill and then chilled till it reached 4 °C. This extract was then microwaved at 700 W for 5 min and 15 s for optimal mucilage extraction. The optimal amount of water to be added was four times the quantity of the raw material used. The authors also reported that pH plays a significant role in mucilage yield. At pH 11, an optimal amount of mucilage yield was reported because of the high degree of dissociation of the acid groups and the high degree of solubility of the mucilage and, thus, water extraction. After microwaving, the extracted mucilage was cooled (4 °C), filtered to remove the pulp and centrifuged at 4000× g for 15 min. The filtrate was precipitated with ethanol (95%, v/v) at 4 °C overnight. The precipitate was also washed with ethanol (75%, v/v) and then subjected to lyophilisation at −55 °C for 12 h. Felkai-Haddache et al. [24] reported an optimal mucilage recovery of 25.6%.

In a procedure patented by Du Toit and De Wit [17], an extraction process for mucilage was described, which also used a microwave-heating step. The extraction process requires the cladodes first to be peeled, removing all of the hard outer layers and fibrous material, with only the 'light green slimy inside' remaining, which is then sliced into manageable sizes and placed into a microwave oven at maximum power for 4 min (or until the cladode pieces are cooked soft). The cladode pieces are macerated by mincing or cutting without the addition of water. The authors used a juicing apparatus typically found in a household kitchen. Lastly, the macerated mucilage pulp is centrifuged for 15 min at 8000 rpm, maintaining a temperature of 4 °C to separate the solids from the mucilage. There was no reference to an ethanol precipitation step. Mucilage yields of ~39–62% per weight of extracted pulp were obtained, while percentages ranging between 10 and 17% per cladode weight were found.

Majdoub et al. [11] used cladodes (6–12 months old) that were shredded and blended. The mixture was then subjected to degreasing with petroleum ether, washing with deionised water, filtering out the solid components, centrifuging and then filtering again. The supernatant was separated according to different molecular weights using ultrafiltration into high and low molecular weight components. The authors suggested that the presence of proteins in the original sample interacted with the polysaccharides and thus affected the purification results [11]. There was no report of the inclusion of an ethanol precipitation step.

Goycoolea and Cárdenas [15] reported on an extraction procedure distinguishing two main mucilage fractions: gelling (GE) pectin and a non-gelling (NE) mucilage fraction. For the extraction of both fractions, cladodes were diced and heated for 20 min at 85 °C, then neutralised with sodium hydroxide (NaOH), liquidised and then centrifuged. The resulting precipitate was considered the GE fraction, and the supernatant the NE fraction. The precipitate and supernatant were subjected to the extraction protocols represented in Figure 2. After these separate procedures were completed (Figure 2), both fractions were again precipitated with 50% v/v ethanol, again centrifuged, then washed with ethanol/water mixtures (70, 80, 90, 95 and 100% v/v) and then dried at room temperature.

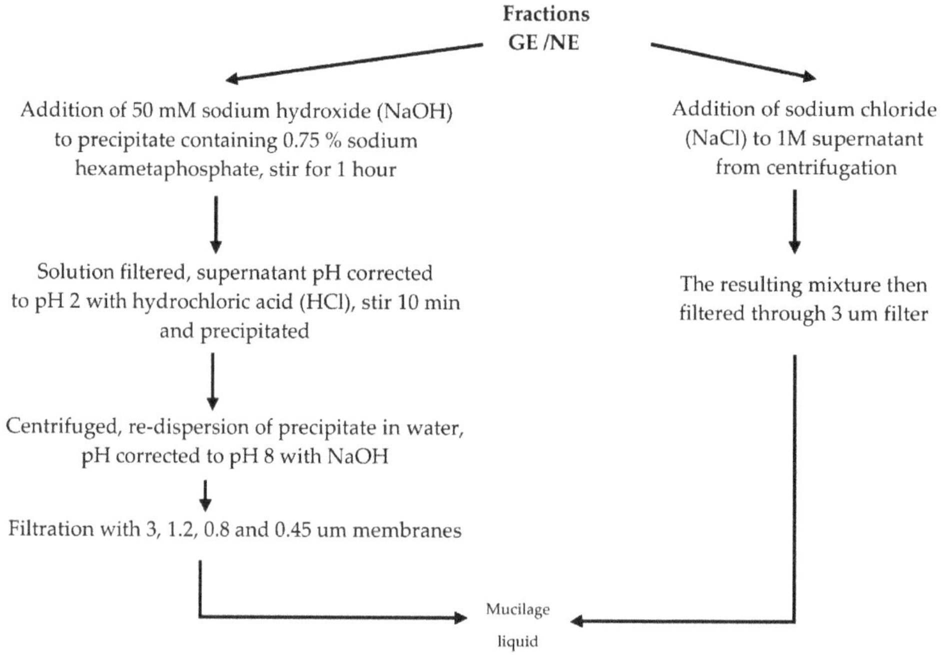

Figure 2. Differences in methods proposed by Goycoolea and Cárdenas [15] to extract and isolate both the gelling fraction (GE) and non-gelling (NE) mucilage/pectin fraction from *Opuntia* spp. cladodes.

4. Functional Properties Associated with Mucilage

Various researchers have explored multiple procedures detailing the extraction and purification of mucilage [13,14,21]. Investigations of the rheological and physico-chemical behaviour of native extracts from cladodes have highlighted the important factors affecting the implementation of mucilage as a functional food ingredient [11,32,38]. Certain dominant factors have been identified that influence the functional behaviour of mucilage in a solution. Specifically, the polymer concentration, pH and presence of a cross-linker in a solution considerably alter mucilage's functional behaviour [19,20,23].

4.1. Rheology-Altering Properties of Mucilage

As a result of mucilage's pseudoplastic nature and unique structural conformation aspects, it has an excellent water-holding capacity. The viscosity-enhancement properties of mucilage in an aqueous solution have been directly linked to the physical entanglements between the polymer chains [11]. Specifically, Medina-Torres [23] found similar viscosity-altering capabilities of a 10% mucilage solution to that of 3% xanthan solutions in aqueous mediums. Monrroy et al. [13] also compared mucilage's thickening ability to that of commercially available hydrocolloids. The authors noted similarities between the ability of mucilage and gum arabic to form low-viscosity solutions [13].

The molecular weight, chain flexibility and surface charge were found to influence the interfacial activity of mucilage (i.e., mucilage's ability to mix with water) [39]. In general, an increase in mucilage concentration, together with high chain flexibility and the presence of uronic acid along its polymer backbone, can promote the adsorption of mucilage onto the liquid phase [11,39]. Water absorption by the polymer is directly influenced by the amount of active water-binding sites, which are affected by the physico-chemical, topological and structural parameters [39].

Generally, viscosity increases by the dispersion of mucilage into aqueous solutions, which has been accounted for by the negatively charged nature of the polymer. These similarly charged, long, flexible mucilage molecules repel themselves in a solution, resulting in the uncoiling and stretching out of the molecule. This is often referred to as electrostatic repulsion. This 'stretching' of a molecule is predominantly responsible for viscosity increases in a solution [11]. A mucilage polymer's thickening ability is further aided by the intermolecular interactions of its polymer side-chain groups via hydrophobic interactions or hydrogen bonding. These viscous solutions formed by mucilage have been shown by Medina-Torres et al. [23] to display non-Newtonian shear-thinning behaviour. This behaviour of increased pseudoplasticity was noted for solutions with increasing amounts of mucilage. Majdoub et al. [11] also confirmed this behaviour by explaining that the zero-shear viscosity value increases with an increase in the concentration of the polymer.

Du Toit et al. [33] investigated the rheological behaviour of mucilage from *Opuntia* spp. Non-Newtonian, pseudoplastic flow properties, indicative of a shear-thinning effect of mucilage, were observed by the authors [33]. They further reported that the viscosity of mucilage solutions are influenced by changes in pH and the presence of charged ions, specifically $CaCl_2$ and $FeCl_3$, highlighting the functional nature of mucilage [33].

The non-Newtonian, pseudoplastic flow behaviour was further reported on by a more recent study conducted by van Rooyen et al. [20]. The authors found that increasing the concentration of mucilage in an aqueous solution resulted in the increased pseudoplastic flow behaviour of the mucilage solution. Interestingly, the authors found that mucilage possesses an overall poorer viscosity-enhancement potential compared to other commercially available biopolymers, such as pectin and alginate, at all the polymer concentrations investigated [20]. Although mucilage has the ability to alter the rheology of a solution, certain factors have been shown to directly influence mucilage's functional behaviour in a solution. Other than polymer concentration, the pH of a solution and the presence of a cross-liker must be considered key influencing factors when evaluating the rheological behaviour of mucilage in a solution [19,20,33,38].

Factors Influencing the Rheology of a Mucilage Solution

Due to the presence of charged sugars, the viscosity of a mucilage solution has been shown to be influenced by the presence of cations, specifically calcium [11,20]. Since the native extract from cactus pear cladodes has been shown to comprise various gelling and non-gelling fractions, the efficacy of the extraction and purification methods used have the potential to greatly influence the functionality displayed by the extracted biopolymer from cactus pear cladodes. Therefore, diverse and often contradictory findings have been reported on the potential of cactus pear (*Opuntia* spp.) extracts to form a gel or display gel-like properties [11,14,15,21].

It has been suggested that mucilage displays elastic properties but is unable to form a gel, regardless of the ionic strength of a solution [11,23]. Majdoub et al. [11] reported on a native polysaccharide extract from cladodes that displayed a slight polyelectrolyte effect, as the molecule's confirmation and viscosity were shown to be influenced by the addition of cations. The authors reported that monovalent cations have less of an effect on viscosity than divalent cations, such as calcium. However, even with the addition of calcium, only slight changes in viscosity were observed, indicative of the limited extension of the polymer in an aqueous solution, which accounted for the low proportion of uronic acids present in the polymer investigated [11]. For mucilage, however, a loss in viscosity is related to a high degree of chain flexibility of the polymer and a low occurrence of changed uronic acid residues. This loss of viscosity with the addition of salt has also been referred to as the salting-out effect. For mucilage, physical gel formation is unable to be occur as expected, as the formation of "inactive loops" prevents the formation of intermolecular junctions between polymers that are required to form a 3D gel network. However, solvent conditions will influence a polymer's chain–chain interactions and configuration [11]. The rheological properties of ionic gels depend on the concentration of cations and the presence of uronic acid in the mucilage. Thus, low uronic acid and few cations have little effect on the viscosity of mucilage in an aqueous system.

However, reports of extracts from cactus pears containing higher contents of charged sugars, displaying increased gel-like properties by the addition of cations, such as Ca^{2+}, have also been reported. The pectin-like fraction present in native mucilage was suggested to be the reason for these observed gel-like properties associated with native mucilage [15,19,20].

The presence of charged sugars has been reported as a significant parameter with which to modulate the structuring capabilities of a mucilage polymer, resulting in mucilage displaying pectin-like behaviour with consequential gel-like properties if certain parameters are met [39]. It has, therefore, also been noted that native mucilage has the potential to be influenced by changes in solution pH. As charged sugars have been associated with native mucilage's chemical structure, changes in pH have been shown to consequently alter the rheological properties of native mucilage. Instances of native mucilage solutions displaying increased gel-like behaviour have been reported under acidic conditions, generally related to a type of acid gelation displayed by native mucilage [19].

4.2. Mucilage in the Development of Biofilms

4.2.1. Homopolymeric Mucilage Biofilms

Cactus pear mucilage has also been considered as an ingredient for biofilms' development. Research on mucilage biofilms and edible coatings has increased considerably, but research remains relativity limited compared to that on other biopolymers, such as pectin and alginate. Biofilm characteristics are generally reported in one of two states, either as 'dried' or 'wet' [25,40,41]. In the case of mucilage biofilms, authors have generally reported on the films when they are in a dried state, as mucilage has been shown to have a poor gelling capacity with cations typically. The mechanical properties of biofilms are of great importance, as these parameters are directly related to their chemical structure and potential for commercial application [41–43]. However, as with most dried polysaccharide-based films, the addition of a plasticiser has been set as the standard for developing mucilage-based biopolymer films. 'Dried' mucilage films have generally displayed brittle and fragile properties if no plasticiser is used in their development. Adding a plasticiser, usually glycerol, allows for more flexible mucilage films that display adequate handling and mechanical properties [44–46]. In research conducted by Gheribi et al. [41], the authors specifically found that using glycerol as a plasticiser at a 40% w/w in the development of mucilage (*Opuntia ficus-indica*) films produced films with superior elongation at break (%E) values, together with adequate film tensile strength. It has been suggested that natural plasticisers, such as glycerol, are essential in developing biopolymer films that show adequate film-forming properties if intended to be used as natural packaging [9,10,41]. As with other commercially available biopolymers used in the development of biofilms, mucilage films have also been developed both with and without the addition of cross-linking agents and at varying pHs. Limited comparative work is available on these factors' influence on mucilage biofilm properties. More recently, publications have explored these factors in more detail [9,10,31,43].

Espino-Díaz et al. [31] were one of the first authors to investigate biofilm formation using mucilage from *Opuntia ficus-indica*. The authors developed these mucilage biofilms together with 50% glycerol, which was added based on the weight (w/w) of the mucilage used in the films [31]. These mucilage films were produced at different pH ranges (3, 4, 5, 6, 7 and 8), with and without the addition of calcium (in the form of $CaCl_2$) as a cross-linker. The pH was adjusted using hydrochloric acid or sodium hydroxide to achieve the desired pH. At the native pH, without the addition of calcium, the films showed the highest TS. Alternatively, the films exposed to calcium displayed enhanced %E and enhanced changes in pH. An important finding regarding the adhesion of the films was further highlighted by the authors; improved film strength and handling was achieved at a higher pH range (pH 5–8) because at a low pH (pH 3), the films were difficult to work with due to the high adhesion potential and elasticity [31].

It has been observed that cross-linked films typically result in a less compact structure due to the intermolecular linking with calcium, resulting in the reduction in and tying of the polymer chains. However, without the addition of calcium, there is a higher chain flexibility, allowing for the formation of more compact films, as there is more contact surface among the molecules [31]. This was supported by Livney et al. [47], who suggested that greater conformation flexibility was possible for open structures than for intermolecular cross-linked structures. Practically, films formed without the addition of calcium allow for reduced permeability as a more organised and compact 3D structure network is formed, retarding movement across the film when compared to films with calcium incorporated into them [31]. Importantly, if mucilage is at a low pH of 4, it is positively charged and may result in reduced intermolecular bonding with calcium [31]. Some authors have reported increased film TS values associated with polymers of increased molecular weight [31,48].

Lira-Vargas et al. [43] are some of the few researchers who have also reported on the potential of using mucilage to form biopolymer films. These authors investigated the non-gelling mucilage fraction extracted from the cladodes of *Opuntia ficus-indica*. Films were formed by drying without the addition of a cross-linker. The films displayed an inconsistent surface morphology, with a lumpy and granular texture. These mucilage biopolymer films resulted in a TS average of 0.49 MPa, similar to that reported by Espino-Díaz et al. [31]. More recent studies have also shown that mucilage can be successfully used in the development of single-polymer films [8,49].

Van Rooyen et al. [10] reported on the mechanical properties of mucilage biofilms. The polymer concentration and changes in pH were shown to have a considerable influence on the biofilm's mechanical properties [10]. It was found that mucilage films developed at a low polymer concentration of 2.5% (*w*/*w*) were unable to produce films displaying satisfactory mechanical properties. However, increasing the polymer concentration of the mucilage in the films to 5% (*w*/*w*) and 7.5% (*w*/*w*) resulted in mucilage films displaying adequate mechanical properties [10]. Furthermore, when decreasing the pH of mucilage films to pH 3–3.5, a noticeable increase in the films' elasticity (when considering the elongation at break % measurements) was reported. This trend of increased film tensile strength at a pH 3–3.5 was ascribed to a type of acid by the authors, indicating the functional nature of mucilage [10].

In addition to changes in pH, Brandon van Rooyen et al. [9] further showed that calcium, in the form of calcium chloride, could alter the physical properties of mucilage biofilms. The authors specifically reported that calcium reduces elasticity and increases the strength of mucilage biofilms [9].

4.2.2. Composite/Blended Mucilage Biofilms

Although the benefits of composite biopolymer films have been well established in the literature, limited research is available on mucilage's interaction with other commercial polymers. It has been suggested that compatible chemical synergetic interactions between ingredients displaying similar mechanical profiles could result in enhanced film mechanical properties. That would be essential for developing low-cost, biodegradable packaging [3,8,29,40,50].

In the recent work of Van Rooyen et al. [3], the authors investigated a blend of mucilage in combination with pectin and mucilage in combination with alginate biofilms. The influence of calcium as a cross-linker was also considered. It was found that, overall, pectin and mucilage displayed a synergist interaction, producing biofilms of increased strength. In contrast, alginate and mucilage showed a poor interaction with each other, reducing film strength. Polymer compatibility accounted for these differences, with pectin and mucilage showing a positive correlation [3].

Andreuccetti et al. [28] investigated the influence that mucilage (*Opuntia ficus-indica*) would have in combination with starch, with glycerol as a plasticiser, to improve film processability, functionality and certain mechanical properties. Composite rice starch and mucilage films produced adequate film mechanical properties, reporting a TS range of 2.80–3.96 MPa and %*E* of 12.07–20.63%. The authors suggested that combining the various biomaterials in the films allowed for a synergistic interaction between starch, mucilage and glycerol. These synergistic interactions resulted in increases in the films' %*E* values, coinciding with decreases in the films' TS values [15].

In another study, Sandoval et al. [8] investigated the effect that various cultivars' mucilage (*Opuntia ficus-indica*) would have on pectin-based film properties for developing biodegradable mucilage films. The authors suggested that determining a film's puncture tests would determine the ability of the film to maintain structural integrity and offer protection to the food product, as it would directly evaluate the film's mechanical strength. It was reported that different *Opuntia* spp. cultivars' mucilage influenced the mechanical properties of composite films to different degrees. The authors also found that decreases in the intermolecular association between polymers in the film matrix consequently resulted in the film's displaying increased flexibility/elasticity and, consequently, decreased film strength. The low polymer concentration used in the development of the films was also suggested to be a reason for this occurrence [8].

Luna-Sosa et al. [50] also investigated the influence mucilage would have on the properties of composite pectin films together with glycerol. The films were developed using the 'casting' method and then dried prior to investigation. The addition of mucilage was shown to directly influence the single-polymer pectin films' morphological and mechanical properties. The authors suggested that mucilage interfered with the pectin film matrix, resulting in less compact film morphological structures. These composite films further showed reduced mechanical properties for both TS and %*E* in comparison to the single-polymer pectin films [50].

Scognamiglio et al. [51] comparatively investigated the tensile tests and microstructures of composite mucilage films developed with thermoplastic starch and high amounts of glycerol. Mucilage was shown to negatively influence the films' TS, showing a more than 50% reduction in film strength in comparison to the single-polymer thermoplastic starch films. A TS range of 0.68–1.64 MPa was reported for composite mucilage films. The films' %*E* values, however, were increased by the addition of mucilage, interestingly, in some instances more than 50%, when compared to the single-polymer thermoplastic starch films. The authors further suggested that although high amounts of Ca and Mg were associated with mucilage, these did not seem to alter the mechanical performance of the films, suggesting that they were not bioavailable to alter the functional behaviour of the films.

4.3. Limitation and Future Perspectives

Variations in extraction methods, coupled with the varying physiological and environmental growth conditions of *Opuntia* spp. have resulted in considerable variation in the functional properties displayed by mucilage. In addition, specific factors have been identified as further altering the functional behaviour of mucilage in a solution and in the development of biofilms, limiting the consistency and expected outcomes of mucilage functional behaviour and applications. The presence of a cross-linker, polymer concentration and alteration in pH have specifically been identified and discussed in this review, with variations observed between different authors [19,20,23].

With regards to mucilage biofilm development, only more recently have specific publications been produced that address mucilage use in the development of biodegradable packaging in the food industry [3]. However, the limited work available on mucilage used in composite film development often shows inconsistent and contradictory findings. Nevertheless, a general consensus indicates that mucilage shows an undoubtable potential for use in the development of composite biofilms to be used as biodegradable packaging in the food industry [8,50].

As knowledge of the interactions between mucilage and other well-established polymers remains limited or non-existent, further investigations of the possibility of various cultivars' mucilage being used in the development of biopolymer films must be strongly considered. Understanding mucilage's functional behaviour relative to that of well-established functional polymers would further prove essential when holistically evaluating mucilage as a functional food ingredient for various applications. These findings could be used to improve the physical properties of packaging further and reduce the gap between low-cost, biodegradable packaging and single-use, petroleum-based packaging.

5. Conclusions

Consumer demands and environmental concerns have been a driving force behind the identification and investigation of novel natural biopolymers, such as cactus pear mucilage. Mucilage has specifically been considered for its unique and desirable functional properties that have been used in an attempt to address the current needs of the food industry. Compared to other commercially available polymers, limited research is available on mucilage's overall functional potential as a natural biopolymer and its commercial feasibility. Specifically, the functional potential of mucilage has found applications from a rheological and biofilm development perspective. It has, however, been seen that mucilage displays a functional potential that can be manipulated by a variety of factors, such as pH, a cross-linker and further influenced by polymer concentration. Variations amongst the different extraction procedures have also been identified as a critical factor for determining the functional behaviour of the resultant mucilage. It is suggested that mucilage's functional properties and applications require additional investigation. A better understanding of the functional behaviour of mucilage will aid in furthering and developing optimal extraction procedures. Although mucilage's rheological behaviour and biofilm development formation have been investigated more recently, certain aspects are still lacking from a research and application perspective. Therefore, further investigation of mucilage will prove critical for making future recommendations on mucilage's functional potential, especially if certain industry requirements and applications must be achieved.

Author Contributions: Conceptualisation: B.V.R. and M.D.W.; methodology: B.V.R., M.D.W. and G.O.; formal analysis: M.D.W. and G.O.; investigation: B.V.R.; resources: M.D.W. and J.V.N.; data curation: B.V.R., G.O. and M.D.W.; writing—original draft preparation: B.V.R. and M.D.W.; writing—review and editing: B.V.R. and J.V.N.; supervision: M.D.W., G.O. and J.V.N.; project administration: B.V.R., M.D.W. and J.V.N. All authors have read and agreed to the published version of the manuscript.

Funding: This research received no external funding.

Institutional Review Board Statement: Not applicable.

Data Availability Statement: Not applicable.

Conflicts of Interest: The authors declare no conflicts of interest.

References

1. Nath, P.C.; Sharma, R.; Debnath, S.; Sharma, M.; Inbaraj, B.S.; Dikkala, P.K.; Nayak, P.K.; Sridhar, K. Recent Trends in Polysaccharide-Based Biodegradable Polymers for Smart Food Packaging Industry. *Int. J. Biol. Macromol.* **2023**, *253*, 127524. [CrossRef] [PubMed]
2. Zena, Y.; Periyasamy, S.; Tesfaye, M.; Tumssa, Z.; Mohamed, B.A.; Karthik, V.; Asaithambi, P.; Getachew, D.; Aminabhavi, T.M. Trends on Barrier Characteristics Improvement of Emerging Biopolymeric Composite Films Using Nanoparticles—A Review. *J. Taiwan Inst. Chem. Eng.* **2024**, 105488. [CrossRef]
3. Van Rooyen, B.; De Wit, M.; Osthoff, G.; Van Niekerk, J.; Hugo, A. Effect of Native Mucilage on the Mechanical Properties of Pectin-Based and Alginate-Based Polymeric Films. *Coatings* **2023**, *13*, 1611. [CrossRef]
4. Kumar, S.; Mukherjee, A.; Dutta, J. Chitosan Based Nanocomposite Films and Coatings: Emerging Antimicrobial Food Packaging Alternatives. *Trends Food Sci. Technol.* **2020**, *97*, 196–209. [CrossRef]
5. Tkaczewska, J. Peptides and Protein Hydrolysates as Food Preservatives and Bioactive Components of Edible Films and Coatings—A Review. *Trends Food Sci. Technol.* **2020**, *106*, 298–311. [CrossRef]

6. Van Rooyen, B.; De Wit, M.; Osthoff, G.; Van Niekerk, J. Cactus Pear (*Opuntia* spp.) Crop Applications and Emerging Biopolymer Innovations. *Acta Hortic.* **2023**, *1380*, 129–134. [CrossRef]
7. Gheribi, R.; Khwaldia, K. Cactus Mucilage for Food Packaging Applications. *Coatings* **2019**, *9*, 655. [CrossRef]
8. Sandoval, D.C.G.; Sosa, B.L.; Martínez-Ávila, G.C.G.; Fuentes, H.R.; Abarca, V.H.A.; Rojas, R. Formulation and Characterization of Edible Films Based on Organic Mucilage from Mexican *Opuntia ficus-indica*. *Coatings* **2019**, *9*, 506. [CrossRef]
9. Van Rooyen, B.; De Wit, M.; Osthoff, G.; Van Niekerk, J.; Hugo, A. Microstructural and Mechanical Properties of Calcium-Treated Cactus Pear Mucilage (*Opuntia* spp.), Pectin and Alginate Single-Biopolymer Films. *Polymers* **2023**, *15*, 4295. [CrossRef]
10. Van Rooyen, B.; De Wit, M.; Osthoff, G.; Van Niekerk, J.; Hugo, A. Effect of pH on the Mechanical Properties of Single-Biopolymer Mucilage (*Opuntia ficus-indica*), Pectin and Alginate Films: Development and Mechanical Characterisation. *Polymers* **2023**, *15*, 4640. [CrossRef]
11. Majdoub, H.; Roudesli, S.; Picton, L.; Le Cerf, D.; Muller, G.; Grisel, M. Prickly Pear Nopals Pectin from *Opuntia ficus-indica* Physico-Chemical Study in Dilute and Semi-Dilute Solutions. *Carbohydr. Polym.* **2001**, *46*, 69–79. [CrossRef]
12. Matsuhiro, B.; Lillo, L.E.; Sáenz, C.; Urzúa, C.C.; Zárate, O. Chemical Characterization of the Mucilage from Fruits of *Opuntia ficus indica*. *Carbohydr. Polym.* **2006**, *63*, 263–267. [CrossRef]
13. Monrroy, M.; García, E.; Ríos, K.; García, J.R. Extraction and Physicochemical Characterization of Mucilage from *Opuntia cochenillifera* (L.) Miller. *J. Chem.* **2017**, *2017*, 4301901. [CrossRef]
14. Rodríguez-González, S.; Martínez-Flores, H.E.; Chávez-Moreno, C.K.; Macías-Rodríguez, L.I.; Zavala-Mendoza, E.; Garnica-Romo, M.G.; Chacõn-García, L. Extraction and Characterization of Mucilage from Wild Species of *Opuntia*. *J. Food Process Eng.* **2014**, *37*, 285–292. [CrossRef]
15. Goycoolea, F.M.; Cárdenas, A. Pectins from *Opuntia* Spp.: A Short Review. *J. Prof. Assoc. Cactus Dev.* **2003**, *5*, 17–29.
16. Cárdenas, A.; Goycoolea, F.M.; Rinaudo, M. On the Gelling Behaviour of "nopal" (*Opuntia ficus indica*) Low Methoxyl Pectin. *Carbohydr. Polym.* **2008**, *73*, 212–222. [CrossRef]
17. Du Toit, A.; De Wit, M. Patent PA153178P A Process for Extracting Mucilage from *Opuntia ficus-indica*, Aloe Barbadensis and Agave Americana. Ph.D. Thesis, University of the Free State, Bloemfontein, South Africa, 2021. [CrossRef]
18. Feugang, J.M. Nutritional and Medicinal Use of Cactus Pear (*Opuntia* spp.) Cladodes and Fruits. *Front. Biosci.* **2006**, *11*, 2574. [CrossRef] [PubMed]
19. Van Rooyen, B.; De Wit, M.; Osthoff, G. Gelling Potential of Native Cactus Pear Mucilage. *Acta Hortic.* **2022**, *1343*, 489–496. [CrossRef]
20. Van Rooyen, B.; De Wit, M.; Osthoff, G. Functionality of Native Mucilage from Cactus Pears as a Potential Functional Food Ingredient at Industrial Scale. *Acta Hortic.* **2022**, *1343*, 481–488. [CrossRef]
21. Sáenz, C.; Sepúlveda, E.; Matsuhiro, B. Opuntia Spp. Mucilage's: A Functional Component with Industrial Perspectives. *J. Arid Environ.* **2004**, *57*, 275–290. [CrossRef]
22. Sepúlveda, E.; Sáenz, C.; Aliaga, E.; Aceituno, C. Extraction and Characterization of Mucilage in *Opuntia* spp. *J. Arid Environ.* **2007**, *68*, 534–545. [CrossRef]
23. Medina-Torres, L.; Brito-De La Fuente, E.; Torrestiana-Sanchez, B.; Katthain, R. Rheological Properties of the Mucilage Gum (*Opuntia ficus indica*). *Food Hydrocoll.* **2000**, *14*, 417–424. [CrossRef]
24. Felkai-Haddache, L.; Dahmoune, F.; Remini, H.; Lefsih, K.; Mouni, L.; Madani, K. Microwave Optimization of Mucilage Extraction from *Opuntia ficus indica* Cladodes. *Int. J. Biol. Macromol.* **2016**, *84*, 24–30. [CrossRef] [PubMed]
25. Barbut, S.; Harper, B.A. Dried Ca-Alginate Films: Effects of Glycerol, Relative Humidity, Soy Fibers, and Carrageenan. *LWT* **2019**, *103*, 260–265. [CrossRef]
26. Bierhalz, A.C.K.; Da Silva, M.A.; Kieckbusch, T.G. Natamycin Release from Alginate/Pectin Films for Food Packaging Applications. *J. Food Eng.* **2012**, *110*, 18–25. [CrossRef]
27. da Silva, M.A.; Bierhalz, A.C.K.; Kieckbusch, T.G. Alginate and Pectin Composite Films Crosslinked with Ca^{2+} Ions: Effect of the Plasticizer Concentration. *Carbohydr. Polym.* **2009**, *77*, 736–742. [CrossRef]
28. Andreuccetti, C.; Galicia-García, T.; Martínez-Bustos, F.; Ferreira Grosso, R.; González-Núñez, R. Effects of Nopal Mucilage (*Opuntia ficus-indica*) as Plasticizer in the Fabrication of Laminated and Tubular Films of Extruded Acetylated Starches. *Int. J. Polym. Sci.* **2021**, *2021*, 6638756. [CrossRef]
29. Tosif, M.M.; Najda, A.; Bains, A.; Zawiślak, G.; Maj, G.; Chawla, P. Starch–Mucilage Composite Films: An Inclusive on Physicochemical and Biological Perspective. *Polymers* **2021**, *13*, 2588. [CrossRef]
30. Garfias Silva, V.; Cordova Aguilar, M.S.; Ascanio, G.; Aguayo, J.P.; Pérez-Salas, K.Y.; Susunaga Notario, A.D.C. Acid Hydrolysis of Pectin and Mucilage from Cactus (*Opuntia ficus*) for Identification and Quantification of Monosaccharides. *Molecules* **2022**, *27*, 5830. [CrossRef]
31. Espino-Díaz, M.; Ornelas-Paz, J.D.J.; Martínez-Téllez, M.A.; Santillán, C.; Barbosa-Cánovas, G.V.; Zamudio-Flores, P.B.; Olivas, G.I. Development and Characterization of Edible Films Based on Mucilage of *Opuntia ficus-indica* (L.). *J. Food Sci.* **2010**, *75*, 347–352. [CrossRef]
32. Du Toit, A.; De Wit, M.; Fouché, H.J.; Taljaard, M.; Venter, S.L.; Hugo, A. Mucilage Powder from Cactus Pears as Functional Ingredient: Influence of Cultivar and Harvest Month on the Physicochemical and Technological Properties. *J. Food Sci. Technol.* **2019**, *56*, 2404–2416. [CrossRef] [PubMed]

33. Du Toit, A.; De Wit, M.; Seroto, K.D.; Fouché, H.J.; Hugo, A.; Venter, S.L. Rheological Characterization of Cactus Pear Mucilage for Application in Nutraceutical Food Products. *Acta Hortic.* **2019**, *1247*, 63–72. [CrossRef]
34. Madera-Santana, T.J.; Vargas-Rodríguez, L.; Núñez-Colín, C.A.; González-García, G.; Peña-Caballero, V.; Núñez-Gastélum, J.A.; Gallegos-Vázquez, C.; Rodríguez-Núñez, J.R. Mucilage from Cladodes of Opuntia Spinulifera Salm-Dyck: Chemical, Morphological, Structural and Thermal Characterization. *CYTA—J. Food* **2018**, *16*, 650–657. [CrossRef]
35. Gawkowska, D.; Cybulska, J.; Zdunek, A. Structure-Related Gelling of Pectins and Linking with Other Natural Compounds: A Review. *Polymers* **2018**, *10*, 762. [CrossRef] [PubMed]
36. Miya, S.; De Wit, M.; Van Biljon, A.; Venter, S.L.; Amonsou, E. *Opuntia ficus-indica* Mill. and *O. robusta* Cladode Mucilage: Carbohydrates. *Acta Hortic.* **2022**, *1343*, 511–518. [CrossRef]
37. Contreras-Padilla, M.; Rodríguez-García, M.E.; Gutiérrez-Cortez, E.; Valderrama-Bravo, M.d.C.; Rojas-Molina, J.I.; Rivera-Muñoz, E.M. Physicochemical and Rheological Characterization of Opuntia Ficus Mucilage at Three Different Maturity Stages of Cladode. *Eur. Polym. J.* **2016**, *78*, 226–234. [CrossRef]
38. De Wit, M.; Du Toit, A.; Fouché, H.J.; Hugo, A.; Venter, S.L. Screening of Cladodes from 42 South African Spineless Cactus Pear Cultivars for Morphology, Mucilage Yield and Mucilage Viscosity. *Acta Hortic.* **2019**, *1247*, 47–55. [CrossRef]
39. Soukoulis, C.; Gaiani, C.; Hoffmann, L. Plant Seed Mucilage as Emerging Biopolymer in Food Industry Applications. *Curr. Opin. Food Sci.* **2018**, *22*, 28–42. [CrossRef]
40. Galus, S.; Lenart, A. Development and Characterization of Composite Edible Films Based on Sodium Alginate and Pectin. *J. Food Eng.* **2013**, *115*, 459–465. [CrossRef]
41. Gheribi, R.; Puchot, L.; Verge, P.; Jaoued-Grayaa, N.; Mezni, M.; Habibi, Y.; Khwaldia, K. Development of Plasticized Edible Films from *Opuntia ficus-indica* Mucilage: A Comparative Study of Various Polyol Plasticizers. *Carbohydr. Polym.* **2018**, *190*, 204–211. [CrossRef]
42. Fabra, M.J.; Talens, P.; Chiralt, A. Influence of Calcium on Tensile, Optical and Water Vapour Permeability Properties of Sodium Caseinate Edible Films. *J. Food Eng.* **2010**, *96*, 356–364. [CrossRef]
43. Lira-Vargas, A.A.; Lira-Vargas, A.A.; Corrales-Garcia, J.J.E.; Valle-Guadarrama, S.; Peña-Valdivia, C.B.; Trejo-Marquez, M.A. Biopolymeric Films Based on Cactus (*Opuntia ficus-indica*) Mucilage Incorporated with Gelatin and Beeswax. *J. Prof. Assoc. Cactus Dev.* **2014**, *16*, 51–70.
44. Allegra, A.; Inglese, P.; Sortino, G.; Settanni, L.; Todaro, A.; Liguori, G. The Influence of *Opuntia ficus-indica* Mucilage Edible Coating on the Quality of "Hayward" Kiwifruit Slices. *Postharvest Biol. Technol.* **2016**, *120*, 45–51. [CrossRef]
45. Damas, M.S.P.; Junior, V.A.P.; Nishihora, R.K.; Quadri, M.G.N. Edible Films from Mucilage of Cereus Hildmannianus Fruits: Development and Characterization. *J. Appl. Polym. Sci.* **2017**, *134*, 1–9. [CrossRef]
46. Zibaei, R.; Hasanvand, S.; Hashami, Z.; Roshandel, Z.; Rouhi, M.; Guimarães, J.d.T.; Mortazavian, A.M.; Sarlak, Z.; Mohammadi, R. Applications of Emerging Botanical Hydrocolloids for Edible Films: A Review. *Carbohydr. Polym.* **2021**, *256*, 117554. [CrossRef] [PubMed]
47. Livney, Y.D.; Schwan, A.L.; Dalgleish, D.G. A Study of β-Casein Tertiary Structure by Intramolecular Crosslinking and Mass Spectrometry. *J. Dairy Sci.* **2004**, *87*, 3638–3647. [CrossRef] [PubMed]
48. Lazaridou, A.; Biliaderis, C.G.; Kontogiorgos, V. Molecular Weight Effects on Solution Rheology of Pullulan and Mechanical Properties of Its Films. *Carbohydr. Polym.* **2003**, *52*, 151–166. [CrossRef]
49. Gheribi, R.; Gharbi, M.A.; Ouni, M.E.; Khwaldia, K. Enhancement of the Physical, Mechanical and Thermal Properties of Cactus Mucilage Films by Blending with Polyvinyl Alcohol. *Food Packag. Shelf Life* **2019**, *22*, 100386. [CrossRef]
50. Luna-Sosa, B.; Martínez-Ávila, G.C.G.; Rodríguez-Fuentes, H.; Azevedo, A.G.; Pastrana, L.M.; Rojas, R.; Cerqueira, M.A. Pectin-Based Films Loaded with Hydroponic Nopal Mucilages: Development and Physicochemical Characterization. *Coatings* **2020**, *10*, 467. [CrossRef]
51. Scognamiglio, F.S.; Gattia, D.M.; Roselli, G.; Persia, F.; De Angelis, U.; Santulli, C. Thermoplastic Starch (TPS) Films Added with Mucilage from *Opuntia ficus indica*: Mechanical, Microstructural and Thermal Characterization. *Materials* **2020**, *13*, 1000. [CrossRef]

polymers

MDPI

Article

Effect of Mixed Particulate Emulsifiers on Spray-Dried Avocado Oil-in-Water Pickering Emulsions

Vicente Espinosa-Solis [1], Yunia Verónica García-Tejeda [2,*], Oscar Manuel Portilla-Rivera [1], Carolina Estefania Chávez-Murillo [3] and Víctor Barrera-Figueroa [4]

[1] Coordinación Académica Región Huasteca Sur, Universidad Autónoma de San Luis Potosí, km 5, Carretera Tamazunchale-San Martín, Tamazunchale 79960, Mexico; vicente.espinosa@uaslp.mx (V.E.-S.); manuel.portilla@uaslp.mx (O.M.P.-R.)

[2] Academia de Ciencias Básicas, UPIITA, Avenida Instituto Politécnico Nacional No. 2580, Col. Barrio la Laguna Ticomán, Gustavo A. Madero, Mexico City 07340, Mexico

[3] Academia de Bioingeniería, UPIIZ, Instituto Politécnico Nacional, Circuito del Gato No. 202, Col. Ciudad Administrativa, Zacatecas 98160, Mexico; cchavezm@ipn.mx

[4] Sección de Estudios de Posgrado e Investigación, UPIITA, Avenida Instituto Politécnico Nacional No. 2580, Col. Barrio la Laguna Ticomán, Gustavo A. Madero, Mexico City 07340, Mexico; vbarreraf@ipn.mx

* Correspondence: ygarciat@ipn.mx; Tel.: +52-555-729-6000 (Ext. 56918)

Abstract: Avocado oil is a very valuable agro-industrial product which can be perishable in a short time if it is not stored in the right conditions. The encapsulation of the oils through the spray drying technique protects them from oxidation and facilitates their incorporation into different pharmaceutical products and food matrices; however, the selection of environmentally friendly emulsifiers is a great challenge. Four formulations of the following solid particles: Gum Arabic, HI-CAP®100 starch, and phosphorylated waxy maize starch, were selected to prepare avocado oil Pickering emulsions. Two of the formulations have the same composition, but one of them was emulsified by rotor-stator homogenization. The rest of the emulsions were emulsified by combining rotor-stator plus ultrasound methods. The protective effect of mixed particle emulsifiers in avocado oil encapsulated by spray drying was based on the efficiency of encapsulation. The best results were achieved when avocado oil was emulsified with a mixture of phosphorylated starch/HI-CAP®100, where it presented the highest encapsulation efficiency.

Keywords: avocado oil; Pickering emulsions; encapsulation; phosphorylated starch

check for updates

Citation: Espinosa-Solis, V.; García-Tejeda, Y.V.; Portilla-Rivera, O.M.; Chávez-Murillo, C.E.; Barrera-Figueroa, V. Effect of Mixed Particulate Emulsifiers on Spray-Dried Avocado Oil-in-Water Pickering Emulsions. *Polymers* **2022**, *14*, 3064. https://doi.org/10.3390/polym14153064

Academic Editor: César Leyva-Pórras

Received: 16 May 2022
Accepted: 25 July 2022
Published: 28 July 2022

1. Introduction

The avocado is a highly valued fruit in the international market, but when it does not have adequate characteristics to the exported to the country of destination, it is sometimes discarded and lost. A sustainable alternative is to use avocado oil to promote new agro-industrial products. Avocado oil is obtained from the mesocarp and seed of the fruit of the avocado tree (*Persea americana*); avocado oil could serve as a food supplement in the diet due to the multiple benefits that its consumption confers on health. The inclusion of avocado oil in the diet improves the skin collagen metabolism [1], postprandial metabolic responses to a hypercaloric-hyperlipidemic meal in overweight subjects [2], the glucose and insulin resistance induced by high sucrose diet in Wistar rats [3]. Eight fatty acids are present in avocado flesh [4], including palmitic (C16:0), palmitoleic (C16:1), stearic (18:0), oleic (C18:1), and linoleic (18:2), myristic (C14:0), and arachidic (C20:0). Moreover, avocado oil contains fat-soluble vitamins, including vitamin A, B, E, and vitamin D precursors [5]. Vitamin D is formed in the skin after exposure to sunlight. Vitamin D2 and D3 are synthesized from precursors such as ergosterol and 7-dehydrocholesterol, respectively [6]. Vitamin D deficiency has been linked to an increased risk and morbidity associated with COVID-19 [7]. In addition, refined oil is used in skincare products since it is rapidly absorbed by the skin and has sunscreen properties [8,9].

It is important to incorporate avocado oil into the diet; however, incorporating it into processed food matrices is limited since avocado oil is degraded dramatically at temperatures above 180 °C, and light accelerates the degradation of avocado oil. Incorporating oil in food and pharmaceutical products that have immiscible phases can be carried out by using an emulsifier to form a homogeneous mixture, that is, an emulsion. Traditional emulsifiers comprise biopolymers and low-molecular-weight surfactants (LMWEs), such as monoacylglycerols and polysorbates; which can be either natural or synthetic origin. LMWEs consist of a hydrophilic head, which can be nonionic, or fully charged; and a hydrophobic tail, usually consisting of at least one acyl chain [10]. However, most surfactants currently used to stabilize food emulsions are ionic molecules, which can induce irritating skin reactions and can cause toxic symptoms in animals and humans [11].

An emulsion that uses solid particles for stabilization instead of chemical surfactants is called a Pickering-type emulsion; this emulsion consists of solid particles absorbed at the oil-water interface, which prevents flocculation and coalescence by forming a densely packed layer that retards the formation of creams or sedimentation [12]. Pickering emulsions are environmentally friendly due to being byproducts of the agri-food industry, and can be used for formulation [13], as well as polysaccharides of some microorganisms [14].

Inorganic particles such as silica (SiO_2), calcium carbonate ($CaCO_3$), and titanium dioxide (TiO_2) have been used as Pickering stabilizers [15]. Bio-based particles, such as starch granules, fat crystals [16], gums [17,18], and proteins like soy glycinin [19], collagen [20], casein micelles, whey protein nanofibers [21,22], salted duck egg white [23,24], and pea proteins [25] as edible and sustainable solid particles for their use in infant formulas [26], edible foams [27], gluten-free rice bread [28], and for the replacement of saturated fat with vegetable oils in sausages [29].

Pickering emulsion technique has become increasingly important as a template for microcapsule formation and relies on solid-stabilized emulsions [30]. Spray drying is the most commonly used technique for microencapsulation of oils [31,32], and the selection of wall materials with good emulsifying properties is crucial to prolong the stability of the encapsulated oil. Among the solid particles synthesized from bio-based particles, *n*-octenyl succinic anhydride starch (HI-CAP®100) derived from waxy maize is widely used for stabilizing emulsions and microencapsulation of bioactive compounds by spray drying, offering advantages such as neutral aroma and taste, low viscosity at high solids concentrations, and good protection against oxidation [33]. Gum Arabic (gum Acacia) is a hydrocolloid produced by the natural exudation of acacia trees and is an effective carrier agent due to its high water solubility and its ability to act as an oil-in-water emulsifier [34]. Maltodextrins are hydrolyzed starches produced via enzymatic or acid hydrolysis of the starch, followed by purification and spray drying [35]. Anion starch phosphate has not been fully exploited in the microencapsulation of oils; it has negatively charged phosphate groups that cause repulsion between the starch chains and, consequently, an increase in its emulsifying and hydration capacity [36].

In recent works, *n*-octenyl succinic anhydride starches [37,38] were evaluated for the microencapsulation of avocado oil by spray drying; these report low encapsulation efficiencies (40–61.7%). The encapsulation efficiency, and the oxidation stability of avocado oil are improved by combining two encapsulating materials, maltodextrins in combination with whey protein isolate [39], or maltodextrins in combination with HI-CAP®100 [40]. Phosphorylated starch has great potential as a wall material for spray drying, but it has lower emulsifying properties than *n*-octenyl succinic anhydride starch [41]. To the best of our knowledge, the combination of phosphorylated starch, Gum Arabic and HI-CAP®100, for the emulsification and microencapsulation of avocado oil, has not been reported.

In this way, the objectives of the present work are: (1) to examine the ability of the combination of phosphorylated starch, Gum Arabic, and HI-CAP®100 to emulsify avocado oil, (2) to evaluate the protection of these different polymers combination in microencapsulation process, and (3) to determine the critical storage conditions of avocado oil microparticles.

2. Materials and Methods

2.1. Plant Materials and Chemical Reagents

Extra Virgin Avocado Oil (EVAO) was extracted from the mesocarp of the avocado fruit cultivar "Hass" *Persea gratissima*, which was purchased from Avocare Oleo Lab (Guadalajara, Mexico).

Gum Arabic 8287 was purchased from Norevo (CDMX, Mexico), HI-CAP®100 starch was purchased from Ingredion (CDMX, Mexico), and phosphorylated waxy maize starch with a degree of substitution of 0.04 was prepared by reactive extrusion, as described in a previous work [41].

2.2. Preparation of Avocado Oil Emulsions

Avocado oil-in-water emulsions were prepared using blends of the following biopolymers as emulsifiers: Gum Arabic, HI-CAP®100 starch, and phosphorylated maize starch. Four different emulsions, AOE1, AOE2, AOE3, and AOE4, were formulated as shown in Table 1. These formulations were selected based on a preliminary experimental design of mixtures (see supplementary material). The ratio between the avocado oil and each biopolymer mixture was 1:4 (*w*/*w*); each polysaccharide suspension was prepared by suspending the solids at 20% (*w*/*w*) in distilled water. Then, avocado oil was slowly incorporated into each polysaccharide suspension by high shear stirring at 11,000 rpm for 5 min, using a rotor-stator blender (Ultra-Turrax IKA T18 basic, Wilmington, USA) to form emulsions. After the homogenization process by high shear, AOE2, AOE3, and AOE4 samples (Table 1) were submitted to ultrasonication at 160 W of nominal power in a Branson DigitalSonifier® Model S-450D (Branson Ultrasonics Corporation, Danbury, CT, USA), 20 kHz (60% amplitude), for 1 min at 4 °C in an ice bath to dissipate heat and prevent overheating of the sample. The avocado-loaded Pickering o/w emulsion AOE1 was prepared by homogenization.

Table 1. Formulation and composition of avocado oil-in-water emulsions by using different biopolymers.

Sample	Composition of Emulsifiers in wt %			Emulsification Method [1]
	Phosphorylated Starch	Gum Arabic	HI-CAP®100	
AOE1	66.66	16.66	16.66	H
AOE2	66.66	16.66	16.66	H + U
AOE3	66.6	0.8	32.5	H + U
AOE4	50	0	50	H + U

[1] Avocado oil-in-water emulsions were prepared by high shear (H) and ultrasound (U) homogenization methods.

2.2.1. Stability of the Kinetics of the Emulsion

The stability of emulsions was investigated on avocado oil emulsions samples by light scattering by a Turbiscan Lab Expert (Formulation, Toulouse, France). The detection head is composed of a pulsed near-infrared light source (λ = 850 nm) and two synchronous detectors. A glass cell was placed in the equipment to analyze the stability index of the emulsion for fifteen days, the measurements were taken every five minutes during the first hour, every twenty minutes during five hours, and every three days during fifteen days at 25 °C. The analysis of stability was performed as a variation of backscattering (BS) profiles as a function of time at the middle and top layer of the samples and then exported as % BS and peak thickness, respectively, by Turbisoft Lab 2.2 software. The curves obtained by subtracting the BS profile at time 0 from the profile at time t (that is, $\Delta BS = BS_t - BS_0$) showed a typical shape that allows a better quantification of creaming, flocculation, and other destabilization processes [42]. The global Turbiscan Stability Index (TSI) was calculated to compare the stability of the different formulations under analysis using the following formulæ

$$BS = 1/\sqrt{\lambda^*}, \lambda^*(\varphi, d) = \frac{2d}{3\varphi(1-g)Qs}, \quad (1)$$

$$TSI = \left(\frac{1}{n-1} \sum_{i=1}^{n} (x_i - x_{BS})^2 \right)^{1/2}, \tag{2}$$

where λ^* is the photon transport mean free path in the analyzed dispersion: φ is the volume fraction of particles; d is the mean diameter of particles; g and Qs are the optical parameters given by Mie's theory; x_i is the average backscattering for each minute of measurement; x_{BS} is the average of x_i and n is the number of scans [42].

2.2.2. Morphology, Droplet Size Distribution, and ζ-Potential Measurements

The interfacial structure of aged emulsions (fifteen days of storage) was analyzed using an Eclipse H550S microscope (Nikon, Chiyoda-ku, Japan) equipped with a Kodak DC 120 digital camera (Servier Country, TN, USA).

The size characterization and the ζ-potential were measured separately using a Zetasizer (NanoZS, Malvern Instruments Ltd., Malvern, UK) by diluting 1 µL of each emulsion sample in 10 mL of type I water. All measurements reported in this paper were made at a temperature of 25 °C. Size measurements were carried out using a process called dynamic light scattering (DLS), which uses a 4 mW He–Ne laser operating at a wavelength of 633 nm and a detection angle of 173°. The size distribution was obtained from the analysis of correlation function in the instrument software.

The ζ-potential on emulsion droplets was determined by the Henry equation, which relates the electrophoretic mobility to ζ-potential. The electrophoretic mobility is obtained by performing an electrophoresis experiment on the sample and measuring the velocity of the particles using a laser doppler velocimetry [43].

2.3. *Spray-Drying of Pickering Emulsions to Produce Microparticles*

The encapsulation was carried out by spray-drying in a Mobile Minor 2000 (GEA Niro, Søborg, Denmark) using a peristaltic pump (Watson-Marlow 520S, Altamonte Springs, FL, USA) with the following drying conditions: inlet air temperature of 170 ± 5 °C; outlet air temperature of 75 ± 5 °C; nozzle diameter of 0.5 mm; and liquid flow rate of 10 mL/min. The equipment's air flow was set at 70 m^3/h.

The product yield (Y) was calculated as the ratio between the mass of the output powders ($M_{recovered}$), recovered from the equipment in the end of the spray-drying process, and the mass of the solid content of the initial solution M_{infeed}, infeed to the spray-dryer chamber.

$$Y(\%) = \frac{M_{recovered}}{M_{infeed}} \times 100, \tag{3}$$

2.3.1. Determination of Moisture Content and Water Activity of Microparticles

The moisture content of spray-dried powders was measured using a gravimetric method (AOAC, 1995). Briefly, 1 g of each powder was placed in an aluminum plate at constant weight and heated at 105 °C until constant weight was reached. Water activity (a_w) was measured employing an Aqualab meter (Decagon Devices, Model 4 TE, Pullman, DC, USA).

2.3.2. Determination of Avocado Oil Content in Microparticles

The determination of avocado oil in the microcapsules was carried out according to the previously reported methodology [41]. The onset (T_o) and peak (T_p) temperatures of crystallization and the enthalpy (ΔH_c) in J/g of both avocado oil and microcapsules were determined in a Differential scanning calorimetry (DSC) analysis by using a TRIOS 5.1.1 Sofware (TA Instruments, New Castle, UK). Approximately 3 mg (dry basis) was weighed directly into aluminum trays. The oil percentage in microcapsules was calculated by dividing the integrated area under the single exothermal peak corresponding to the oil in microcapsules by the crystallization enthalpy of the pure avocado oil,

$$\text{EVAO} = \frac{\Delta H_{\text{Microcapsules}}}{\Delta H_{\text{EVAO}}} \times 100, \quad (4)$$

where EVAO is the percentage of extra virgin avocado oil in microcapsules; $\Delta H_{\text{Microcapsules}}$ is the enthalpy of crystallization of EVAO in microcapsules; and ΔH_{EVAO} is the enthalpy of crystallization of pure EVAO.

Encapsulation efficiency (*EE*) is the percentage of total avocado oil in spray-dried product with reference to the corresponding avocado oil infeed in the emulsion,

$$EE = \frac{\Delta H_{\text{Microcapsules}}}{1.48} \times 100, \quad (5)$$

where 1.48 is the enthalpy of crystallization in J/g of EVAO.

2.3.3. Morphology of Microparticles

The external morphology of the microparticles was observed by scanning electron microscopy (SEM), consisting of a FEI-Sirion S4800 JEOL 7401F instrument operated a 5 kV with secondary electrons (Hitachi High-Technologies Corporation, Instruments Co., Ltd., Tokyo, Japan).

2.4. Determining Storage Conditions for Avocado Oil Microparticles

2.4.1. Moisture Adsorption Isotherm and Its Modeling

On the basis to determine critical storage conditions, samples of M-AOE4 (1 g) were put into dishes; the dishes were placed into hermetically sealed desiccators at 25 °C, each containing one of the following saturated solutions: LiCl, CH_3CO_2K, KCl, K_2CO_3, $Mg(NO_3)_2$, NaCl, KCl, BaCl. In this way, the water activity ranges from 0.11 to 0.94 a_w. The samples were stored until reaching equilibrium moisture content, that is, when the differences between two consecutive weights were within 0.001 g.

Guggenheim-Anderson-De Boer (GAB) Model

Water sorption isotherm was fitted by GAB model [44], and is described by the formula

$$M = \frac{M_0 C K a_w}{(1 - K a_w)(1 - K a_w + C K a_w)}, \quad (6)$$

where M is the moisture content in the sample (g water per 100 g of dry solids) at a_w, M_0 is the monolayer moisture content (g water per 100 g of dry solids), C is the Guggenheim's parameter, and K is a dimensionless parameter.

LSF-Polynomials of 6-th Order

In LSF-polynomials (Least squares fitting-polynomials), the sum of the squares of the vertical offsets between the data points and the polynomial is minimized [45]. The coefficients of an LSF-polynomial were determined by the instruction `Fit[]` of Wolfram Mathematica® (Champaign, IL, USA). A 6-th order LSF-polynomial is employed to minimize the fitting error by increasing the number of degrees of freedom instead of increasing the number of data points for modelling moisture sorption isotherm.

$$M_6(a_w) = b_0 + b_1 a_w + b_2 a_w^2 + b_3 a_w^3 + b_4 a_w^4 + b_5 a_w^5 + b_6 a_w^6, \quad (7)$$

In contrast to the use of polynomials as interpolating functions, where the degree of the interpolating polynomial depends on the number of data points, LSF-polynomials do not necessarily pass through the data points as interpolating polynomials do. As a result, the degree of a LSF-polynomial can be much lower than the number of data points [45].

2.4.2. Determination of the Critical Water Activity (RHc)
Determination of Inflection Points

From the experimental data of M obtained from sorption isotherm of avocado oil microparticles (M-AOE4) at 25 °C, LSF-polynomials of 6-th order were denoted by $M_6(a_w)$. The inflection points were calculated from the zeros of the polynomial equations $M_i(a_w) = 0$, where $i = 6$. Given a polynomial of degree 6, its inflection points were calculated from its second derivative, which is indeed another polynomial of degree $6 - 2$.

Determination of T_g of M-AOE4 Microparticles

The values of T_g of M-AOE4 samples stored at different a_w at 25 °C were determined by DSC. Runs were performed within a temperature range of 25 to −85 °C, finally cooling from −85 to 80 °C; the heating and cooling rates were set at 10 °C/min. The values of T_g were determined as the midpoint in the proximity of change of the specific heat (ΔCp). The plasticizing effect of water on T_g is described by the Gordon–Taylor model [46], which is described by the formula

$$T_g = \frac{w_1 T_{g1} + k w_2 T_{g2}}{w_1 + k w_2}, \tag{8}$$

where T_g is the glass transition temperature of a mixture of solids and water; w_1 is the anhydrous fraction having a glass transition T_{g1}; w_2 is the water fraction having a glass transition T_{g2}, which is often taken as −135 °C, corresponding to pure water; and k is a parameter [47]. If $k = 1$, the relation between T_g and the anhydrous fraction is linear, whose plot is a straight line in the $w_1 - T_g$ plane. If $k > 1$ the resulting plot is concave, while if $k < 1$ the plot is convex [48].

2.5. Fitting of Models

The fitting of the above models was evaluated through the mean relative deviation error (P), which is defined as follows

$$P = \frac{100}{N} \sum_{i=1}^{N} \left| \frac{X_i - X_i'}{X_i} \right|, \tag{9}$$

where N is the number of data points, X_i denotes the experimental data, and X_i' is the forecast value calculated by a model.

2.6. Statistical Analysis

The results obtained were conducted in triplicate and results are reported as mean ± standard deviation. Data were analyzed statistically by one-way analysis of variance (ANOVA) using Minitab 19 Statistical Software (Minitab, Inc., State College, Pennsylvania, PA, USA). The Tukey's test was used to determine differences in the mean values ($p \leq 0.05$).

3. Results

3.1. Droplet Size Distribution and Morphology

Droplet size distributions of Pickering emulsions containing avocado oil is shown in Figure 1. The mean particle size of avocado oil emulsions was between a range of 363.8–858.5 nm. The samples AOE2 and AOE1 contain the same composition of biopolymers, the difference lies in the emulsification method. Coarse emulsion AOE1 was not sonicated and showed the largest emulsion droplets, with an average droplet size of 858.5 nm. On the other hand, the smallest emulsion droplets with an average droplet size of 363.8 nm (AOE2) were obtained by using rotor-stator homogenizer, followed by ultrasonic emulsification. According to [49], ultrasonic emulsification decreases the median diameter of oil droplets from 1.141 to 0.891 µm, then droplet size reduction is attributed to the combined emulsification methods.

The emulsifier type and the emulsification method influenced the control of the size of the droplets, and Figure 1 shows the droplet size distribution observed in all the samples. AOE1 and AOE4 show a trimodal distribution, however, samples AOE2 and AOE3 show a bimodal distribution; this phenomenon is attributed to the coalescence and rupture of the droplets [50]. It can be seen with the naked eye that emulsions with a trimodal droplet distribution have larger droplets than emulsions with a bimodal droplet distribution. The values obtained in the present work are comparable to those reported for olive oil emulsions [41] homogenized by microfluidization (345–996 nm) and by using starch derivatives.

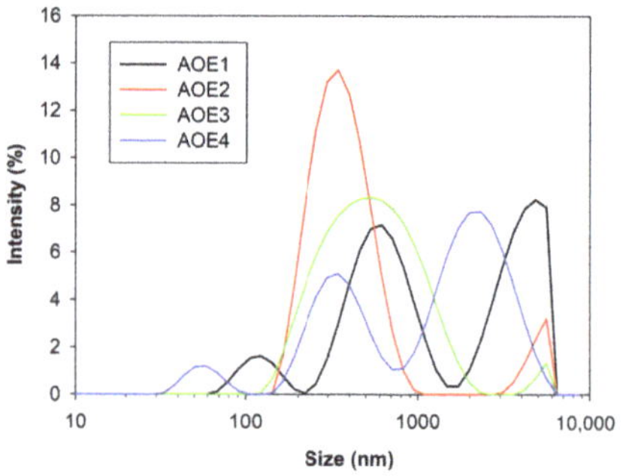

Figure 1. Droplet size distribution of O/W emulsions stabilized by mixed particulate emulsifiers (AOE1, AOE2, AOE3, AOE4).

3.2. ζ-Potential and Stability of Emulsions

Table 2 presents the ζ-potential values. This parameter characterizes the surface charge of the droplets and reflects the repulsive force between the emulsion droplets. According to results (Table 2), ζ-potential values are within the range $-0.34 < \zeta < -27.5$; it is considered in systems with $\zeta > 25$ mV and $\zeta < -25$ mV to have a high degree of stability [51]. Stabilized emulsions with Gum Arabic show more electronegative values than AOE4 emulsion, particularly attributed to the negative ζ-potential of carboxylic groups in Gum Arabic (−28.97 mV). Less negative values were obtained in Phosphorylated (−18.1 mV) and HI-CAP®100 (−19.20 mV) starches.

Table 2. Influence of mixed particle/emulsifier on stability and physical characteristics of emulsions

Sample	Average Size (nm)	Droplet Size (nm) Peak 1	Peak 2	Peak 3 [1]	ζ-Potential (mV)	TSI (6 h, 15 days)
AOE1	858.5	639 (48.9)	3991 (43.9)	120.9 (7.2)	−27.5 ± 1.08	1.92, 19.93
AOE2	363.8	372.1 (92.6)	4907.00 (7.4)	0.00 (0.0)	−22.4 ± 0.11	1.85, 13.02
AOE3	453.7	609.4 (97.8)	5196.00 (2.2)	0.00 (0.0)	−27.6 ± 0.46	1.83, 10.23
AOE4	605.7	2305 (62.4)	363.6 (32.5)	57.31 (5.1)	−0.34 ± 0.23	1.92, 17.10

[1] Droplet size average of each peak in a tri-modal droplet size distribution (% Intensity), and mean value of ζ-potential ± standard error of fresh avocado oil emulsions. Turbiscan stability index (TSI) by dynamic light scattering.

Samples AOE1 and AOE2 have the same composition, but they present different values of ζ-potential; this can be attributed to the emulsification method. The type of emulsification, either rotor-stator or microfluidization, impacts the distribution of the functional groups of mixed biopolymers on the surface composition of spray dried emulsions [52]. By using a rotor-stator mixer, micelle formation occurs through shear forces. In another way by ultra-sonication, ultrasound waves are propagated through the emulsion and denaturation of the secondary structure of macromolecules is carried out [53]. Carbonyl, sulfhydryl, hydroxyl groups, etc., could be exposed due to molecular unfolding and stretching of proteins of Gum Arabic in sonicated samples; thus, they can affect electrokinetic potential in emulsions.

Not only ζ-potential, but also steric hindrance among droplets, is another mechanism to prevent the coagulation or flocculation in emulsions [18], as it is the main stabilization mechanism [54], particularly in pickering-type emulsions. Previous studies showed that the union of hydrophilic polysaccharides on the surface of oil droplets reinforces steric repulsion, preventing droplet aggregation [55,56]. The AOE4 shows a ζ-potential close to zero. This formulation is composed of starch derivatives and it suggests that starch particles confer a charge shielding effect.

Samples AOE1 and AOE4 show a lower stability of the droplets to flow compared to samples AOE2 and AOE3, which is consistent with the TSI of the emulsions (Table 2). It is important to mention that the lowest values of TSI obtained for AOE2 and AOE3 imply better stability during storage, and that in those samples Gum Arabic helped to stabilize the emulsions since the hydrophobic and protein rich backbone at Gum Arabic adsorbs onto the O/W emulsion interface [57]. Moreover, a yellowish color can be seen in samples containing Gum Arabic (Figure 1).

The visual appearance of the vials containing emulsions after fifteen days of storage can be seen in Figure 2. The photograph shows that sedimentation takes place after the coalescence of droplets, it is appreciated that the AOE1 sample shows more sediment than AOE2, AOE3, and AOE4 samples. Optical images of samples AOE1 and AOE2 were taken from the bottom of the vials and confirm the flocculation and coalescence phenomenon, and the green circles enclose the largest droplets that originated from the collision of two or more droplets. Optical images of AOE2 and AOE3 samples were taken from the supernatant of the vials, which show droplets of a uniform size that maintain a certain distance between each one. The destabilization mechanisms associated with the coalescence and flocculation of the droplets in all the samples could be observed, and both phenomena were observed during their analysis by optical microscopy in aged samples. No appreciable changes were observed after fifteen days of storage.

The backscattering (BS) profiles monitored during fifteen days are shown in Figure 3. No noticeable destabilization of the emulsions was observed during the first six hours of storage, therefore, the emulsions can be spray-dried during that time. After three to fifteen days of storage, it is shown in AOE1 sample (Figure 3) that sedimentation and creaming occurred. It should be noted that the AOE2 sample is more stable than the AOE1, both containing the same composition of biopolymers, but AOE2 was processed by rotor-stator mixer and ultrasonication homogenization.

A direct relationship was observed between droplet size and creaming, and the formation of a cream layer in AOE1 and AOE4 resulted from the generation of big lipid droplets due to weak steric repulsion [58]. The AOE2 sample that showed the smallest droplet size (363.8 nm) did not show destabilization by creaming. The presence of Gum Arabic in AOE2 and AOE3 samples provided a better stability by immobilizing the emulsion droplets into a network stabilized by electrostatic repulsion or steric effects between the droplets.

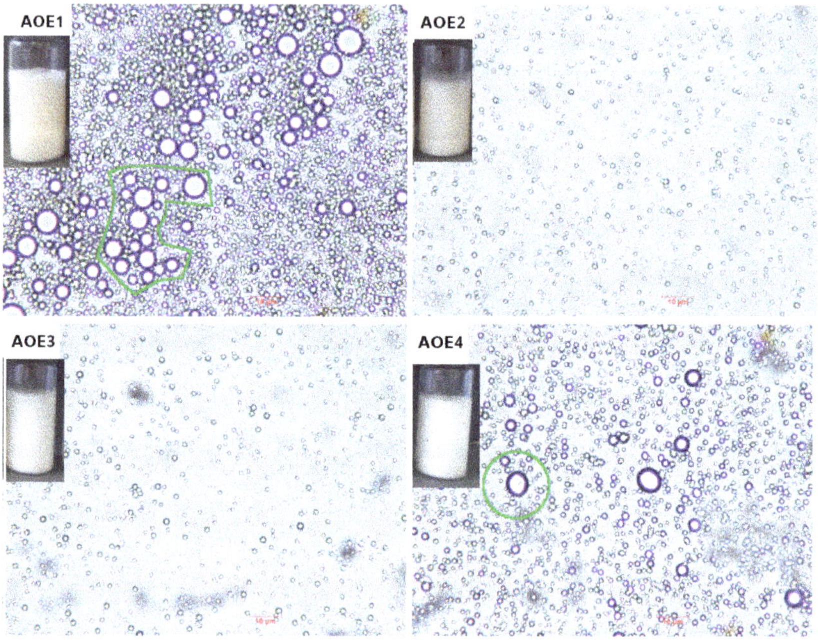

Figure 2. Visual appearance of measuring vials containing avocado oil emulsions stabilized by particulate emulsifiers after fifteen days of storage and optical microscopy (40× magnification) of avocado oil emulsions: AOE1, AOE2, AOE3, and AOE4. Optical microscopy images of AOE1 and AOE4 samples were taken from the supernatant of the vials, and AOE2 and AOE3 samples were taken from the sediment in the vials. The green circles enclose droplet coalescence. Scale bar of 10 µm (—).

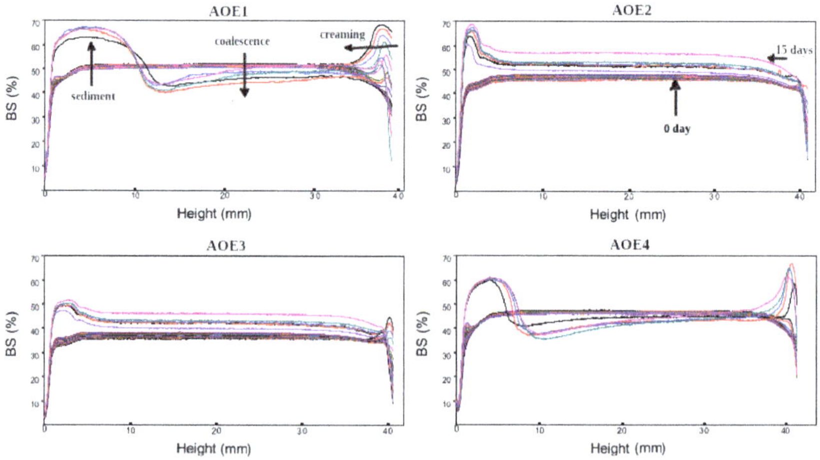

Figure 3. Backscattered light intensity as a function of the height of a measuring cell of O/W emulsions stabilized by particulate emulsifiers through fifteen days of storage at 25 °C.

3.3. Characterization of Spray-Dried Microparticles from Emulsions

The moisture content of the spray-dried powders and a_w are shown in Table 3. Moisture content ranged from 1.334 to 3.156% and water activity ranged from 0.11 to 0.15 a_w. Those results are similar to those reported for microparticles of avocado oil [39] by using mixtures of whey protein and maltodextrin. The authors report a maximum moisture content of 2.89% of moisture content [39]. It should be noted that low moisture content and water activity are desirable for greater product stability during storage.

Table 3. Influence of mixed particulate emulsifiers (AOE1, AOE2, AOE3, AOE4) on the yield, moisture content, and a_w of spray dried microcapsules.

Sample	Yield (%)	Moisture Content	a_w
M-AOE1	70.34 ± 1.574 [c]	1.344 ± 0.188 [a]	0.11 ± 0.00 [a]
M-AOE2	66.00 ± 0.818 [b]	3.156 ± 0.147 [c]	0.15 ± 0.00 [b]
M-AOE3	76.10 ± 0.561 [d]	2.386 ± 0.163 [b]	0.13 ± 0.01 [c]
M-AOE4	85.92 ± 2.513 [a]	1.725 ± 0.321 [a]	0.11± 0.00 [a]

Results show the mean value ± standard error from three samples. Significant differences ($p < 0.05$) were labeled with different lowercase letters within a column.

Product yield of avocado oil microparticles ranged from 66 to 85.92%, being the highest value allied to M-AOE4 sample, while the lowest value corresponds to M-AOE2 sample. The quantity of product yield after spray drying is influenced by the stickiness of the solution fed to the equipment [59], which is mainly attributed to the glass transition temperature of the wall materials [60]. For the case of samples which present the same composition but different processing (M-AOE1 and M-AOE2), it can be seen that the sample homogenized by using high shear plus ultrasound (M-AOE1) presented a higher yield after spray drying than the sample homogenized by using high shear (M-AOE2). Different studies have previously shown that emulsification with ultrasound improves the performance in the recovery of microparticles during spray drying process [61], compared with high shear homogenization.

Moreover, the composition of the encapsulating materials in the emulsions influenced the performance of spray drying. By comparing the samples AOE2, AOE3, and AOE4 with different composition of biopolymers (Table 1), but processed in the same way, it can been seen that sample AOE4 shows the highest product yield (85.92%). This result can be attributed to the absence of Gum Arabic in M-AOE4 sample: at the higher Gum Arabic content, the lower the product yield. In addition to the type of emulsion and processing, product yield performance depends on the type of equipment used and its operating parameters. In the present study, the emulsions were spray-dried in the same way.

3.4. Morphology of Spray Dried Microparticles

The effect of the composition on the morphology of spray dried powders from emulsions is observed in Figure 4. Particles with different sizes and wrinkled morphology can be seen in all treatments, and the rough surface of the microparticles is attributed to rapid formation of spherical structures and rapid evaporation of moisture from inside during spray-drying process [62]. It should be noted that M-AOE1, M-AOE2, and M-AOE3 samples show agglomeration of particles; on the other hand, sample M-AOE4 that does not contain Gum Arabic is in free-flowing powder. This can be attributed to the presence of Gum Arabic in these samples (see fomulation in Table 1).

Gum arabic is a highly branched and complex polysaccharide composed of galactose, arabinose, rhamnose, glucoronic acid, and a protein fraction (1.5–2.6%) [63]. Those molecules contain carboxylate anion groups and confer a negative ζ-potential value of −28.97 mV, thus chemical structure could contribute to the electrostatic interaction between the microparticles and adhesiveness between them. Fusion of microparticles have been reported by [39,45] to be related to glass transition and crystallization of amorphous polymer matrix of wall materials during spray drying, however in the present study this phenomenon is not observed. Then, the agglomeration of the microparticles can be due to electrostatic interactions.

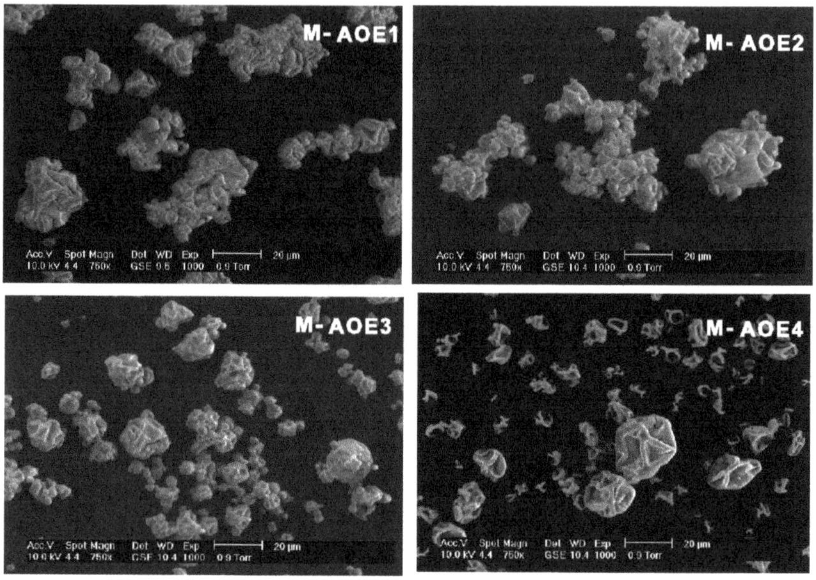

Figure 4. Microstructure of spray dried microparticles from avocado oil-in-water emulsions observed by SEM (at 750× magnification). M-AOE1, M-AOE2, and M-AOE3 contain phosphorylated starch, Gum Arabic, and HI-CAP®100 starch. Sample M-AOE4 contains phosphorylated starch and HI-CAP®100 starch.

3.5. Thermal Analysis and Encapsulation Efficiency of Avocado Oil

Thermogram profiles of avocado oil and spray dried powders are shown in Figure 5A,B respectively. The observed thermal transitions depend on the formed crystals of TAG, which could present three typical polymorphs based on fatty acid moieties; parallel (β), perpendicular (β'), and random (α) in the order of the melting point [64]. This packing is defined as the type of cross-sectional packing of aliphatic chains of the oil. Avocado oil is characterized by showing a high proportion of monounsaturated fatty acids (65.29–71.31%) and saturated fatty acids (13.41–19.25%), followed by a low proportion of polyunsaturated fatty acid (11.30–16.41%) [65]. The crystallization profile of avocado oil (Figure 5A) shows two exothermic peaks during cooling, the first peak (1) detected at −20.3 °C and a second peak (2) detected at −43.87 °C. The melt curve profile of avocado oil shows an exothermic peak (3) at −72.19 °C, which can be attributed to the crystallization of the avocado oil fraction that did not solidify during cooling [66].

Three endothermic peaks (4, 5, and 6) are observed in Figure 5A; these are attributed to melting of monounsaturated fatty acids and polyunsaturated fatty acid fractions. According to [67], melting temperature at ≈−25 °C is attributed to stearin fraction, however no melting peaks are observed at temperatures greater than 0 °C in the present work. It should be noted that avocado oil was extracted by cold pressing using mechanical methods and subsequently filtered to remove fats that solidify at room temperature.

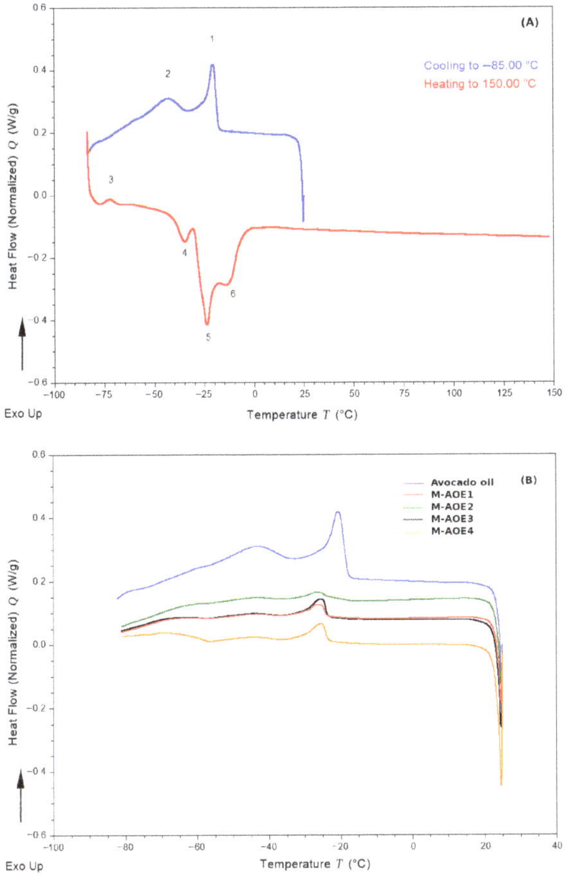

Figure 5. DSC thermogram of avocado oil (**A**) and avocado oil microcapsules after spray drying (**B**).

DSC analysis is sensitive to phase transitions of edible oils present in food matrices. Next, crystallization curves of avocado oil in powders M-AOE1, M-AOE2, M-AOE3, and M-AOE4 were determined and illustrated in Figure 5B in order to quantify the oil content in the microparticles. In all powders, a shift of the crystallization peak towards lower temperatures was observed, consistent with a previous study on olive oil [41]. The heat of crystallization of avocado oil (peak 1) is 5.92 J/g of avocado oil, as the spray dried powders were formulated in a ratio of 1:4 *w*/*w* (avocado oil to wall material). The corresponding value in infeed emulsion is 1.48/g of powder.

As can be seen in Table 4, the type of particulate emulsifier shows a significant effect on the avocado oil content ($p < 0.05$). M-AOE4 sample shows the highest *EE* for encapsulation of avocado oil, and its composition consists of 50% HI-CAP®100 starch and 50% phosphorylated starch. In a previous study [41] it was shown that phosphorylated starch provided a greater protective effect against the oxidation of olive oil microcapsules than octenyl succinylated starch, however it presented lower *EE* than octenyl succinylated starch. In the present work, by combining both types of starch derivatives, HI-CAP®100 starch and phosphorylated starch, it was possible to obtain a higher *EE* (95.4%) than the values reported for olive oil microcapsules produced by using phosphorylated starch (71.8%).

Table 4. Thermal properties of avocado oil powders.

Sample	Thermal Properties				
	T_o (°C)	T_p (°C)	ΔHc (J/g)	AO (%)	EE (%)
Avocado oil	−18.288 ± 0.009 [a]	−20.300 ± 0.000 [a]	5.920 ± 0.088 [a]	100.000	
M-AOE1	−23.719 ± 0.204 [b]	−26.721 ± 0.114 [bd]	0.662 ± 0.133 [c]	11.180	44.721
M-AOE2	−23.683 ± 0.531 [b]	−26.484 ± 0.005 [d]	1.053 ± 0.034 [de]	17.783	71.134
M-AOE3	−23.717 ± 0.104 [b]	−25.181 ± 0.047 [c]	1.366 ± 0.209 [be]	23.070	92.278
M-AOE4	−23.636 ± 0.057 [b]	−25.162 ± 0.212 [c]	1.412 ± 0.089 [b]	23.981	95.386

Results show the mean value ± standard error from two samples. T_o = onset temperature; T_p = melting temperature; ΔH = crystallization enthalpy; AO = avocado oil percent. Significant differences ($p < 0.05$) were labeled with different lowercase letters within a column.

Critical Storage Conditions of Avocado Oil Microparticles

According to the best performance of the powders, M-AOE4 sample was selected to analyze its water adsorption isotherm and to determine its T_g as a function of moisture content. Figure 6A shows adsorption isotherm of avocado oil microparticles at 25 °C, and shows the typical sigmoidal shape of type II isotherm (Brunauer–Emmett–Teller classification). This type of isotherm is consistent with those previously published for microcapsules of octenyl succinylated starch [68], lauroylated starch [69], and acetylated starch [70].

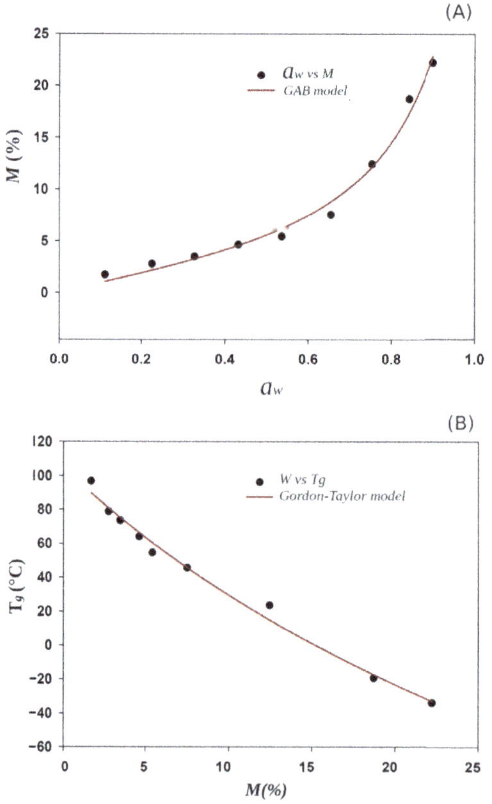

Figure 6. Predicted models for avocado oil microcapsules: (**A**) GAB model (—) sorption isotherm and M (•) in function of a_w; (**B**) Gordon–Taylor model (—) and experimental data of T_g (•) in function of moisture content M.

BET isotherm (Brunauer–Emmett–Teller) was developed to model a multilayer adsorption system. Although the sorption isotherm of avocado oil powder can be estimated from BET, if it is stored at $a_w > 0.5$ the isotherm suddenly grows and loses the fitting. The GAB model (Guggenheim, Anderson, and De Boer) is applicable in a wider range $0.1 \leq a_w \leq 0.9$ because it introduces a second well-differentiated adsorption stage, resulting in the addition of an extra degree of freedom (the K parameter) in its formula [71]. It has been established that M_0 value of the GAB model is the saturation of polar groups of the materials with adsorbed water molecules in the most active sites and that in the obtained M_0 value the product should be stable against microbial deterioration [72].

The M_0 value of M-AOE4 corresponds to 0.047 g H_2O/g of powder sorbed at 0.453 a_w, which is slightly less than the value reported for coffee oil microcapsules [73] prepared by using a mixture of HI-CAP®100 and maltodextrin (MD), as can be seen in the Table 5. The combination of octenyl succinylated commercial starch (HI-CAP®100) with phosphorylated starch in present work led to decrease in moisture sorption, compared to the mixture octenyl succinylated starch with maltodextrin for coffee oil encapsulation [73]. On the other hand, the use of octenyl succinylated taro starch as encapsulating material for avocado oil [37] without mixing originates more hygroscopic avocado oil microparticles (M_0 = 0.1416 g H_2O/g of powder) than the value obtained in the present work (M_0 = 0.047 g H_2O/g of powder).

Table 5. Estimated parameters for selected sorption isotherms models, and prediction of T_g by using Gordon–Taylor model.

Models		Carrier Agents [1]	
GAB	Parameters	M-AOE4	HI-CAP/MD
	C	2.279	2.88
	K_{GAB}	0.905	0.95
	M_0 (H_2O/g)	0.047	0.042
	R^2	0.995	0.996
	P (%)	12.393	6.09
Gordon–Taylor			
	T_{g1} (°C)	105.109	56.3
	k_{GT}	3.174	2.57
	R^2	0.989	0.984
	P (%)	3.06	0.96

[1] Parameters of the GAB and Gordon–Taylor models for M-AOE4, HI-CAP/MD are reported in [73] for coffee oil microcapsules.

The parameter K_{GAB} determines the rate of growth of the isotherm curve for the higher values of a_w. $K_{GAB} = 0.905$ (Table 5) lies in the allowed interval of $0.24 \leq K \leq 0.1$ in order to obtain an error value (P) lower than 15.5%. In the present work, a 6-th oder LSF polynomial model provided a better fit to the experimental data ($P = 1.079\%$) than GAB model ($P = 12.393\%$). Moreover, its inflection point I_3 (0.477 a_w) is allied to the moisture content in the M_0. According to the sorption isotherm, 0.477 a_w corresponds to 0.049 g H_2O/g of powder, a value very close to M_0 (0.047 g H_2O/g). Inflection point I_4 (0.818 a_w) is related to the solubilization of M-AOE4 powder, characterized by a dramatic increase in moisture content.

The T_g is an important parameter to predict powder stability against caking, which is a phenomenon caused by the adsorption of water from powders, and often causes loss of fluidity of the powder due to agglomeration of particles [45]. As T_g value indicates the transition from the vitreous state into the rubber state, thus the T_g of the mixed particulate emulsifiers should be above room temperature to promote greater stability during storage. The M-AOE4 powder obtained by spray drying is in the vitreous state, in which the mobility of polysaccharides chains is restricted, limiting the diffusion phenomena of water molecules [74]. Figure 5B shows that T_g decreases with the increase of the moisture content in avocado oil microparticles due to the plasticizing effect of water [45].

The Gordon–Taylor model shows an state diagram of the T_g as a function of a_w, and shows that $T_g(RHc) = 25$ °C occurs when M-AOE4 contains 0.107 g H_2O/g of powder and 0.73 a_w. It means that at room temperature, samples should not be stored at relative humidities greater than 73%. If the powder is above the T_g, the oxidation reactions of the avocado oil will be accelerated and its structural properties will be lost.

According to the Gordon–Taylor model, anhydrous fraction of M-AOE4 shows a $T_{g1} = 105$ °C, which is higher than these reported for coffee oil microcapsules made with mixtures of the following materials: Hi-Cap/MD (56.3 °C), Capsul/MD (65 °C), N-lock/MD (60.4 °C), GA/MD (62.9 °C). It should be noted that cross-linked phosphorylated starch is characterized by its resistance to high temperatures and high hygroscopicity. Phosphorylated starch has negatively charged phosphate groups, which allowed greater water penetration and swelling [41,75]. In the present work, the combination of phosphorylated starch with Hi-CAP starch overcomes the drawbacks of of the both types of derivatization, succinylation, and phosphorylation. The critical storage conditions determined are suitable for delivery system in skim milk powders with values ranging from 0.100 to 0.380 a_w [76] and powder care cosmetics [77].

According to the 6-th order LSF polynomial model, it provides a better fit (P = 1.08) than the GAB model (P = 12.39):

$$M_6(a_w) = -0.0603597 + 1.48479a_w - 10.608a_w^2 + 38.9682a_w^3 - 74.0669a_w^4 + 69.409a_w^5 - 24.9497a_w^6.$$

Then the following four inflection points can be determined from the polynomial model with high precision:

$$I_1 = 0.204976a_w, I_2 = 0.353877a_w, I_3 = 0.477614a_w, I_4 = 0.81817a_w;$$

where, according to a previous work [45], the physico-chemical sense is related to the moisture sorption properties. I_3 represents the moisture content in monolayer, which corresponds to 0.049 H_2O/g of powder, very close to the monolayer of GAB model (M_0 = 0.047 H_2O/g). I_4 is attributed to the caking phenomenon, which is accompanied with a dramatic moisture sorption of 0.16 g H_2O/g. Therefore, powders should be stored at $a_w < 0.73$ to preserve their physical characteristics and free flowing properties.

4. Conclusions

The formulation of four avocado oil-in-water emulsions influenced the oil content of the spray-dried powders. Two formulations with the same composition of emulsifying particles were processed differently; the first one was homogenized using rotor/stator, and the second one by combining rotor/stator and ultrasound methods. The combination of both emulsification methods favored the decrease in droplet size of the emulsions and the increase in oil encapsulation efficiency, compared to the sample processed by rotor-stator method.

Three formulations with different particle composition were homogenized by combining rotor-stator plus ultrasound methods, two of them with Gum Arabic showed greater stability and smaller droplet size than the sample that only contained modified starches. The composite emulsion, formulated with phosphorylated and HI-CAP®100 starches, showed the highest encapsulation efficiency of avocado oil compared to the rest of the emulsions that contained Gum Arabic. On the other hand, it was the most unstable, as it did not contain Gum Arabic.

The results of this work showed that phosphorylated starch in a mixture of HI-CAP®100 is a sustainable alternative to replace the use of Gum Arabic, since starch is a tunable biopolymer. This methodology allowed us to obtain microparticles with high efficiency of avocado oil encapsulation and good physical stability up to 73% of relative humidity.

Supplementary Materials: The following supporting information can be downloaded at: https://www.mdpi.com/article/10.3390/polym14153064/s1. Table S1: Composition of the feed emulsions and response variables, Table S2: ANOVA analysis for zeta potential, droplet size and TSI, Table S3: Optimization design for zeta potential, droplet size, and Turbiscan Stability Index (TSI) for emulsions prepared from the 22 experimental runs of the D-optimal design, Figure S1: Graphic response model for droplet size (nm), Figure S2: Graphic response model for TSI The visual appearance of the vials containing emulsions can be seen in Figure S3, Figure S3: Photograph of the prepared emulsions.

Author Contributions: Conceptualization, Y.V.G.-T.; methodology, V.E.-S., Y.V.G.-T., O.M.P.-R. and C.E.C.-M.; software, Y.V.G.-T.; validation, Y.V.G.-T. and V.E.-S.; formal analysis, V.B.-F.; investigation, Y.V.G.-T. and V.E.-S.; resources, V.E.-S., Y.V.G.-T. and V.B.-F.; data curation, V.E.-S. and C.E.C.-M.; writing–original draft preparation, Y.V.G.-T., V.E.-S. and V.B.-F.; writing–review and editing, Y.V.G.-T. and V.B.-F.; visualization, V.B.-F.; supervision, Y.V.G.-T.; project administration, Y.V.G.-T.; funding acquisition, Y.V.G.-T. and V.B.-F. All authors have read and agreed to the published version of the manuscript.

Funding: This research was funded b SIP projects 20220843 (V.B.F.), 20222121 (Y.V.G.T.).

Institutional Review Board Statement: Not applicable.

Informed Consent Statement: Not applicable.

Data Availability Statement: Not applicable.

Acknowledgments: The authors acknowledge the experimental support of the CNMN-IPN in the realization of the present work.

Conflicts of Interest: The authors declare no conflict of interest.

Abbreviations

The following abbreviations are used in this manuscript:

MD	Maltodextrin
LSF	Least squares fitting
EVAO	Extra virgin avocado oil
AOE	Avocado oil emulsion
M-AOE	Microparticles of avocado oil

References

1. Werman, M.J.; Mokady, S.; Ntmni, M.E.; Neeman, I. The Effect of Various Avocado Oils on Skin Collagen Metabolism. *Connect. Tissue Res.* **1991**, *26*, 1–10. [CrossRef] [PubMed]
2. Furlan, C.P.B.; Valle, S.C.; Östman, E.; Maróstica, M.R.; Tovar, J. Inclusion of Hass avocado-oil improves postprandial metabolic responses to a hypercaloric-hyperlipidic meal in overweight subjects. *J. Funct. Foods* **2017**, *38*, 349–354. [CrossRef]
3. Del Toro-Equihua, M.; Velasco-Rodríguez, R.; López-Ascencio, R.; Vásquez, C. Effect of an avocado oil-enhanced diet (Persea americana) on sucrose-induced insulin resistance in Wistar rats. *J. Food. Drug. Anal.* **2016**, *24*, 350–357. [CrossRef] [PubMed]
4. Ge, Y.; Ma, F.; Wu, B.; Tan, L. Morphological and Chemical Analysis of 16 Avocado Accessions (Persea americana) From China by Principal Component Analysis and Cluster Analysis. *J. Agric. Sci.* **2018** *10*, 80. [CrossRef]
5. Baur, A.C.; Brandsch, C.; König, B.; Hirche, F.; Stangl, G.I. Plant Oils as Potential Sources of Vitamin D. *Front Nutr.* **2016** *3*, 29. [CrossRef]
6. Engelsen, O. The relationship between ultraviolet radiation exposure and vitamin D status. *Nutrients* **2010**, *2*, 482–495. [CrossRef]
7. Hutchings, N.; Babalyan, V.; Baghdasaryan, S.; Qefoyan M.; Sargsyants, N.; Aghajanova, E.; Martirosyan, A.; Harutyunyan, R; Lesnyak, O.; Formenti, A.M.; et al. Patients hospitalized with COVID-19 have low levels of 25-hydroxyvitamin D. *Endocrine* **2021**, *71*, 267–269. [CrossRef]
8. Woolf, A.; Wong, M.; Eyres, L.; McGhie, T.; Lund, C.; Olsson, S.; Wang, Y.; Bulley, C.; Wang, M.; Friel, E.; et al. 2-Avocado oil. In *Gourmet and Health-Promoting Specialty Oils*; Moreau, R.A., Kamal-Eldin, A., Eds.; AOCS Press: Amsterdam, The Netherlands , 2009; pp. 73–125.
9. Mota, A.H.; Silva, C.O.; Nicolai, M.; Baby, A.; Palma, L.; Rijo, P.; Ascensão, L.; Reis, C.P. Design and evaluation of novel topical formulation with olive oil as natural functional active. *Pharm. Dev. Technol.* **2018**, *23*, 794–805. [CrossRef]
10. Berton-Carabin, C.; Schroën, K. Pickering Emulsions for Food Applications: Background, Trends, and Challenges. *Annu. Rev. Food Sci. Technol.* **2015**, 6, 263–297. [CrossRef]

11. Bouyer, E.; Mekhloufi, G.; Rosilio, V.; Grossiord, J.; Agnely, F. Proteins, polysaccharides, and their complexes used as stabilizers for emulsions: Alternatives to synthetic surfactants in the pharmaceutical field? *Int. J. Pharm.* **2012**, *436*, 359–378. [CrossRef]
12. Angkuratipakorn, T.; Sriprai, A.; Tantrawong, S.; Chaiyasit, W.; Singkhonrat, J. Fabrication and characterization of rice bran oil-in-water Pickering emulsion stabilized by cellulose nanocrystals. *Colloids Surf. Physicochem. Eng. Asp.* **2017**, *522*, 310–319. [CrossRef]
13. Joseph, C.; Savoire, R.; Harscoat-Schiavo, C.; Pintori, D.; Monteil, J.; Faure, C.; Leal-Calderon, F. Pickering Emulsions Stabilized by various Plant Materials: Cocoa, Rapeseed Press Cake and Lupin Hulls. *LWT* **2020**, *130*, 109621. [CrossRef]
14. Jiang, B.; Wang, L.; Zhu, M.; Wu, S.; Wang, X.; Li, D.; Liu, C.; Feng, Z.; Tian, B. Separation, Structural Characteristics and Biological Activity of Lactic Acid Bacteria Exopolysaccharides Separated by Aqueous Two-Phase System. *LWT* **2021**, *147*, 111617. [CrossRef]
15. Bouhoute, M.; Taarji, N.; de Oliveira Felipe, L.; Habibi, Y.; Kobayashi, I.; Zahar, M.; Isoda, H.; Nakajima, M.; Neves, M. A. Microfibrillated cellulose from Argania spinosa shells as sustainable solid particles for O/W Pickering emulsions. *Carbohydr. Polym.* **2021**, *251*, 116990. [CrossRef]
16. Schröder, A.; Sprakel, J.; Boerkamp, W.; Schroën, K.; Berton-Carabin, C.C. Can we Prevent Lipid Oxidation in Emulsions by using Fat-Based Pickering Particles? *Food Res. Int.* **2019**, 120, 352–363. [CrossRef]
17. Cai, X.; Wang, Y.; Du, X.; Xing, X.; Zhu, G. Stability of pH-responsive Pickering emulsion stabilized by carboxymethyl starch/xanthan gum combinations. *Food Hydrocoll.* **2020**, *109*, 106093. [CrossRef]
18. Li, Q.; Huang, Y.; Du, Y.; Chen, Y.; Wu, Y.; Zhong, K.; Huang, Y.; Gao, H. Food-grade olive oil Pickering emulsions stabilized by starch/β-cyclodextrin complex nanoparticles: Improved storage stability and regulatory effects on gut microbiota. *LWT* **2021**, *155*, 112950. [CrossRef]
19. Liu, F.; Tang, C. Soy glycinin as food-grade Pickering stabilizers: Part. III. Fabrication of gel-like emulsions and their potential as sustained-release delivery systems for β-carotene. *Food Hydrocoll.* **2016**, *56*, 434–444. [CrossRef]
20. Zhu, Q.; Li, Y.; Li, S.; Wang, W. Fabrication and characterization of acid soluble collagen stabilized Pickering emulsions. *Food Hydrocoll.* **2020**, *106*, 105875 . [CrossRef]
21. Yang, Y.; Jiao, Q.; Wang, L.; Zhang, Y.; Jiang, B.; Li, D.; Feng, Z.; Liu, C. Preparation and Evaluation of a Novel High Internal Phase Pickering Emulsion Based on Whey Protein Isolate Nanofibrils Derived by Hydrothermal Method. *Food Hydrocoll.* **2022**, *123*, 107180. [CrossRef]
22. Jiao, Q.; Liu, Z.; Li, B.; Tian, B.; Zhang, N.; Liu, C.; Feng, Z.; Jiang, B. Development of Antioxidant and Stable Conjugated Linoleic Acid Pickering Emulsion with Protein Nanofibers by Microwave-Assisted Self-Assembly. *Foods* **2021**, *10*, 1892. [CrossRef] [PubMed]
23. Du, M.; Sun, Z.; Liu, Z.; Yang, Y.; Liu, Z.; Wang, Y.; Jiang, B.; Feng, Z.; Liu, C. High Efficiency Desalination of Wasted Salted Duck Egg White and Processing into Food-Grade Pickering Emulsion Stabilizer. *LWT* **2022**, *161*, 113337. [CrossRef]
24. Chen, Z.; Cui, B.; Guo, X.; Zhou, B.; Wang, S.; Pei, Y.; Li, B.; Liang, H. Fabrication and Characterization of Pickering Emulsions Stabilized by Desalted Duck Egg White Nanogels and Sodium Alginate. *J. Sci. Food Agric.* **2022**, *102*, 949–956. [CrossRef] [PubMed]
25. Hinderink, E.B.A.; Berton-Carabin, C.C.; Schroën, K.; Riaublanc, A.; Houinsou-Houssou, B.; Boire, A.; Genot, C. Conformational Changes of Whey and Pea Proteins upon Emulsification Approached by Front-Surface Fluorescence. *J. Agric. Food Chem.* **2021**, *69*, 6601–6612. [CrossRef]
26. Chen, M.; Sun, Q. Current Knowledge in the stabilization/destabilization of Infant Formula Emulsions during Processing as Affected by Formulations. *Trends Food Sci. Technol.* **2021**, *109*, 435–447. [CrossRef]
27. Yucel Falco, C.; Geng, X.; Cárdenas, M.; Risbo, J. Edible Foam Based on Pickering Effect of Probiotic Bacteria and Milk Proteins. *Food Hydrocoll.* **2017**, *70*, 211–218. [CrossRef]
28. Yano, H.; Fukui, A.; Kajiwara, K.; Kobayashi, I.; Yoza, K.; Satake, A.; Villeneuve, M. Development of Gluten-Free Rice Bread: Pickering Stabilization as a Possible Batter-Swelling Mechanism. *LWT* **2017**, *79*, 632–639. [CrossRef]
29. Monteiro, G.M.; Souza, X.R.; Costa, D.P.B.; Faria, P.B.; Vicente, J. Partial substitution of pork fat with canola oil in Toscana sausage. *Innov. Food Sci. Emerg. Technol.* **2017**, *44*, 2–8. [CrossRef]
30. Jalali, E.; Maghsoudi, S.; Noroozian, E. Ultraviolet protection of Bacillus thuringiensis through microencapsulation with Pickering emulsion method. *Sci. Rep.* **2020**, *10*, 20633. [CrossRef]
31. Koç, M.; Güngör, Ö.; Zungur, A.; Yalçın, B.; Selek, İ.; Ertekin, F.K.; Ötles, S. Microencapsulation of Extra Virgin Olive Oil by Spray Drying: Effect of Wall Materials Composition, Process Conditions, and Emulsification Method. *Food Biop. Tech.* **2015**,*8*, 301–318. [CrossRef]
32. Karrar, E.; Mahdi, A.A.; Sheth, S.; Mohamed Ahmed, I.A.; Manzoor, M.F.; Wei, W.; Wang, X. Effect of maltodextrin combination with gum arabic and whey protein isolate on the microencapsulation of gurum seed oil using a spray-drying method. *Int. J. Biol. Macromol.* **2021**, *171*, 208–216. [CrossRef]
33. Carneiro, H.C.F.; Tonon, R.V.; Grosso, C.R.F.; Hubinger, M.D. Encapsulation efficiency and oxidative stability of flaxseed oil microencapsulated by spray drying using different combinations of wall materials. *J. Food Eng.* **2013**, *115*, 443–451. [CrossRef]
34. McNamee, B.F.; O'Riorda, E.D.; O'Sullivan, M. Emulsification and Microencapsulation Properties of Gum Arabic. *J. Agric. Food Chem.* **1998**, *46*, 4551–4555. [CrossRef]
35. Hofman, D.L.; van Buul, V.J.; Brouns, F.J.P.H. Nutrition, Health, and Regulatory Aspects of Digestible Maltodextrins. *Crit. Rev. Food Sci. Nutr.* **2016**, *56*, 2091–2100. [CrossRef]

36. Nabeshima, E.H.; Bustos, F.M.; Hashimoto, J.M.; El Dash, A.A. Improving Functional Properties of Rice Flours Through Phosphorylation. *Int. J. Food Prop.* **2010**, *13*, 921–930. [CrossRef]
37. Romero-Hernandez, H.A.; Sánchez-Rivera, M.M.; Alvarez-Ramirez, J.; Yee-Madeira, H.; Yañez-Fernandez, J.; Bello-Pérez, L.A. Avocado oil encapsulation with OSA-esterified taro starch as wall material: Physicochemical and morphology characteristics. *LWT* **2021**, *138*, 110629. [CrossRef]
38. Sotelo-Bautista, M.; Bello-Perez, L.; Gonzalez-Soto, R.; Yañez-Fernandez, J.; Alvarez-Ramirez, J. OSA-maltodextrin as wall material for encapsulation of essential avocado oil by spray drying. *J. Dispers. Sci. Technol.* **2020**, *41*, 235–242. [CrossRef]
39. Bae, E.K.; Lee, S.J. Microencapsulation of avocado oil by spray drying using whey protein and maltodextrin. *J. Microencapsul.* **2008**, *25*, 549–560. [CrossRef]
40. Chimsook, T. Microwave Assisted Extraction of Avocado Oil from Avocado Skin and Encapsulation Using Spray Drying. *Key Eng. Mater.* **2017**, *737*, 341–346. [CrossRef]
41. Espinosa-Solís, V.; García-Tejeda, Y.V.; Portilla-Rivera, O.; Barrera-Figueroa, V. Tailoring Olive Oil Microcapsules Via Microfluidization of Pickering o/w Emulsions. *Food Bioprocess Technol.* **2021**, *14*, 1835–1843. [CrossRef]
42. Raikos, V. Encapsulation of vitamin E in edible orange oil-in-water emulsion beverages: Influence of heating temperature on physicochemical stability during chilled storage. *Food Hydrocoll.* **2017**, *72*, 155–162. [CrossRef]
43. Kaszuba, M.; Corbett, J.; Watson, F.M.; Jones, A. High-concentration zeta potential measurements using light-scattering techniques. *Phil. Trans. R. Soc. A* **2010**, *368* 3684439–4451. [CrossRef]
44. Timmermann, E.O.; Chirife, J.; Iglesias, H.A. Water sorption isotherms of foods and foodstuffs: Bet or gab parameters? *J. Food Eng.* **2001**, *48*, 19. [CrossRef]
45. García-Tejeda, Y.V.; García-Armenta, E.; Martínez-Audelo, J.M.; Barrera-Figueroa, V. Determination of the structural stability of a premix powder through the critical water activity. *J. Food Meas. Charact.* **2019**, *13*, 1323–1332. [CrossRef]
46. Gordon, M.; Taylor, J.S. Ideal Copolymers and the Second-Order Transitions of Synthetic Rubbers. i. Non-Crystalline Copolymers. *J. Appl. Chem.* **1952**, *2*, 493–500. [CrossRef]
47. Tonon, R.V.; Baroni, A.F.; Brabet, C.; Gibert, O.; Pallet, D.; Hubinger, M.D. Water Sorption and Glass Transition Temperature of Spray Dried Açai (Euterpe Oleracea Mart.) Juice. *J. Food Eng.* **2009**, *94*, 215–221. [CrossRef]
48. Sablani, S.S.; Syamaladevi, R.M.; Swanson, B.G. A Review of Methods, Data and Applications of State Diagrams of Food Systems.*Food Eng. Rev.* **2010**, *2*, 168–203. [CrossRef]
49. Kaltsa, O.; Gatsi, I.; Yanniotis, S.; Mandala, I. Influence of Ultrasonication Parameters on Physical Characteristics of Olive Oil Model Emulsions Containing Xanthan.*Food Bioprocess Technol.* **2014**, *7*, 2038–2049. [CrossRef]
50. Leiva, J.M.; Geffroy, E. Evolution of the Size Distribution of an Emulsion under a Simple Shear Flow. *Fluids* **2018**, *3*, 46. [CrossRef]
51. Shnoudeh, A.J.; Hamad, I.; Abdo, R.W.; Qadumii, L.; Jaber, A.Y.; Surchi, H.S.; Alkelany, S.Z. Synthesis, characterization, and applications of metal nanoparticles. In *Biomaterials and Bionanotechnology*; Tekade, R.K., Ed.; Academic Press: Cambridge, MA, USA, 2019; pp. 527–612.
52. Villalobos-Castillejos, F.; Lartundo-Rojas, L.; Leyva-Daniel, D.E.; Porras-Saavedra, J.; Pereyra-Castro, S.; Gutiérrez-López, G.F.; Alamilla-Beltrán, L. Effect of Emulsification Techniques on the Distribution of Components on the Surface of Microparticles obtained by Spray Drying. *Food Bioprod. Process.* **2021**, *129*, 115–123. [CrossRef]
53. Duerkop, M.; Berger, E.; Dürauer, A.; Jungbauer, A. Influence of Cavitation and High Shear Stress on HSA Aggregation Behavior. *Eng. Life Sci.* **2018**, *18*, 169–178. [CrossRef] [PubMed]
54. García-Tejeda, Y.V.; Leal-Castañeda, E.J.; Espinosa-Solis, V.; Barrera-Figueroa, V. Synthesis and Characterization of Rice Starch Laurate as Food-Grade Emulsifier for Canola Oil-in-Water Emulsions. *Carbohydr. Polym.* **2018**, *194*, 177–183. [CrossRef] [PubMed]
55. Pan, Y.; Wu, Z.; Xie, Q.; Li, X.; Meng, R.; Zhang, B.; Jin, Z. Insight into the Stabilization Mechanism of Emulsions Stabilized by Maillard Conjugates: Protein Hydrolysates-Dextrin with Different Degree of Polymerization. *Food Hydrocoll.* **2020**, *99*, 105347. [CrossRef]
56. Wang, C.; Li, J.; Sun, Y.; Wang, C.; Guo, M. Fabrication and Characterization of a Cannabidiol-Loaded Emulsion Stabilized by a Whey Protein-Maltodextrin Conjugate and Rosmarinic Acid Complex. *J. Dairy Sci.* **2022**, *105*, 4631–6446 . [CrossRef]
57. Jayme, M.L.; Dunstan, D.E.; Gee, M.L. Zeta Potentials of Gum Arabic Stabilised Oil in Water Emulsions. *Food Hydrocoll.* **1999**, *13*, 459–465. [CrossRef]
58. Niknam, S.M.; Escudero, I.; Benito, J.M. Formulation and Preparation of Water-in-Oil-in-Water Emulsions Loaded with a Phenolic-Rich Inner Aqueous Phase by Application of High Energy Emulsification Methods. *Foods* **2020**, *9*, 1411. [CrossRef]
59. Ribeiro, A.M.; Shahgol, M.; Estevinho, B.N.; Rocha, F. Microencapsulation of Vitamin A by Spray-Drying, using Binary and Ternary Blends of Gum Arabic, Starch and Maltodextrin. *Food Hydrocoll.* **2020**, *108*, 106029. [CrossRef]
60. Fitzpatrick, J.J.; Hodnett, M.; Twomey, M.; Cerqueira, P.S.M.; O'Flynn, J.; Roos, Y.H. Glass Transition and the Flowability and Caking of Powders Containing Amorphous Lactose. *Powder Technol.* **2007**, *178*, 119–128. [CrossRef]
61. Tontul, I.; Topuz, A. Mixture Design Approach in Wall Material Selection and Evaluation of Ultrasonic Emulsification in Flaxseed Oil Microencapsulation. *Dry. Technol.* **2013**, *31*, 1362–1373. [CrossRef]
62. Akram, S.; Bao, Y.; Butt, M.S.; Shukat, R.; Afzal, A.; Huang, J. Fabrication and Characterization of Gum Arabic- and Maltodextrin-Based Microcapsules Containing Polyunsaturated Oils. *J. Sci. Food Agric.* **2021**, *101*, 6384–6394. [CrossRef]
63. Sarika, P.R.; Pavithran, A.; James, N.R. Cationized gelatin/gum Arabic Polyelectrolyte Complex: Study of Electrostatic Interactions. *Food Hydrocoll.* **2015**, *49*, 176–182. [CrossRef]

64. Takeguchi, S.; Sato, A.; Hondoh, H.; Aoki, M.; Uehara, H.; Ueno, S. Multiple β Forms of Saturated Monoacid Triacylglycerol Crystals. *Molecules* **2020**, *25*, 5086. [CrossRef]
65. Tan, C.X. Virgin avocado oil: An emerging source of functional fruit oil. *J. Funct. Foods* **2019**, *54*, 381–392. [CrossRef]
66. Barba, L.; Arrighetti, G.; Calligaris, S. Crystallization and melting properties of extra virgin olive oil studied by synchrotron XRD and DSC. *Eur. J. Lipid Sci. Technol.* **2013**, *115*, 322–329. [CrossRef]
67. Tan, C.P.; Che Man, Y.B. Differential scanning calorimetric analysis of edible oils: Comparison of thermal properties and chemical composition. *J. Am. Oil Chem. Soc.* **2000**, *77*, 143–155. [CrossRef]
68. García-Tejeda, Y.V.; Salinas-Moreno, Y.; Barrera-Figueroa, V.; Martínez-Bustos, F. Preparation and characterization of octenyl succinylated normal and waxy starches of maize as encapsulating agents for anthocyanins by spray-drying. *J. Food Sci. Technol.* **2018**, *55*, 2279–2287. [CrossRef]
69. Espinosa-Solis, V.; García-Tejeda, Y.V.; Leal-Castañeda, E.J.; Barrera-Figueroa, V. Effect of the Degree of Substitution on the Hydrophobicity, Crystallinity, and Thermal Properties of Lauroylated Amaranth Starch. *Polymers* **2020**, *12*, 2548. [CrossRef]
70. García-Tejeda, Y.V.; Salinas-Moreno, Y.; Martínez-Bustos, F. Acetylation of normal and waxy maize starches as encapsulating agents for maize anthocyanins microencapsulation. *Food Bioprod. Process.* **2015**, *94*, 717–726. [CrossRef]
71. García-Tejeda, Y.V.; Barrera-Figueroa, V. Least Squares Fitting-Polynomials for Determining Inflection Points in Adsorption Isotherms of Spray-Dried Açaí Juice (Euterpe Oleracea Mart.) and Soy Sauce Powders. *Powder Technol.* **2019**, *342*, 829–839. [CrossRef]
72. Bonilla, E.; Azuara, E.; Beristain, C.I.; Vernon-Carter, E.J. Predicting suitable storage conditions for spray-dried microcapsules formed with different biopolymer matrices. *Food Hydrocoll.* **2010**, *24*, 633–640. [CrossRef]
73. Silva, V.M.; Vieira, G.S.; Hubinger, M.D. Influence of different combinations of wall materials and homogenisation pressure on the microencapsulation of green coffee oil by spray drying. *Food Res. Int.* **2014**, *61*, 132–143. [CrossRef]
74. Saavedra-Leos, M.Z.; Román-Aguirre, M.; Toxqui-Terán, A.; Espinosa-Solís, V.; Franco-Vega, A.; Leyva-Porras, C. Blends of Carbohydrate Polymers for the Co-Microencapsulation of Bacillus clausii and Quercetin as Active Ingredients of a Functional Food. *Polymers* **2022**, *14*, 236. [CrossRef] [PubMed]
75. García-Tejeda, Y.V.; Salinas-Moreno, Y.; Hernández-Martínez, Á.R.; Martínez-Bustos, F. Encapsulation of Purple Maize Anthocyanins in Phosphorylated Starch by Spray Drying. *Cereal Chem.* **2016**, *93*, 130–137. [CrossRef]
76. Pugliese, A.; Cabassi, G.; Chiavaro, E.; Paciulli, M.; Carini, E.; Mucchetti, G. Physical characterization of whole and skim dried milk powders. *J. Food Sci. Technol.* **2017**, *54*, 3433–3442. [CrossRef]
77. Costa, R.; Santos, L. Delivery systems for cosmetics—From manufacturing to the skin of natural antioxidants. *Powder Technol.* **2017**, *322*, 402–416. [CrossRef]

polymers

Article

Evaluation of Two Active System Encapsulant Matrices with Quercetin and *Bacillus clausii* for Functional Foods

Hector Alfonso Enciso-Huerta [1], Miguel Angel Ruiz-Cabrera [1], Laura Araceli Lopez-Martinez [2], Raul Gonzalez-Garcia [1], Fidel Martinez-Gutierrez [1] and Maria Zenaida Saavedra-Leos [3,*]

1 Facultad de Ciencias Químicas, Universidad Autónoma de San Luis Potosí, Av. Dr. Manuel Nava 6, San Luis Potosí 78210, Mexico
2 Coordinación Académica Región Altiplano Oeste, Universidad Autónoma de San Luis Potosí, Salinas de Hidalgo 78600, Mexico
3 Coordinación Académica Región Altiplano, Universidad Autónoma de San Luis Potosí, 11 Carretera Cedral Km, 5+600 Ejido San José de las Trojes, Matehuala 78700, Mexico
* Correspondence: zenaida.saavedra@uaslp.mx

Abstract: Currently, demand for functional foods is increasing in the public interest in order to improve life expectations and general health. Food matrices containing probiotic microorganisms and active compounds encapsulated into carrier agents are essential in this context. Encapsulation via the lyophilisation method is widely used because oxidation reactions that affect physicochemical and nutritional food properties are usually avoided. Encapsulated functional ingredients, such as quercetin and *Bacillus clausii*, using two carrier agents' matrices—I [inulin (IN), lactose (L) and maltodextrin (MX)] and II [arabic (A), guar (G), and xanthan (X) gums)]—are presented in this work. A D-optimal procedure involving 59 experiments was designed to evaluate each matrix's yield, viability, and antioxidant activity (AA). Matrix I (33.3 IN:33.3 L:33.3 MX) and matrix II (33.3 A:33.3 G:33.3 X) exhibited the best yield; viability of 9.7 log10 CFU/g and 9.73 log10 CFU/g was found in matrix I (using a ratio of 33.3 IN:33.3 L:33.3 MX) and matrix II (50 G:50 X), respectively. Results for the antioxidant capacity of matrix I (100 IN:0 L:0M X) and matrix II (0 A:50 G:50 X) were 58.75 and 55.54 (DPPH* scavenging activity (10 μg/mL)), respectively. Synergy between matrices I and II with use of 100IN:0L:OMX and 0A:50G:50X resulted in 55.4 log10 CFU/g viability values; the antioxidant capacity was 9. 52 (DPPH* scavenging activity (10 μg/mL). The present work proposes use of a carrier agent mixture to produce a functional ingredient with antioxidant and probiotic properties that exceed the minimum viability, 6.0 log10 CFU/g, recommended by the FAO/WHO (2002) to be probiotic, and that contributes to the recommended daily quercetin intake of 10–16 mg/day or inulin intake of 10–20 g/day and dietary fibre intake of 25–38 g per day.

Keywords: functional; food; inulin; lactose; *Bacillus clausii*

check for updates

Citation: Enciso-Huerta, H.A.; Ruiz-Cabrera, M.A.; Lopez-Martinez, L.A.; Gonzalez-Garcia, R.; Martinez-Gutierrez, F.; Saavedra-Leos, M.Z. Evaluation of Two Active System Encapsulant Matrices with Quercetin and *Bacillus clausii* for Functional Foods. *Polymers* **2022**, *14*, 5225. https://doi.org/10.3390/polym14235225

Academic Editor: Alfredo Cassano

Received: 26 October 2022
Accepted: 22 November 2022
Published: 1 December 2022

Publisher's Note: MDPI stays neutral with regard to jurisdictional claims in published maps and institutional affiliations.

1. Introduction

A functional food (FF) is defined by the European Society for Clinical Nutrition and Metabolism guide as an enriched food with ingredients, nutrients, or additional compounds intended to manifest specific benefits to health. In the last decade, FF production has become an important biotechnology industry, given growing consumer interest in improving life expectancy and healthy due to raising of awareness about prevention of certain diseases such as diabetes, cancer, and Alzheimer's [1,2]. Antioxidants are commonly compounds added to other nutritional particles to promote synergism: e.g., vitamin C to regenerate the vitamin E tocopheryl radical after its oxidation [3]. Additionally, antioxidants are added to suppress lipid oxidation, increase products' shelf life, and reduce free radical concentrations generated in organisms [4]. Flavonoids such as quercetin are antioxidants found in apples, grapes, beans, broccoli, red onion, tomatoes, oilseeds, flowers, tea leaves, and Ginkgo biloba. However, though recommended ingestion of quercetin is 1 g per day, the average consumption is about 10–16 mg [5]. Different molecular mechanisms have been reported in treatment of various diseases. For example, in allergic asthma, the compound showed inhibition of *MUC5AC* gene expression in NCI-H292 cells, triggering human nasal mucosa anti-secretory agents that prevented mucosa secretion in epithelial cells while maintaining a normal ciliary movement [6]. Ingestion of 150–730 mg of quercetin per day over four weeks resulted in antihypertensive action, reducing systolic and diastolic pressure in patients in the first stage of hypertension [7]. Similarly, patients with metabolic syndrome who consumed a daily dose of 150 mg of quercetin over five weeks significantly reduced their systolic pressure. An in vitro study performed by Reyes-Farias and Carrasco-Pozo [8] showed that quercetin acts as an antiviral agent against HIV, inhibiting integrase, protease, and inverse transcriptase enzymes.

Other important compounds found in FFs are probiotics, which confer benefits to health through production of biliary enzymes, organic acid, satiety hormones, and immune system modulation. These microorganisms also improve antibody response, improve substrate competence against pathogenic organisms, and interact with microbiota [9]. An important probiotic employed in food enrichment is *Bacillus clausii* (*B. clausii*): an anaerobic gram-positive bacterium capable of generating spores and intestinal colonisers [10]. Moreover, *B. clausii* is resistant to heat, gastric pH conditions, and antibiotics. Nevertheless, its optimal growth conditions are 40 °C and a pH of 9.0. De Castro et al. [11] treated acute infant diarrhoea (by viral cause or associated with antibiotics) with *B. clausii*, showing that its consumption over seven days reduced disease duration, gastrointestinal symptoms, and evacuation frequency. Plomer et al. [12] used *B. clausii* to reduce adverse effects of treatment of *Helicobacter pylori*: a pathology, usually treated with antibiotics, that causes nausea, inflammation, vomit, and diarrhoea, triggering treatment failure and bacterial resistance.

A microencapsulation process is employed to conserve active ingredient properties susceptible to suffering damage under processing or environmental conditions. Major environmental conditions that could affect ingredient activity include atmospheric oxygen, pH, humidity, light irradiation, and high temperature exposure. The microencapsulation technique involves use of an encapsulating material that maintains its microstructural integrity in aggressive environments in which active ingredients may lose their functions. Nutraceutical and functional ingredients such as antioxidants, vitamins, minerals, lipids, and probiotics have been microencapsulated via different methodologies [13–15]. Polysaccharides, lipids, and proteins are examples of different compounds employed as wall materials in microencapsulation [16]. Inulin (IN) is a non-digestible polymer, presenting fructose linear chains with a terminal group consisting of glucose molecules joined through β-(2,1) bonds. This molecule is found in many vegetables, fruits, and cereals; is structurally considered a short-chain carbohydrate with a polymerisation rate between 1 and 60 repetitive units per molecule; and is a water-soluble biopolymer [17]. On the other hand, maltodextrin (MX) is a polysaccharide derived from starch acidic hydrolysis, presenting a nutritional contribution of only 4 calories per gram. This molecule is commercialised in a wide range of molecular weight distributions (MWDs), each with different thermal properties and potential applications. Recently, Saavedra-Leos et al. [18] reported use of a set of four MXs as carrying agents in spray-drying of blueberry juice–maltodextrin (BJ–MX); they set application limits of the maltodextrins based on the MWDs. Lactose in its monohydrated form has been widely used as an excipient in order to facilitate administration of drugs, especially those that target the lungs [19]. Gum arabic (GA) is a naturally occurring polysaccharide, obtained from resin of certain varieties of Acacia (*Mimosoidae* subfamily), of low viscosity, solubility, and emulsion formation; it can act as an encapsulating agent in combination with other agents, such as xanthan gum, MX, or modified starch. GA is a carbohydrate extracted from the plant *Cyamopsis tetragonoloba*, whose main characteristics are to hydrate rapidly in cold water and produce highly viscous solutions [20]. It is used as a thickener and a viscosity modifier in a wide variety of processed foods, such as ice cream, cheese, bread, meats, dressings, and sauces, or pharmaceuticals and cosmetics [21]. Xanthan gum (XG) is a natural polysaccharide produced through fermentation of *Xan-*

thomonas campestris. It is highly soluble in water, producing high viscosity; stable in alkaline or acidic conditions; and widely used as a stabiliser in foods such as creams, artificial juices, sauces, syrups, ice cream toppings, meat, poultry and fish. An investigation performed by Lombardo and Villares [22] demonstrated that cellulose, MX, IN, and starch polysaccharides have been used as carrier agents to improve rigidity of microencapsulations. Other investigations [14,15] performed freeze-drying encapsulation of ethanolic *Elsholtzia ciliata* and *Lactobacillus plantarum* extracts using different combinations of gum A, MX, lactose, and skimmed milk; these investigations obtained yields in the range of 90–100%.

Production of foods that contain probiotics usually employs spray-drying and lyophilisation techniques. The latter is considered by the food, pharmaceutical, and biotechnological industries as a drying process capable of stabilizing and preserving products through reduction of loss of unstable compounds. Consequently, it is the preferred methodology of preserving aromas, flavours, and nutritional compounds [23]. In contrast to the spray-drying process, because of the low temperature at which lyophilisation is carried out, oxidation reactions are not catalysed, thus preventing physicochemical and nutritional food damage [24]. The freeze-drying process consists of freezing the product at -40 °C and sublimating the ice at sub-atmospheric pressures [25]. A study performed by Gümüşay et al. [26] compared different drying methods, with the aim to obtain higher content of phenolic compounds, ascorbic acid and antioxidant activity from tomatoes. These authors observed that freeze-drying resulted in about double phenolic compounds compared to those yielded by other drying methods. Rockinger et al. [27] reviewed current approaches to cell preservation through freeze-drying and found that stability of cells is achieved by cryopreservation at sub-zero temperatures (-130 °C). Solid-state water is removed through sublimation, and no residual moisture remaining in the solid is enough to allow molecular movement and biochemical reactions; this, in turn, may preserve the food product and promote longer storage periods.

Taking into consideration the importance of antioxidants and probiotics in health, the objective of this work is to evaluate a combination of three carrier agents (inulin, lactose and maltodextrin) and three gums (xanthan, arabic and guar) in co-encapsulation of *Bacillus clausii* and quercetin in a functional food prepared via freeze-drying. A special cubic design of experiments employing the Scheffe mix model was implemented to achieve this purpose and to compare the extent of the response variables (viability of *B. clausii* and antioxidant activity of quercetin).

2. Materials and Methods

2.1. Materials

Commercial maltodextrin (MX) extracted from maize starch was acquired from INGREDION Mexico (Guadalajara, Mexico). The dextrose (DE) equivalent of MX was 10, with a molecular weight of 1625 g/mol and a polymerisation grade (DP) of 2–16 glucose units. Inulin (IN) was purchased from INGREDION Mexico (Guadalajara, Mexico). α-lactose monohydrate (L) (Lα·H2O, purity $\geq$ 99.9%) was purchased from Sigma-Aldrich Chemical Co (Toluca, Mexico).; methanol (MeOH, purity $\geq$ 99.8) was obtained from J.T. Baker (Guadalajara, Mexico). Gums arabic (A), guar (G) and xanthan (X) were obtained from INGREDION Mexico (Guadalajara, Mexico). The *Bacillus* strain (*B. clausii*) in sinuberase solution was purchased from Sanofi-Aventis Mexico, S.A. de C.V. (Coyocan, Mexico City, Mexico). Quercetin 3-D-Galactose (purity $\geq$ 99%) was acquired from Química Farmacéutica Esteroidal S.A de C.V. (Tlahuac, Mexico City, Mexico). Trypticase Soy Agar was obtained from Dickinson de México S.A. de C.V. (Mexico City, Mexico). Finally, analytical-grade 2,2-diphenyl-1-picrilhidrazile (DPPH) was obtained from Sigma–Aldrich Chemical Co (Toluca, Mexico).

2.2. Lyophilisation Preparation

In Table 1—the experimental design of two matrices for a special cubic x special cubic model—the resulting 59 tests performed in the laboratory were carried out in a random order. For each test, 100 g samples (*w*/*w*) were prepared.

Table 1. Experimental design of two matrices for a special cubic x special cubic model.

		Matrix I			Matrix II			Yield	DPPH * Scavenging Activity			Bc
No	Run	IN	L	MX	A	G	X	(%)	5 µg/mL	10 µg/mL	30 µg/mL	(Log10 CFU/g)
1	44	100.0	0.0	0.0	0.0	50.0	50.0	82.7	52.35	55.7	89	9.7
2	15	0.0	100.0	0.0	100.0	0.0	0.0	82.6	52.32	61.7	76	9.67
3	51	0.0	100.0	0.0	0.0	100.0	0.0	87.6	52.45	56.6	80	9.6
4	3	0.0	100.0	0.0	0.0	0.0	100.0	86.1	51.91	55	72	9.52
5	20	0.0	100.0	0.0	50.0	50.0	0.0	87.8	56.05	58.8	84	9.48
6	57	0.0	100.0	0.0	50.0	0.0	50.0	86.6	51.45	55.4	71	9.56
7	48	0.0	100.0	0.0	0.0	50.0	50.0	84.4	48.36	52.1	82	9.3
8	32	0.0	0.0	100.0	100.0	0.0	0.0	83.9	53.74	55.5	79	9.3
9	35	0.0	0.0	100.0	0.0	100.0	0.0	87.6	52.6	56.6	71	9.3
10	41	0.0	0.0	100.0	0.0	0.0	100.0	89.2	52.3	55.3	70	9.48
11	11	0.0	0.0	100.0	50.0	50.0	0.0	86.4	49.86	60.2	77	9.43
12	18	0.0	0.0	100.0	50.0	0.0	50.0	86.7	50.18	53.1	71	9.52
13	39	0.0	0.0	100.0	0.0	50.0	50.0	87.2	54.85	55.4	67	9.48
14	25	50.0	50.0	0.0	100.0	0.0	0.0	84.6	48.55	51.2	75	9.64
15	43	50.0	50.0	0.0	0.0	100.0	0.0	86.3	51.8	58	78	9.6
16	47	50.0	50.0	0.0	0.0	0.0	100.0	88.1	41.95	42.9	63	9.75
17	8	50.0	50.0	0.0	50.0	50.0	0.0	93.5	50.06	55.1	72	9.73
18	40	50.0	50.0	0.0	50.0	0.0	50.0	79.9	49.69	51.2	77	9.37
19	27	50.0	50.0	0.0	0.0	50.0	50.0	87.9	48.36	52.1	82	9.87
20	19	50.0	0.0	50.0	100.0	0.0	0.0	82.2	51.63	59.6	68	9.82
21	1	50.0	0.0	50.0	0.0	100.0	0.0	76.5	51.22	52.8	63	9.56
22	23	50.0	0.0	50.0	0.0	0.0	100.0	86.8	49.2	51.2	71	9.75
23	7	50.0	0.0	50.0	50.0	50.0	0.0	74.6	50.68	53.3	63	9.37
24	34	50.0	0.0	50.0	50.0	0.0	50.0	77.2	38.58	43.3	63	9.67
25	55	50.0	0.0	50.0	0.0	50.0	50.0	89.1	50.68	58.5	76	9.6
26	45	0.0	50.0	50.0	100.0	0.0	0.0	82.9	49.41	54.9	75	9.37
27	37	0.0	50.0	50.0	0.0	100.0	0.0	81.1	49.32	54.3	76	9.52
28	2	0.0	50.0	50.0	0.0	0.0	100.0	85.7	42.19	49.8	59	9.56
29	30	0.0	50.0	50.0	50.0	50.0	0.0	84.5	39.42	45.4	61	9.43
30	54	0.0	50.0	50.0	50.0	0.0	50.0	85.6	51.21	53.6	81	9.52
31	58	0.0	50.0	50.0	0.0	50.0	50.0	88.3	38.58	44.1	62	9.7
32	16	33.3	33.3	33.3	33.3	33.3	33.3	85.8	48.18	57.9	74	9.56
33	5	100.0	0.0	0.0	33.3	33.3	33.3	81	51.13	56	68	9.75
34	56	0.0	100.0	0.0	33.3	33.3	33.3	81.8	53.27	54.5	72	9.6
35	24	0.0	0.0	100.0	33.3	33.3	33.3	79.8	51.82	58.7	78	9.48
36	6	50.0	50.0	0.0	33.3	33.3	33.3	86.5	54.85	55.4	67	9.52
37	28	50.0	0.0	50.0	33.3	33.3	33.3	81.2	50.78	57.3	79	9.37
38	29	0.0	50.0	50.0	33.3	33.3	33.3	78.9	50.7	54.5	65	9.85
39	31	33.3	33.3	33.3	100.0	0.0	0.0	82.5	51.76	58.3	71	9.67
40	38	33.3	33.3	33.3	0.0	100.0	0.0	86	52.05	57.3	83	9.48
41	26	33.3	33.3	33.3	0.0	0.0	100.0	84.2	48.45	54.3	78	9.7
42	14	33.3	33.3	33.3	50.0	50.0	0.0	84.1	54.85	55.4	67	9.73
43	33	33.3	33.3	33.3	50.0	0.0	50.0	83.4	52.88	58	82	9.43
44	42	33.3	33.3	33.3	0.0	50.0	50.0	86.4	49.05	54.3	72	9.73
45	52	100.0	0.0	0.0	100.0	0.0	0.0	82.8	53.06	56.9	80	9.67
46	36	100.0	0.0	0.0	0.0	100.0	0.0	81.2	55.24	57.6	74	9.48
47	53	100.0	0.0	0.0	0.0	0.0	100.0	84	52.88	58	82	9.67
48	59	100.0	0.0	0.0	50.0	50.0	0.0	85.2	45.12	54.7	79	9.78
49	13	100.0	0.0	0.0	50.0	0.0	50.0	81.2	53.87	57	72	9.7
50	22	66.7	16.7	16.7	66.7	16.7	16.7	84.5	54.36	56.8	69	9.64

Table 1. *Cont.*

		Matrix I			Matrix II			Yield	DPPH * Scavenging Activity			Bc
No	Run	IN	L	MX	A	G	X	(%)	5 μg/mL	10 μg/mL	30 μg/mL	(Log10 CFU/g)
51	12	66.7	16.7	16.7	16.7	66.7	16.7	83.4	40.24	52.4	70	9.88
52	17	66.7	16.7	16.7	16.7	16.7	66.7	86.2	51.91	52.8	63	9.67
53	4	16.7	66.7	16.7	66.7	16.7	16.7	81.1	52.01	52.7	70	9.73
54	50	16.7	66.7	16.7	16.7	66.7	16.7	85.7	43.39	43.9	60	9.6
55	21	0.0	0.0	100.0	0.0	0.0	100.0	88.5	53.18	55.4	65	9.37
56	10	0.0	0.0	100.0	100.0	0.0	0.0	86.4	40.26	42.3	62	9.8
57	46	0.0	0.0	100.0	0.0	100.0	0.0	87.7	46.75	53.9	75	9.6
58	9	0.0	0.0	100.0	50.0	0.0	50.0	85.2	40.38	42.6	62	9.9
59	49	0.0	0.0	100.0	0.0	50.0	50.0	86.8	53.38	56.8	71	9.56

* Free radical.

Each mass fraction for matrices I and II was set according to the experimental design. The compounds of matrix I (10 g) and matrix II (1 g) were passed through a 1 mm sieve. Subsequently, 87 g deionised water was added and magnetically stirred at 35 °C for 5 min. Next, 1 g quercetin and 1 g *B. clausii* were added. The samples were stored in the dark at −80 °C. The microencapsulation process was carried out by sublimation in a freeze dryer (Ilshin Bio Base® Model TFD8501, Gyeonggi-do, South Korea) under a vacuum pressure of 5 mTorr −65 °C for approximately 120 h. Yield was determined using Equation (1):

$$Yield = \left[\frac{SL}{SI}\right] * 100 \quad (1)$$

SL = solids recovered at the end of freeze-drying
SI = Initial solids (10 g matrix I + 1 g matrix II + 1 g quercetin + 1 g *B. clausii*)

2.3. Determination of Microbial Viability

Viability of *B. clausii* before and after the encapsulation process was determined through resuspension of 1 g of the microparticles obtained in 9 mL of saline solution (NaCl, 0.9% *w*/*v*). To break microcapsules, the suspension was agitated for 10 min with a vortex and incubated in a water bath for 10 min at 50 °C. Viable cells were analysed according to the method described by Miles et al. [28]. Briefly, dilutions of 1×10^{-3} a 1×10^{-9} performed in saline solution were sown on trypticase soy agar and incubated at 35 °C for 24 h. The evaluation was performed in triplicate and reported in colony-forming units per gram (CFU/g), using Equation (2):

$$Viability = \left[\frac{Number\ of\ colonies\ in\ box * dilution\ factor}{mL\ of\ sample\ sown}\right] \quad (2)$$

2.4. Antioxidant Activity (AA)

Quercetin antioxidant capacity was determined according to the method described by Brand-Williams et al. [29]. Briefly, 1.7 mL of alcoholic solution of *DPPH* (0.1 mmol DPPH/L) was mixed with 1.7 mL of microencapsulated suspension in which concentration of microencapsulation varied from 2.5 to 5 or 15 μg/mL. The mixture was left to stand in darkness for 30 min, and absorbance at 537 nm was measured using a spectrophotometer UV-Vis Evolution 220 (Thermo Scientific, Walthman, MA. USA). The sweep percentage was calculated using equation 3:

$$AA\ (\%DPPH) = \frac{A0 - A30}{A0} \times 100 \quad (3)$$

where *A*0 represents absorbance of blank solution (*DPPH* mixture and ethanol without microencapsulates) and *A*30 represents absorbance of *DPPH* solution and ethanol with microencapsulates after 30 min. Sweep activity was determined in triplicate for each sample.

2.5. Design of Experiments and Statistical Analysis

Two independent mixtures were tested. Matrix I consisted of inulin (IN), lactose (L), and maltodextrin (MX), while matrix II consisted of gums arabic (A), guar (G), and xanthan (X). The lower and upper levels of these variables were between 0 and 100 (wt %), and the sum of the components in each mixture was 100% for each trial. The response variables were yield (%), *Bc* (Log10 CFU/g), and antioxidant activity (DPPH at concentrations of 5, 10, and 30 (µg/g)). In this manner, a combined experimental design of two matrices for a special cubic x special cubic model was selected to evaluate the effect of each factor for each response variable. Table 1 shows the resulting 59 trials performed at the laboratory in a random order.

An analysis of variance (ANOVA) was performed for each response (yield, *Bc*, and antioxidant activity) at the significance level of 0.05, using Design-Expert® Version 12 Software (trial version). The analysed Scheffe model (special cubic x special cubic) was written as Equation (4):

$$Y = (\alpha_1 A + \alpha_2 B + \alpha_3 C + \alpha_4 AB + \alpha_5 AC + \alpha_6 BC + \alpha_7 ABC) \times (\kappa_1 D + \kappa_2 E + \kappa_3 F + \kappa_4 DE + \kappa_5 DF + \kappa_6 EF + \kappa_7 DEF) \quad (4)$$

which is an expanded method that results in 49 adjustable parameters. In Table 2: ANOVA statistical analyse for each response observed

3. Results and Discussion

The ANOVA for capacity antioxidant and *B. clausii* response variables is discussed herein.

Table 2. Statistical ANOVA details for each response analysed.

Response	SST	SSR	SSE	DFT	DFR	DFE	F	P(F)	R^2
Antioxidant Capacity for 2,2 Difenil-1-Picrilhidrazil (DPPH) 5 µg/mL	1177.97	1053.83	124.14	58	48	10	1.77	0.1665	0.8946
Antioxidant Capacity for 2,2 Difenil-1-Picrilhidrazil (DPPH) 10 µg/mL	1218.56	1117.41	101.15	58	48	10	2.3	0.0778	0.9170
Antioxidant Capacity for 2,2 Difenil-1-Picrilhidrazil (DPPH) 30 µg/mL	2914.83	2738.77	176.06	58	48	10	3.24	0.0246	0.9396
B. clausii	1.38	1.3	0.076	58	48	10	3.55	0.0177	0.9445

SST: Sum of Squares Total
SSR: Sum of Squares Regression
SSE: Sum of Squares Error
DFT: Degrees of Freedom Total
DFR: Degrees of Freedom of Regression
DFE: Degrees of Freedom of Error
F: Fisher's Statistic

3.1. Microencapsulation Performance

The microencapsulation technique comprised coating small particles to form capsules with unique properties and different morphologies, each of which could reach diameters from nanometres to millimetres, protect bioactive ingredients from adverse reactions, and improve functionality and bioavailability [30]. Lyophilisation is a microencapsulation process in which a previously frozen product (−40 °C) is lyophilised to ice sublimation at sub-atmospheric pressures. This work evaluated two different encapsulating matrices for two active systems (antioxidant and microorganism) in order to produce functional lyophilised food. Matrix I, containing IN, L, and MX, as shown in Figure 1a, featured an efficiency of 93.7% when IN was present at 66.71% in example experiment 54 (66.7 IN:16.7 L:16.7 MX); it featured an efficiency of 87.6% for IN at 100% in the case of experiment 49 (100 IN:0 L:0 MX). Matrix II, containing arabic (A), guar (G), and xanthan (X) gums, showed better efficiency, corresponding to 88.08% when A, G, and X gums were present in the same proportion: e.g., experiment 32 (33.3 A:33.3 G:33.3 X), as shown in Figure 1b. Our results are consistent with the report of Enache et al. [31], who employed the freeze-drying method of co-microencapsulation of black-currant-extract anthocyanins and lactic-acid bacteria, using inulin and chitosan as carrier agents. They reported a recovery efficiency of 95.46% ± 1.30% for inulin and 87.38% ± 0.48% for chitosan. Pudziuvelyte et al. [14] reported microencapsulate lyophilisation yield of the ethanolic extract of *Elsholtzia ciliata*, with six carrier agents at 20% concentration and mixes at 10% concentration; they employed arabic gum (GUM_E), maltodextrin (MALTO_E), resistant maltodextrin (RES_E), skimmed milk (SKIM_E), sodium caseinate (SOD_CAS_E), and beta-cyclodextrin (BETA_CYCL_E). These authors indicated that a higher yield was observable when they employed SKIM_E and MALTO_E, which showed 100% and 95% efficiency, respectively, for mixtures of an observed 100% yield in two situations: use of SKIM_E with MALTO_E and of GUM_E with BETA_CYCL_E. Sharifi et al. [15] performed analysis of co-microencapsulate Lactobacillus plantarum and phytosterol mixtures formed with β-sitosterol (49.54%), campesterol (26.12%), stigmasterol (19.1%), and brassicasterol (1.48%). Researchers used gum arabic (GA) (2.25% w/v) and whey protein isolate (WPI) (5% w/v) as encapsulating agents. They formed coacervates followed by two dry processing techniques—aspersion drying and lyophilisation—obtaining 58.62 ± 2.01% and 65.23 ± 0.51% yields, respectively. These results demonstrated improved performance with use of a dry freeze-drying process. The yield results contributed to a range of responses; for example, higher yields were obtained when the carrier agent used in matrix I had an IN value closer to 100 or when A:G:X demonstrated the same ratio for matrix II, and lower yields were obtained with use of only MX at 100 for matrix I, or, in the case of matrix II, with use of G 100.

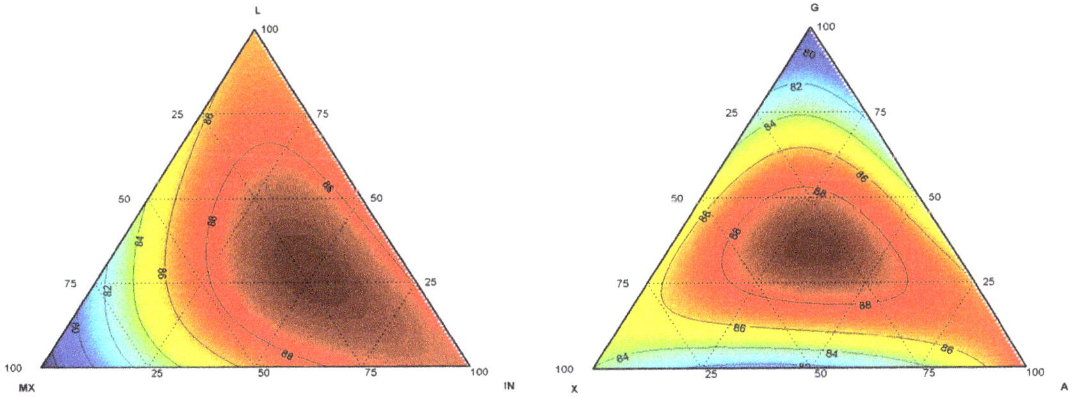

(a) (b)

Figure 1. Yield surface graphic: (**a**) matrix I, IN:L:MX; (**b**) matrix II, A:G:X.

3.2. *Viability of B. clausii Microencapsulated*

Viability of *B. clausii* was measured with the colony forming unit method (CFU/g). The results shown in Figure 2a were determined concerning the viability in matrix I. Experiments 1 to 59 had viability diminution at a 9.3 to 9.9 log10 CFU/g rate against the control at 11.30 log10 CFU/g. As an example, for experiment 40, with the proportion of 33.3IN:33.3L:33.3MX, a viability value of 9.85 log10 CFU/g was determined. Notwithstanding, we observed higher viability of *B. clausii* (9.8 log10 CFU/g) when we used L closer to the unit, as experiment 34 showed (0 IN:100 L:0 MX). Using a 50 L:50 MX ratio, *B. clausii* viability was 9.37 log10 CFU/g, corresponding to experiment 27 (0 IN:50 L:50 MX). In matrix I, when carrier agents were present in the same proportion and matched with experiment 32 (33.3 IN:33.3 L:33.3 MX), viability was 9.75 log10 CFU/g. Results showed that use of more than an encapsulating agent improved *B. clausii* survival. It is worth mentioning that the standards presented in these experimental designs, particularly the three components of the Scheffe special cubic x special cubic model, surpassed the 6.0 log10 CFU/g minimum value recommended by the FAO/OMS (2002) to be considered probiotic, and correlated with previous reports from [31], who performed experiments with co-microencapsulate *Lactobacillus casei* and black-currant (Ribes nigrum)-extract anthocyanin dried via lyophilisation. These authors employed whey protein isolate (WPI), chitosan, and inulin at a 2:1:1 rate as carrier agents. They reported that viability of the powder was 11 log10 UFC/g as a starting value; after storage for 90 days at 4 °C, it reduced to 8.13-6.35 log10 UFC/g. Showing stability of carrier agent mixtures, Milea et al. [32] reported viability of co-microencapsulates via flavonoid lyophilisation, obtained from yellow onion peelings (Allium cepa) and *Lactobacillus casei* and employing whey protein isolate (WPI), inulin (I), and maltodextrin (MD) as carrier agents (2:1:1 proportion). These samples were encapsulated by lyophilisation at 1% and 2% concentrations probed into food one (cream cheese). The researchers reported results after storage of 21 days at 4 °C, recording a 6.6 and 7.41 log10 UFC/g viability at the concentrations mentioned above. Cayra et al. [33] suggested protection provided by L, IN, and MX materials for cellular structures of microorganisms via formation of crystals and water-molecule replacement in polar groups of cellular membrane lipids. For matrix II, we found a higher B. clausii viability value in two conditions: with use of a unit of G and in employment of a G and X mixture. This was seen, for example, in experiment 31 (0A:50G:50X composition) and experiment 3 (0 A:100 G:0 X). Obtained results showed that: I) Higher viability was obtained when IN and L and/or IN, L and MX at the same proportion were employed as carrier agents in matrix I, or, in matrix II, when G and/or G and X at the same ratio were used closer to the unit; II) Lower viability was observed with use of L and MX at the same rate in matrix I and with use of X closer to the unit in matrix II; III) The better combination of carrier agents and gums allowed better viability preservation; IV) We obtained one functional food, since the minimum viability value of 6.0 log10 CFU/g recommended by the FAO/OMS (2002) to be considered as probiotic was determined across the values in all experiments in this research.

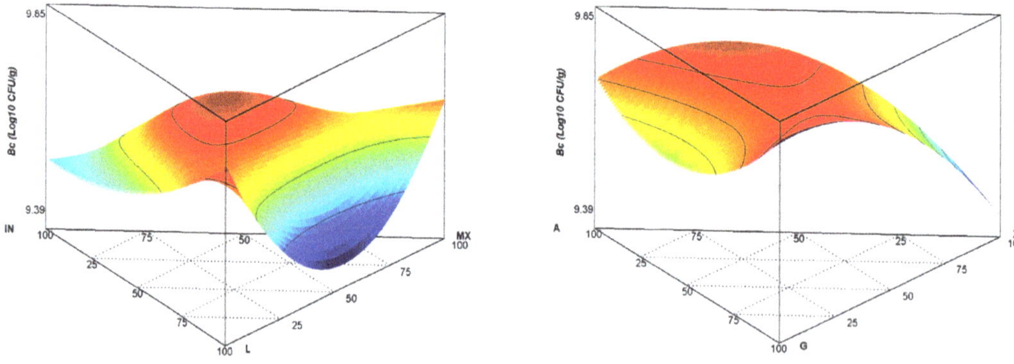

(**a**) (**b**)

Figure 2. Graphics of surface responses to viability of *B. clausii* dried by lyophilisation, expressed as log10 (CFU/g). (**a**) Matrix I, IN:L:MX; (**b**) matrix II, A:G:X.

3.3. Antioxidant Capacity Determination

Antioxidant capacity was determined with 2,2 diphenyl-1-picrilhidrazil (DPPH) radical inhibition using 5, 10 and 30 μg/mL (sample) as concentrations; as can be observed in Figure 3a–c, for matrix I, formed by IN, L, and MX, higher antioxidant activity (AA) was determined when the carrier agent was closer to the inulin unit and corresponded to experiment 33 (100 IN:0 L:0 MX), presenting concentrations of 56.06 AA to 5 μg/mL concentration, 58.75 to 10 μg/mL, and 84.35 to 30 μg/mL. Nevertheless, when the matrix was compounded by 50 IN:50 L, corresponding to experiment 15 (50 IN:50 L:0 MX), AA was 52.88 (5 μg/mL), 57.95 (10 μg/mL) and 81.51 (30 μg/mL), respectively. Samples that presented less AA were closer to the MX unit, as shown in Figure 3a–c, with activity of 38.58 at 5 μg/mL, 43.27 at 10 μg/mL, and 63.1 to 30μg/mL. These results correlate with observations reported by other authors, such as Martins et al. [34], who performed a lemongrass (Cymbopogon citratus (DC.) Stapf) essential oil micro-encapsulation study; the object of that work was to evaluate development, characterisation and production of particle antioxidant potential. Three different formulations were generated for essential oil encapsulates: M1 (5% of essential oil), M2 (10% of essential oil), and M3 (15% of essential oil). Each mixture used maltodextrin MD (DE 20) and gelatine (GEL) at a 4:1 (*w*/*w*) ratio as encapsulating agents. Result emulsions were lyophilised under 0.011 mbar and −60 °C conditions for 48 h. Concerning antioxidant activity, the authors reported that generally, due to a variety of presences of bioactive compounds, functional groups, and polarities as parts of the essential oil, the antioxidant effect had a starting antioxidant capacity value of 22.16 ± 0.04 mg TE/g, measured by the DPPH method. After lyophilisation, samples presented antioxidant potential of 2.46 ± 0.12 mg TE/g for M1, 7.74 ± 0.05 mg TE/g for M2, and 12.10 ± 0.30 mg TE/g for M3. Results showed that MD (DE 20) use influenced antioxidant capacity. Azarpazhooh et al. [35] evaluated pomegranate (Punica granatum L.) grinds extracted via the DPPH method for antioxidant capacity (RSA). The process consisted of use of maltodextrin (MDX) in three proportions—5, 10, and 15%—as a carrier agent, as well as calcium alginate at 0.1/(*w*/*w*) at a 1:5 proportion. Researchers reported lower inhibitory concentration (IC50), at 0.56 mg/mL of MDX for a 15% sample. Against the 0.86 mg/mL IC50 MDX observed in a 5% sample, results indicated the influence of MDX concentration on RSA increase. Notwithstanding, the obtained results could have been influenced by anthocyanins and polyphenols present in the sample.

For matrix II, consisting of G:X:A, we show the results in Figure 4a–c. At the concentrations of 5, 10, and 30μg/mL, more favourable results were obtained using G and X at the same proportions without presence of A; e.g., for experiment 49 (0 A:50 G:50 X). Notwithstanding, when A was closer to the unit, AA was lower; e.g., for experiment 20 (100 A:0 G:0 X), corresponding to AA of 48.36 (5 μg/mL), 52.06 (10 μg/mL), and 81.93 (30 μg/mL), as shown in Table 1. Mansour et al. [36] micro-encapsulated anthocyanin (AC) extracts obtained from raspberries via lyophilisation (Rubus idaeus L.). Evaluation revealed three different anthocyanin concentrations (0.025%, 0.05%, and 0.075%), two encapsulating agents (soy protein isolate (SPI) and gum arabic (GA) at 5% concentration *w*/*v*), and SPI and GA at 2.5:2.5 % *w*/*v* concentration. Antioxidant capacities observed for these compounds at 0.025% were 25% for SPI, 45% for GA and 35% for the SPI+GA mixture.

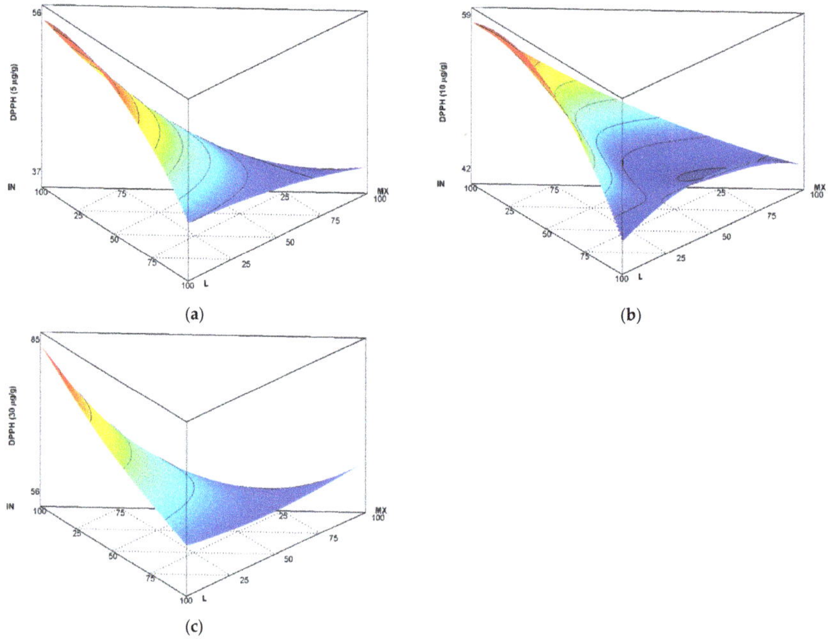

Figure 3. Antioxidant capacity with 2,2 difenil-1-picrilhidrazil (DPPH) radical inhibition for (**a**) 5 μg/mL, (**b**) 10 μg/mL, and (**c**) 30 μg/mL (sample), in matrix I (formed by IN, L, and MX).

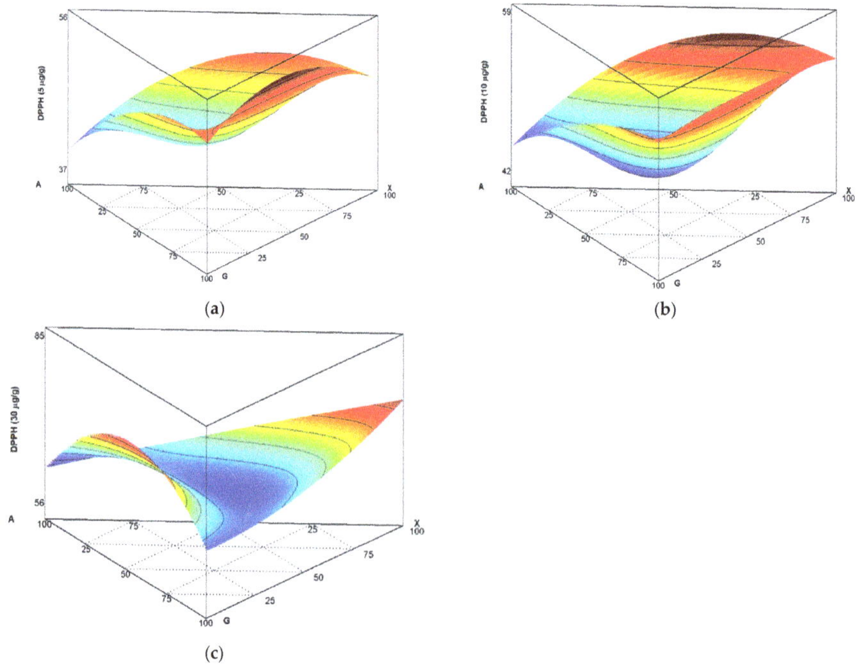

Figure 4. Antioxidant capacity of 2,2 difenil-1-picrilhidrazil (DPPH) radical inhibition: (**a**) 5 μg/mL, (**b**) 10 μg/mL, and (**c**) 30 μg/mL (sample), for matrix II, formed by A, G and X.

Rezende et al. [37] elaborated on industrial waste and steelyard pulp (Malpighia emarginata DC) micro-encapsulates. They employed gum arabic (GA) and maltodextrin (MD) mixture in the same proportion (1:1; w/w) as carrier agents. These researchers used the DPPH method to contrast antioxidant activity for industrial waste and steelyard pulp through the dry aspersion and lyophilisation processes. Antioxidant activity after the sample encapsulation process was 129.16 μM TE/g for lyophilised samples and 155.24 μM TE/g for samples dried via aspersion. Obtained results allowed identification of (I) carrier agent combinations and gums that allowed the best antioxidant activity percentage; (II) higher antioxidant activity obtained when the carrier agent used in matrix I was inulin, or G and X at the same proportion for the 5 and 10 μg/mL concentrations in matrix II; and (III) lower antioxidant activity with use of only maltodextrin in matrix I and with G closer to the unit in the matrix II. This last evidence is in line with results reported by researchers.

4. Conclusions

Through experimental D-optimal designs, we prepared a functional ingredient formed by *B. clausii* and quercetin for two matrices: matrix I, formed by IN:L:MX; and matrix II, formed by A:G:X. With regard to yield, using the three compounds with the same proportion for both matrices, we obtained a higher value (88.08%) for experiment 32 (33.3 IN:33.3 L:33.3 MX) and (33.3 A:33.3 G:33.3 X). For matrix II, we observed higher viability (9.8 log10 CFU/gto) with use of L closer to the unit corresponding to experiment 34 (0 IN; 100 L:0 MX). We obtained higher AA when inulin was used closer to the unit corresponding to experiment 33 (100 IN:0 L:0 MX), presenting as 56.05 at a 5 μg/mL concentration, 8.75 at 10 μg/mL and 84.35 at 30 μg/mL. Matrix II presented higher viability values in two cases: with use of G closer to the unit and in employment of a G and X mixture, e.g., 9.9 log10 CFU/g in experiment 31 (0 A:50 G:50 X) and 9.8 log10 CFU/g in 3 experiment 3 (0 A:100 G:0 X). Higher AA was yielded with use of G and X in equal proportions, corresponding to experiment 49 (0 A:50 G:50 X), with AA of 54.85 at a concentration of 5 μg/mL, 55.44 at 10 μg/mL, and 67.09 at 30 μg/mL. Synergism between two matrices occurred with use of a higher I and G proportion, corresponding to experiment 46, which featured a yield of 86.5%; viability of 9.52 log10 CFU/g9.52 log10 UFC/g; and AA of 54.85 at a 5 μg/mL concentration, 55.44 at 10 μg/mL, and 67.09 at 30 μg/mL.

Author Contributions: Conceptualisation, M.A.R.-C., R.G.-G. and M.Z.S.-L.; data curation, R.G.-G.; formal analysis, M.A.R.-C. and F.M.-G.; investigation, M.A.R.-C., L.A.L.-M. and F.M.-G.; methodology, H.A.E.-H., L.A.L.-M. and F.M.-G.; software, R.G.-G.; writing—original draft, M.Z.S.-L.; writing—review and editing, M.Z.S.-L. All authors have read and agreed to the published version of the manuscript.

Funding: This research received no external funding.

Institutional Review Board Statement: Not acceptable.

Data Availability Statement: Not acceptable.

Acknowledgments: Héctor Alfonso Enciso-Huerta is grateful to the National Council of Science and Technology (CONACYT) in Mexico for financial support provided during his studies through scholarship No. 784106.

Conflicts of Interest: The authors declare no conflict of interest.

References

1. Codex Alimentarius Commission; World Health Organization; Joint FAO/WHO Food Standards Programme. *Codex Alimentarius: Cereals, Pulses, Legumes and Vegetable Proteins*, 1st ed.; Food & Agriculture Org.: Rome, Italy, 2007.
2. Sarao, L.K.; Arora, M. Probiotics, Prebiotics, and Microencapsulation: A Review. *Crit. Rev. Food Sci. Nutr.* **2017**, *57*, 344–371. [CrossRef] [PubMed]
3. Addor, F.A.S. Antioxidants in Dermatology. *An. Bras. Dermatol.* **2017**, *92*, 356–362. [CrossRef] [PubMed]
4. Ciriminna, R.; Meneguzzo, F.; Delisi, R.; Pagliaro, M. Olive Biophenols as New Antioxidant Additives in Food and Beverage. *ChemistrySelect* **2017**, *2*, 1360–1365. [CrossRef]

5. Khan, H.; Ullah, H.; Aschner, M.; Cheang, W.S.; Akkol, E.K. Neuroprotective Effects of Quercetin in Alzheimer's Disease. *Biomolecules* **2020**, *10*, 59. [CrossRef] [PubMed]
6. Jafarinia, M.; Hosseini, M.S.; Kasiri, N.; Fazel, N.; Fathi, F.; Hakemi, M.G.; Eskandari, N. Quercetin with the Potential Effect on Allergic Diseases. *Allergy Asthma Clin. Immunol.* **2020**, *16*, 36. [CrossRef]
7. Marunaka, Y.; Marunaka, R.; Sun, H.; Yamamoto, T.; Kanamura, N.; Inui, T.; Taruno, A. Actions of Quercetin, a Polyphenol, on Blood Pressure. *Molecules* **2017**, *22*, 209. [CrossRef]
8. Reyes-Farias, M.; Carrasco-Pozo, C. The Anti-Cancer Effect of Quercetin: Molecular Implications in Cancer Metabolism. *Int. J. Mol. Sci.* **2019**, *20*, 3177. [CrossRef]
9. Sanders, M.E.; Merenstein, D.J.; Reid, G.; Gibson, G.R.; Rastall, R.A. Probiotics and Prebiotics in Intestinal Health and Disease: From Biology to the Clinic. *Nat. Rev. Gastroenterol. Hepatol.* **2019**, *16*, 605–616. [CrossRef]
10. Paparo, L.; Tripodi, L.; Bruno, C.; Pisapia, L.; Damiano, C.; Pastore, L.; Canani, R.B. Protective Action of *Bacillus clausii* Probiotic Strains in an in Vitro Model of Rotavirus Infection. *Sci. Rep.* **2020**, *10*, 12636. [CrossRef]
11. De Castro, J.A.A.; Guno, M.J.V.R.; Perez, M.O. *Bacillus clausii* as Adjunctive Treatment for Acute Community-Acquired Diarrhea among Filipino Children: A Large-Scale, Multicenter, Open-Label Study (CODDLE). *Trop. Dis. Travel Med. Vaccines* **2019**, *5*, 14. [CrossRef]
12. Plomer, M.; Perez, M.; Greifenberg, D.M. Effect of *Bacillus clausii* Capsules in Reducing Adverse Effects Associated with Helicobacter Pylori Eradication Therapy: A Randomized, Double-Blind, Controlled Trial. *Infect. Dis. Ther.* **2020**, *9*, 867–878. [CrossRef] [PubMed]
13. Calinoiu, L.F.; Ştefanescu, B.E.; Pop, I.D.; Muntean, L.; Vodnar, D.C. Chitosan coating applications in probiotic microencapsulation. *Coatings* **2019**, *9*, 194. [CrossRef]
14. Pudziuvelyte, L.; Marksa, M.; Sosnowska, K.; Winnicka, K.; Morkuniene, R.; Bernatoniene, J. Freeze-Drying Technique for Microencapsulation of *Elsholtzia ciliata* Ethanolic Extract Using Different Coating Materials. *Molecules* **2020**, *25*, 2237. [CrossRef] [PubMed]
15. Sharifi, S.; Rezazad-Bari, M.; Alizadeh, M.; Almasi, H.; Amiri, S. Use of Whey Protein Isolate and Gum Arabic for the Co-Encapsulation of Probiotic *Lactobacillus plantarum* and Phytosterols by Complex Coacervation: Enhanced Viability of Probiotic in Iranian White Cheese. *Food Hydrocoll.* **2021**, *113*, 106496. [CrossRef]
16. Machado, N.D.; Fernández, M.A.; Díaz, D.D. Recent Strategies in Resveratrol Delivery Systems. *ChemPluschem* **2019**, *84*, 951–973. [CrossRef]
17. Apolinário, A.C.; de Carvalho, E.M.; de Lima Damasceno, B.P.G.; da Silva, P.C.D.; Converti, A.; Pessoa, A.; da Silva, J.A. Extraction, isolation and characterization of inulin from Agave sisalana boles. *Ind. Crops Prod.* **2017**, *108*, 355–362. [CrossRef]
18. Saavedra-Leos, M.Z.; Leyva-Porras, C.; López-Martínez, L.A.; González-García, R.; Martínez, J.O.; Martínez, I.C.; Toxqui-Terán, A. Evaluation of the Spray Drying Conditions of Blueberry Juice-Maltodextrin on the Yield, Content, and Retention of Quercetin 3-d-Galactoside. *Polymers* **2019**, *11*, 312. [CrossRef]
19. Lara-Mota, E.E.; Nicolás–Vázquez, M.I.; López-Martínez, L.A.; Espinosa-Solis, V.; Cruz-Alcantar, P.; Toxqui-Teran, A.; Saavedra-Leos, M.Z. Phenomenological study of the synthesis of pure anhydrous β-lactose in alcoholic solution. *Food Chem.* **2021**, *340*, 128054. [CrossRef]
20. Dabeek, W.M.; Marra, M.V. Dietary Quercetin and Kaempferol: Bioavailability and Potential Cardiovascular-Related Bioactivity in Humans. *Nutrients* **2019**, *11*, 2288. [CrossRef]
21. Castañeda-Ovando, A.; González-Aguilar, L.A.; Granados-Delgadillo, M.A.; Chávez-Gómez, U.J. Goma guar: Un aliado en la industria alimentaria. *Pädi Bol. Cient. Cienc. Básic. Ing. ICBI* **2020**, *7*, 107–111. [CrossRef]
22. Lombardo, S.; Villares, A. Engineered Multilayer Microcapsules Based on Polysaccharides Nanomaterials. *Molecules* **2020**, *25*, 4420. [CrossRef] [PubMed]
23. Muñoz-López, C.; Urrea-García, G.R.; Jiménez-Fernández, M.; Rodríguez-Jiménes, G.d.C.; Luna-Solano, G. Efecto de Las Condiciones de Liofilización En Propiedades Fisicoquímicas, Contenido de Pectina y Capacidad de Rehidratación de Rodajas de Ciruela (*Spondias Purpurea* L.). *Agrociencia* **2018**, *52*, 1–13.
24. Caballero, B.L.; Márquez, C.J.; Betancur, M.I. Efecto de La Liofilización Sobre Las Caracteristicas Físico-Químicas Del Ají Rocoto (*Capsicum pubescens* R&P) Con o Sin Semilla. *Bioagro* **2017**, *29*, 225–234.
25. Bhatta, S.; Janezic, T.S.; Ratti, C. Freeze-Drying of Plant-Based Foods. *Foods* **2020**, *9*, 87. [CrossRef]
26. Gümüşay, Ö.; Borazan, A.; Ercal, N.; Demirkol, O. Drying effects on the antioxidant properties of tomatoes and ginger. *Food Chem.* **2015**, *173*, 156–162. [CrossRef]
27. Rockinger, U.; Funk, M.; Winter, G. Current Approaches of Preservation of Cells During (Freeze) Drying. *J. Pharm. Sci.* **2021**, *110*, 2873–2893. [CrossRef]
28. Miles, A.A.; Misra, S.S.; Irwin, J.O. The Estimation of the Bactericidal Power of the Blood. *J. Hyg.* **1938**, *38*, 732–749. [CrossRef]
29. Brand-Williams, W.; Cuvelier, M.E.; Berset, C. Use of a Free Radical Method to Evaluate Antioxidant Activity. *LWT Food Sci. Technol.* **1995**, *28*, 25–30. [CrossRef]
30. Zambrano, V.; Bustos, R.; Mahn, A. Insights about Stabilization of Sulforaphane through Microencapsulation. *Heliyon* **2019**, *5*, e02951. [CrossRef]

31. Enache, I.M.; Vasile, A.M.; Enachi, E.; Barbu, V.; Stanciuc, N.; Vizireanu, C. Co-Microencapsulation of Anthocyanins from Cornelian Cherry Fruits and Lactic Acid Bacteria in Biopolymeric Matrices by Freeze-Drying: Evidences on Functional Properties and Applications in Food. *Polymers* **2020**, *12*, 906. [CrossRef]
32. Milea, S.A.; Vasile, M.A.; Crăciunescu, O.; Prelipcean, A.M.; Bahrim, G.E.; Râpeanu, G.; Oancea, A.; Stănciuc, N. Co-Microencapsulation of Flavonoids from Yellow Onion Skins and Lactic Acid Bacteria Lead to Multifunctional Ingredient for Nutraceutical and Pharmaceutics Applications. *Pharmaceutics* **2020**, *12*, 1053. [CrossRef] [PubMed]
33. Cayra, E.; Dávila, J.H.; Villalta, J.M.; Rosales, Y. Evaluation of the Stability and Viability of Two Probiotic Strains Microencapsulated by Fluidized Bed [Evaluación de La Estabilidad y Viabilidad de Dos Cepas Probióticas Microencapsuladas Por Lecho Fluidizado]. *Inf. Tecnol.* **2017**, *28*, 35–44. [CrossRef]
34. da Silva Martins, W.; de Araújo, J.S.F.; Feitosa, B.F.; Oliveira, J.R.; Kotzebue, L.R.V.; da Silva Agostini, D.L.; de Oliveira, D.L.V.; Mazzetto, S.E.; Cavalcanti, M.T.; da Silva, A.L. Lemongrass (*Cymbopogon citratus* DC. *Stapf*) Essential Oil Microparticles: Development, Characterization, and Antioxidant Potential. *Food Chem.* **2021**, *355*, 1. [CrossRef]
35. Azarpazhooh, E.; Sharayei, P.; Zomorodi, S.; Ramaswamy, H.S. Physicochemical and Phytochemical Characterization and Storage Stability of Freeze-Dried Encapsulated Pomegranate Peel Anthocyanin and In Vitro Evaluation of Its Antioxidant Activity. *Food Bioprocess Technol.* **2019**, *12*, 199–210. [CrossRef]
36. Mansour, M.; Salah, M.; Xu, X. Effect of Microencapsulation Using Soy Protein Isolate and Gum Arabic as Wall Material on Red Raspberry Anthocyanin Stability, Characterization, and Simulated Gastrointestinal Conditions. *Ultrason. Sonochem.* **2020**, *63*, 104927. [CrossRef] [PubMed]
37. Rezende, Y.R.R.S.; Nogueira, J.P.; Narain, N. Microencapsulation of Extracts of Bioactive Compounds Obtained from Acerola (*Malpighia Emarginata* DC.) Pulp and Residue by Spray and Freeze Drying: Chemical, Morphological and Chemometric Characterization. *Food Chem.* **2018**, *254*, 281–291. [CrossRef] [PubMed]

polymers

MDPI

Article

An Equilibrium State Diagram for Storage Stability and Conservation of Active Ingredients in a Functional Food Based on Polysaccharides Blends

César Leyva-Porras [1,*], Zenaida Saavedra-Leos [2,*], Manuel Román-Aguirre [1], Carlos Arzate-Quintana [3], Alva R. Castillo-González [3], Andrés I. González-Jácquez [1] and Fernanda Gómez-Loya [3]

[1] Centro de Investigación en Materiales Avanzados S.C. (CIMAV), Miguel de Cervantes No. 120, Complejo Industrial Chihuahua, Chihuahua 31136, Mexico
[2] Coordinación Académica Región Altiplano (COARA), Universidad Autónoma de San Luis Potosí, Matehuala, San Luis Potosí 78700, Mexico
[3] Facultad de Medicina y Ciencias Biomédicas, Universidad Autónoma de Chihuahua (UACH), Circuito Universitario 31109, Campus UACH II, Chihuahua 31125, Mexico
* Correspondence: cesar.leyva@cimav.edu.mx (C.L.-P.); zenaida.saavedra@uaslp.mx (Z.S.-L.); Tel.: +52-(614)4391106 (C.L.-P.); +52-(488)1250150 (Z.S.-L.)

Abstract: A functional food as a matrix based on a blend of carbohydrate polymers (25% maltodextrin and 75% inulin) with quercetin and *Bacillus claussi* to supply antioxidant and probiotic properties was prepared by spray drying. The powders were characterized physiochemically, including by moisture adsorption isotherms, X-ray diffraction (XRD), scanning electron microscopy (SEM), and modulated differential scanning calorimetry (MDSC). The type III adsorption isotherm developed at 35 °C presented a monolayer content of 2.79 g of water for every 100 g of dry sample. The microstructure determined by XRD presented three regions identified as amorphous, semicrystalline, and crystalline-rubbery states. SEM micrographs showed variations in the morphology according to the microstructural regions as (i) spherical particles with smooth surfaces, (ii) a mixture of spherical particles and irregular particles with heterogeneous surfaces, and (iii) agglomerated irregular-shape particles. The blend's functional performance demonstrated antioxidant activities of approximately 50% of DPPH scavenging capacity and viability values of 6.5 Log10 CFU/g. These results demonstrated that the blend displayed functional food behavior over the complete interval of water activities. The equilibrium state diagram was significant for identifying the storage conditions that promote the preservation of functional food properties and those where the collapse of the microstructure occurs.

Keywords: equilibrium state diagram; functional food; polysaccharide blends; antioxidant activity; bacteria viabililty (*Bacillus clausii*); co-microencapsulation

Citation: Leyva-Porras, C.; Saavedra-Leos, Z.; Román-Aguirre, M.; Arzate-Quintana, C.; Castillo-González, A.R.; González-Jácquez, A.I.; Gómez-Loya, F. An Equilibrium State Diagram for Storage Stability and Conservation of Active Ingredients in a Functional Food Based on Polysaccharides Blends. *Polymers* **2023**, *15*, 367. https://doi.org/10.3390/polym15020367

Academic Editor: Dan Cristian Vodnar

Received: 13 December 2022
Revised: 4 January 2023
Accepted: 8 January 2023
Published: 10 January 2023

1. Introduction

Carbohydrate polymers are molecules found in nature with comparable properties to synthetic polymers, for instance, glass transition temperature (Tg), melting temperature (Tm), and molecular weight distribution (MWD) [1,2]. Commonly, polysaccharide molecules include glucose, sucrose, dextrose, arabinose, and galactose. In the food and pharmaceutical industries, polysaccharides (e.g., starch, chitosan, maltodextrins (MX), inulin (IN), and gum Arabic (GA)) are used as carrier agents in the conservation and microencapsulation of active ingredients [3]. Clearly, the adequate selection of the carrier agents in the blend can affect the microencapsulation efficiency of the active ingredient [4]. For example, Meena et al. [5] microencapsulated curcumin, employing blends of whey protein (WP), MX, and GA. They found that the content of the active ingredient remained at values of 92% and 90% at storage temperatures of 25 and 35 °C, respectively, for a period of 6 months. These carrier agents were selected to increase the microencapsulation capability of WP. With the addition of GA, the charge interactions between the positively and negatively charged fractions of WP and GA, respectively, increased the microencapsulation capacity [6,7]. Likewise, MX suppresses the hygroscopicity of GA [8]. Haladyn et al. [9] found that the stability of polyphenolic compounds in microspheres after 14 and 28 days of storage was increased by adding guar gum and sodium alginate, while decreasing with the addition of chitosan. Recently, Saavedra-Leos et al. [10] informed about the microencapsulating properties of a functional food based on IN-MX blends. They found that the antioxidant activity (AA) of quercetin improved with the incorporation of IN; conversely, the addition of MX was beneficial for the microorganisms (*Bacillus clausii*) since it promoted their viability in the presence of IN.

By definition, a functional food is any healthy food consumed in a regular diet that has health-promoting or disease-preventing properties and physiological benefits [11]. A regular consumption of functional foods provides additional health benefits beyond those traditionally associated with nutrients [12], decreasing the risk of suffering chronic diseases [13]. Kaur and Kas [11] classified the functional foods as: (i) fortified food products with high-quality ingredients. (ii) Foods released to respond to anti-nutritional compounds (iii) Improvement of food raw materials by increasing specific components, e.g., the alimentation of animals; (iv) novel foods produced by genetic manipulation or selection of new varieties; (v) probiotics and prebiotics, which are functional foods containing living organisms. Overall, functional foods classified in (i) and (v) commonly include the incorporation of probiotics and antioxidants as active compounds [14,15]. The term "probiotics" refers to live strains of selected microorganisms that confer health benefits on the host after adequate administration in suitable amounts [16]. The benefits of consuming probiotics can vary with the dose, the type of strain, and the diverse components employed in the formulation of the food. On the other hand, antioxidants are another type of bioactive compound with the ability to strengthen the immune system while also delaying cellular aging [17,18]. A natural source of antioxidants is phenolic compounds, which involve a large group of metabolites that exist in plants [19]. These can be classified into two main groups: flavonoids and non-flavonoids, such as quercetin and resveratrol, respectively. Antioxidants interact with unstable free radicals, inhibiting chain reactions that cause cellular aging and further chronic degenerative diseases [20]. Nevertheless, its consumption is limited, first by the high dose necessary to inhibit all oxidative processes within the body and, second, by instability due to external factors such as decomposition with food processing temperature, reactivity with atmospheric oxygen, and degradation upon exposure to ultraviolet light. Microencapsulation is a strategy implemented to extend the shelf life of bioactive ingredients contained in foods [21,22]. This process consists of trapping and keeping the bioactive ingredients within a protective barrier of a material that can withstand higher processing temperatures without degrading. Consequently, spray drying is a process frequently used in the pharmaceutical and food industries for obtaining dry powder products with a low degradation level and great stability.

A state diagram is a graphic depiction of the behavior of a system in equilibrium or non-equilibrium, subjected to different conditions such as variations in composition, temperature, humidity, and pressure [23,24]. The complexity depends on the number of properties included. In the food area, these diagrams are commonly used to represent the microstructural stability of foods and the preservation of their properties during storage for a period sufficient to reach moisture adsorption equilibrium [25,26]. As a result, state diagrams of pure compounds have been used as model systems for understanding storage behavior and further development for inclusion in complex food systems [27,28]. For instance, stability state diagrams of pure inulin systems differing in the degree of polymerization (DP) showed differences in the microstructural transformation from amorphous to crystalline [29]. A single state change was shown by the highest DP inulin at an activity of water (a_w) of 0.5. Meanwhile, the lowest DP inulin showed an intermediate semicrystalline state at a_w between 0.3–0.5. Adding low-molecular-weight sugars from orange juice also increased system complexity, revealing that the intermediate semicrystalline interval spans the range from 0.2 to 0.53 [24]. With the aid of a state diagram, Roos [30] found the critical storage parameters for lactose to be a water content of 7.8 g of water for every 100 g of dry mass, an a_w of 3.8, and a Tg of 23 °C. Exceeding these parameters during storage, dramatic fluctuations in the physical state of lactose may be observed, including time-dependent crystallization and flow properties. Higl et al. [31] plotted the inactivation constant of microorganisms in lactose at different storage temperatures. They found that as long as the samples remain in the glassy state, moisture adsorption is low and independent of storage temperature. However, when passing to the non-glassy state, the adsorption increases rapidly, as does the inactivation of the microorganisms. Lara-Mota et al. [32] reported the state diagram for β-lactose in an extensive a_w interval (0.07–0.972) at storage temperatures of 15, 25 and 35 °C. They discovered that the stability is conserved up to an a_w value of 0.742, where mutarotation to α-lactose begins. The wide range of stability corresponds to a low adsorbed water content that remained below the monolayer value (0.0219–0.0409 g of water for every g of dry mass).

Evidently, the development of a state diagram is desirable for both the scientific and technological fields because it allows for the prediction of the behavior of physicochemical properties during storage for a compound containing active ingredients, such as a functional food.

Therefore, the goal of this work is to obtain an equilibrium state diagram for a functional food based on a blend of carbohydrate polymers. These carbohydrate polymers were chosen not only for their performance as carrier agents, but also for the properties they provide for the functional food. For example, MX is a sweet-tasting polysaccharide with an energy intake of 4 kcal/g, while IN has a moderately sweet flavor and provides only 1.5 kcal/g [24,33]. Besides, IN is resistant to the pH conditions found in the stomach and duodenum, reaching the small intestine almost undigested, which helps to metabolize some of the intestinal microorganisms, supplying prebiotic activity. For this purpose, blends of MX and IN in a concentration range of 0–100% by weight were first prepared by spray drying. The food was made functional by adding two active ingredients: quercetin for the antioxidant activity and *B. claussi* microorganisms for its performance as a probiotic. Specifically, this product can be described as a powdered functional food with a sweet taste and low calorie intake, as well as prebiotic, probiotic, and antioxidant properties. The powders were characterized physiochemically, and the antioxidant activity of quercetin and the viability of *B. claussi* microorganisms were determined. The blend with the highest antioxidant activity was selected for moisture adsorption tests at 35 °C. From the adsorption isotherm obtained and the subsequent physicochemical characterizations, the equilibrium state diagram was elaborated.

2. Materials and Methods

2.1. Materials

Maltodextrin (MX) and inulin (IN) were employed as carrying agents (Ingredion, Mexico City, Mexico). MX was extracted from cornstarch with a dextrose equivalent (DE) of 10, a DP of 2–16 units of glucose, and a molecular weight of 1625 g/mol. IN was derived from agave. The strain was contained in a solution with bacillus bacteria (*Bacillus clausii*, Bc) purchased from Sanofi-Aventis de Mexico (Mexico City, Mexico). Quercetin 3-D-Galactoside (99%) employed as an antioxidant was acquired from Química Farmacéutica Esteroidal (Mexico City, Mexico). Analytical-grade 2,2-diphenyl-1-picrylhydrazyl (DPPH), (±)-6-hydroxy-2,5,7,8-tetramethylchromane-2-carboxylic acid (Trolox), gallic acid, sodium carbonate (Na_2CO_3), and Folin–Ciocalteu reagent were purchased from Sigma–Aldrich Chemical Co (Toluca, Mexico). Inorganic salts with a purity $\geq$ 90% were purchased from PQM Fermont (Monterrey, Mexico) as microenvironments for subjecting the functional food powders to different conditions of moisture: Sodium hydroxide (NaOH), potassium acetate (CH_3COOK), magnesium chloride ($MgCl_2$), potassium carbonate (K_2CO_3), magnesium nitrate [$Mg(NO_3)_2$], sodium nitrate ($NaNO_3$), potassium chloride (KCl), and potassium sulfate (K_2SO_4).

2.2. Preparation of Spray-Dried Powders

For the microencapsulation of the functional food, spray drying was employed. Typically, 20 g of the corresponding carrying agent (inulin, maltodextrin, or a mixture), 1 mg of quercetin, 5 mL of the commercial solution with bacteria (equivalent to a concentration of 2×10^{12} CFU), and distilled water for a total volume of 100 mL of solution were mixed in the preparation of the feeding solution. Microencapsulation was carried out in a Mini Spray Dryer B290 (BÜCHI, Labortechnik AG, Flawil, Switzerland) at the following conditions: feed temperature of 40 °C, feeding flow of 7 cm^3/min, hot airflow of 28 m^3/h, aspiration of 70%, and pressure of 1.5 bar. Based on the conditions reported in [10], an inlet temperature of 210 °C was selected to obtain the highest antioxidant activity.

The powders, packed in dark, airtight bags made of low-density polyethylene (LDPE), were kept at 4 °C in darkness for one week until their characterization. Samples were labeled as MX-IN:x-y, where MX and IN stand for the carbohydrate polymer employed as a carrying agent in the blend, and *x* and *y* stand for the concentration of each polysaccharide in the blend in percent by weight.

2.3. Water Adsorption Isotherm

The static gravimetric method was employed for constructing the equilibrium sorption isotherms [34] at a temperature of 35 °C. Initially, the drying method determined the moisture content after spray drying. The isotherm points were obtained by subjecting the functional food powder (sample identified as MX-IN:25-75) to different conditions of relative humidity in the interval a_w of 0.07–0.972. Microenvironments contained 2 g of powder and 100 g of inorganic salt. The systems were equilibrated for 30 days at the storage temperature. The samples were weighed every 24 h until reaching a constant weight with a difference of ± 0.001 g.

After the incubation lapse, water activity (a_w) was determined with an Aqualab Series 3 Water Activity Meter (Decagon Devices, Inc., Pullman, WA, USA). The water content measured according to the AOAC method, requires drying the sample in an oven at 110 °C for 2 h. Each experiment was performed in triplicate.

2.4. Phisicochemical Characterization

The physicochemical characterizations described below were exerted on the powders from sample MX-IN:25-75 stored at different moisture conditions.

2.4.1. Scanning Electron Microscopy (SEM)

Morphological characterization was conducted using a scanning electron microscope (SEM) (JEOL JSM-7401F, Tokyo, Japan) operated at an accelerating voltage of 2 kV. Powder samples were first dispersed on a double-sided copper conductive tape, then covered with a thin layer of gold utilizing sputtering to reduce charging effects (Denton Desk II sputter coater, Denton, TX, USA). For each sample, several images were acquired at different magnifications (500X, 1000X, 2500X, and 5000X).

2.4.2. X-ray Diffraction (XRD)

An x-ray diffraction (XRD) analysis was carried out to characterize the microstructure. A D8 Advance ECO diffractometer (Bruker, Karlsruhe, Germany) equipped with Cu-K radiation (λ = 1.5406 Å) that operated at 45 kV, 40 mA, and a detector in a Bragg-Brentano geometry was used. Scans were performed in the 2θ range of 5–50°, with step size of 0.016° and 20 s per step.

2.4.3. Thermal Analysis

For determining the glass transition temperature (Tg), a modulated differential scanning calorimeter (MDSC) Q200 (TA Instruments, New Castle, DE, USA) equipped with an RCS90 cooling system was employed. The instrument was calibrated with indium for melting temperature and enthalpy and sapphire for heat capacity (Cp). About 10 mg of the sample were encapsulated in Tzero® aluminum pans. Thermograms were acquired in the temperature range of −50 to 250 °C with a modulation period of 40 s and amplitude of 1.5 °C. These analyses were done in duplicate.

2.5. Antioxidant Activity (AA)

The antioxidant activity of phenolic compounds is usually determined by a rapid assay [35]. The most popular tests are based on photometric assays for evaluating the antioxidant activity with reagents such as 2,2′-azino-bis(3-ethylbenzothiazoline-6-sulfonic acid) (ABTS), 2,2-diphenyl-1-picrylhydrazyl (DPPH), oxygen radical absorbance capacity (ORAC), and Folin-Ciocalteu (FC). For the specific case of determining the antioxidant activity of quercetin 3-D-galactose, the ABTS and DPPH assays are very similar.

The extract samples were measured in terms of hydrogen-donating or radical-scavenging activity using the stable DPPH radical. Briefly, 1.7 mL of an alcoholic solution of DPPH (0.1 mmol DPPH/L) were mixed with 1.7 mL of microencapsulated suspension, whose concentration was 30 μg/mL. The mixture was left to stand in darkness for 30 min, and the absorbance at 537 nm was measured using a spectrophotometer, the UV-Vis Evolution 220 (Thermo Scientific, Walthman, MA, USA). The measurements were done in duplicate after the spray-drying process or from storage at different moisture conditions.

2.6. Viability of Bacillus Clausii in the Microencapsulated

The number of available Bc bacteria cells was evaluated by means of the plate extension technique with Trypticase-Soy Agar (TSA) (Beckton Dickinson, Germany), using serial dilutions of the encapsulated samples from 1×10^{-1} to 1×10^{-7}. Aerobic growing conditions and an incubation period of 48 h at 37 °C were applied in a Novatech incubator (Guadalajara, Jalisco, Mexico). For the determination of the number of colony-forming units per gram (CFU/g), the concentrations exhibiting between 300 and 30 CFU (1×10^{-4} and 1×10^{-5}) were selected. For the quantification of cultivability, Equation (1) was used. All experiments were repeated five times, and the reported values represent the average of the calculated values.

$$\frac{CFU}{g} = \left[\frac{N^{\circ}\ plate\ colonies \times dilution\ factor}{mL\ sample\ seeded}\right] \quad (1)$$

2.7. Statistical Analysis

The differences between the mean values of antioxidant activity and viability measurements were determined by one-way analysis of variance (ANOVA). For this purpose, Origin software version 8.5 (OriginLab, Northampton, MA, USA) was employed. A significance value of 0.05 was used in all the calculations.

3. Results

3.1. Antioxidant Activity of Functional Food Blends

MX-IN concentrations ranging from 0 to 100% were obtained in the prepared spray-dried functional food. In all cases, fine powder materials were obtained, well dispersed, and without signs of microstructural collapse. The physicochemical characterization of these mixtures is beyond the objectives of this work, and only the results of the AA are presented herein. This property was considered suitable to indicate whether the antioxidant ingredient was actually preserved during the spray-drying microencapsulation process. In Figure 1 are shown the results of the AA for the functional food in the entire range of MX-IN concentrations tested. For all the blends, the AA values varied in the range of 53–57%; the one-way ANOVA showed that the population means were not significantly different at the 0.05 level. Although statistically the AA values may be similar, the blend MX-IN:25-75 showed a relatively higher average value (57.03%). Likewise, Saavedra-Leos et al. [10] reported a synergistic effect upon the mixing of MX and IN and found that the MX-IN 1:1 blend presented higher AA values than the pure carrying agents. This agrees with the results presented herein, where the addition of inulin is beneficial for quercetin co-microencapsulation because it promotes higher antioxidant activity.

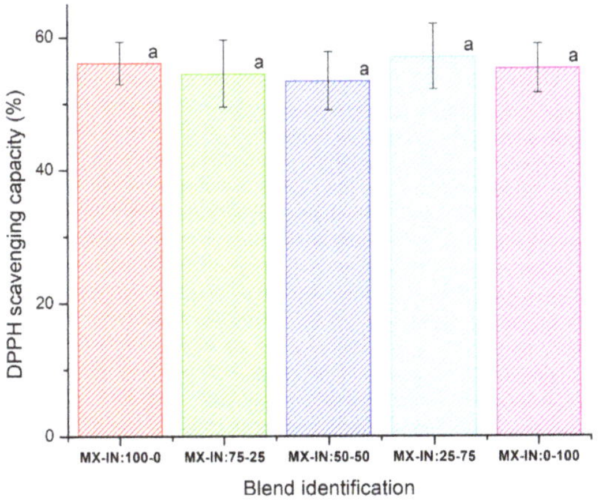

Figure 1. Antioxidant activity of the functional food in all the range of MX-IN concentrations. The lowercase letters indicate mean differences at 0.05 of significance value.

Based on these observations, sample MX-IN:25-75 was selected for storage under different moisture conditions.

3.2. Adsorption Isotherm

Sample MX-IN:25-75 was subjected to the adsorption of water in equilibrium at different conditions of humidity. Figure 2 shows the water adsorption isotherm at 35 °C for sample MX-IN:25-75. From the distinct adsorption conditions, the microenvironment created with K_2SO_4 produced the saturation of the powder and the subsequent collapse of the microstructure. This was observed as a continuous-phase material (not powdered), sticky, and with a remarkable yellow color. Because of this, the corresponding data for a_w (0.98) was not included in the plot nor considered for the physicochemical characterizations.

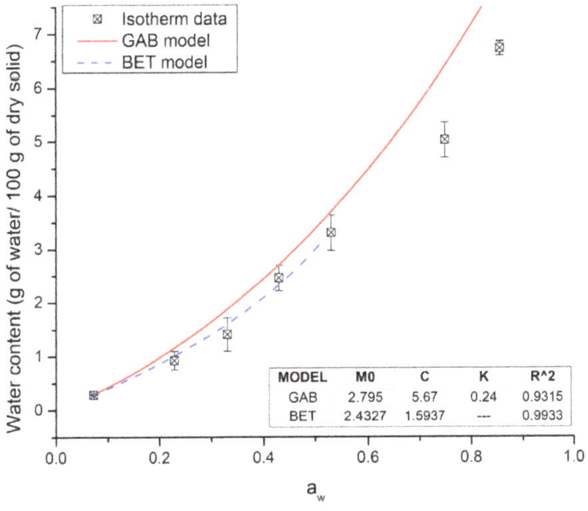

Figure 2. Water adsorption isotherm at 35 °C for the functional food blend MX-IN:25-75.

The adsorbed water content varied in the range of 0.29–6.72 g of water per 100 g of dry sample in the complete interval of a_w (0.073–0.85). In Figure 2, the first observable characteristic is the shape of the isotherm, which presented a low moisture adsorption at low water activities and then a pronounced increase at medium and high a_w values. According to IUPAC, the shape of the isotherm was identified as the Flory-Huggins isotherm (type III). This isotherm accounts for a solvent or plasticizer above the glass transition temperature [36] and is typical of soluble, low-molecular-weight substances like sugars [37]. Several mathematical models for describing sorption isotherms were fitted to the experimental data. The evaluation of these models and their estimated parameters are reported in Figure S1 and Table S1 from the Supplementary Material. From the tested models, some did not fit the experimental data correctly, while others gave unrealistic results. Only those with consistent results are discussed herein. In the low-medium range of a_w, the Brunauer-Emmett-Teller (BET) model was fitted. The BET monolayer value was 2.43 g of water per 100 g of dry sample, and the energy constant (C) related to the heat of sorption was 1.59. However, this model is limited to the linear range of water activity (0.073–0.531), and beyond this range, the BET model underestimates the predicted values. The Guggenheim-Anderson-de Boer (GAB) model has been used to fit isotherms across a wide range of a_w (0–0-9) and provides a more accurate description of sorption behavior for almost every food product [36]. The calculated GAB monolayer content (M_0) was 2.795 g of water per 100 g of dry sample, and the model constants were 5.67 and 0.24, for C and K, respectively. The values of M_0 calculated from the BET and GAB models were very similar and indicated the maximum amount of adsorbed solvent in the form of a monolayer of water molecules on the sample, which is considered the value at which the food product is the most stable [36]. Beyond this value, there are more water molecules available, acting as plasticizers, triggering unwanted microstructure changes, and acting as media for biological reactions such as bacterial growth. Therefore, to maintain the stability during storage of the functional food, it may be kept at a moisture condition lower than the value of the monolayer.

The C and K constants from the GAB model are related to the interaction energy between the first and further adsorbed layers of molecules, respectively. According to Lewicki [38], the estimated value of the monolayer will vary 15% when the values of the constants are within the following limits: $5.67 < C < \infty$, and $0.24 < K < 1$. The results obtained from the fit of the data indicated a good estimation of the monolayer value.

3.3. Microstructural Analysis

The samples stored at different moisture conditions were analyzed by X-ray diffraction to observe possible changes in the microstructure. Figure 3A shows the X-ray diffractograms of the functional food blend MX-IN:25-75 at the different water activities. Microstructural behavior was similar in samples with an a_w of 0.073–0.331, with a broad peak near an angle of 18° and a relatively low intensity diffraction peak at 26.5°. These properties indicated that the microstructure remained amorphous across the entire range of water conditions. For a_w values of 0.43 and 0.531, the intensity of the broad peak decreased and small diffraction peaks appeared, suggesting that the amorphous microstructure is crystallizing due to moisture adsorption. At high a_w values of 0.73 and 0.856, the diffractograms showed a completely collapsed microstructure, which can be understood as a mixture of crystallized material and material in the rubbery state.

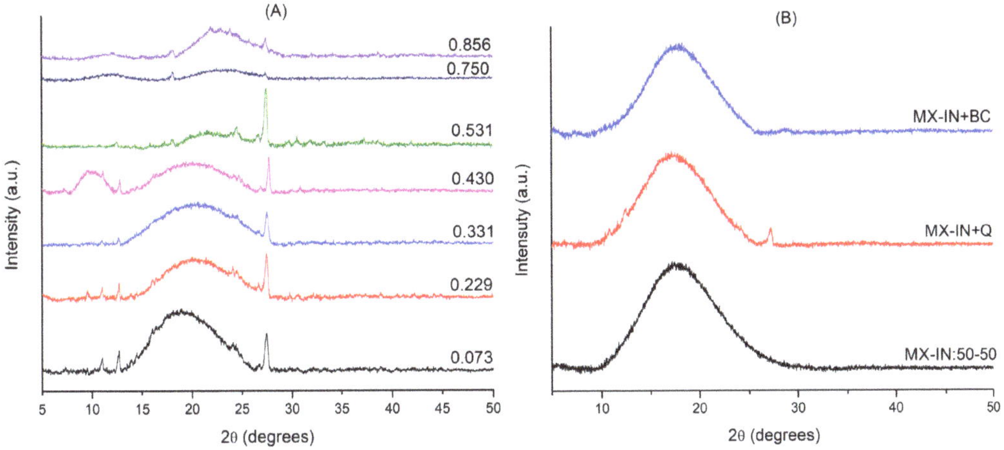

Figure 3. XRD diffractograms: (**A**) functional food blend MX-IN:25-75 at the different a_W; and (**B**) blank samples.

According to previously reported studies, the amorphous microstructure of low molecular weight MXs is maintained with the adsorption of moisture, and only the change from a glassy solid into a rubbery state is observed a_W greater than 0.750 [39]. Meanwhile, with water adsorption, INs present a microstructural transition from amorphous to crystalline, observed as the appearance of diffraction peaks on a broad peak at a_W value of 0.32. The intensity of the diffraction peaks increases in a_W increments until the broad peak completely disappears [29]. Certainly, the blending of these two carbohydrate polymers with contrasting behaviors upon water adsorption causes the microstructural changes reported herein.

On the other hand, the diffraction peak observed at 26.5° suggested the crystallization of some of the active ingredients, such as quercetin or the microorganism *B. cluassi.* To verify the above, three blank samples containing only the mixture of carrier agents (MX-IN:50-50), the mixture plus quercetin (MX-IN+Q), and the mixture plus *B. claussi* (MX-IN+BC) were prepared and characterized by X-rays. Figure 3B shows the diffractograms for the blank samples. In all of these, the same broad peak around 18° was observed. The MX-IN+Q sample presented a diffraction peak at 26.5°, while the MX-IN+BC sample did not show any diffraction peak but only a small, broad peak around 28°.

Clearly, these analyses help to understand the microstructural behavior of the functional food of the MX-IN:25-75 blend during storage at the different moisture conditions tested

3.4. Morphological Characterization

The result of storing information on the morphology of functional food particles was studied by SEM. Figure 4 depicts representative 1000X SEM micrographs of the functional food blend MX-IN:25-75 in various moisture environments. At a_W values of 0.073–0.531, the powder particles presented a quasi-spherical morphology, with sizes ranging from 2–15 microns. Particles with smooth surfaces were observed; however, other particles with an irregular (wrinkled) surface were also present. Likewise, some particles with elongated shapes were also presented, which may correspond to microencapsulated *B. claussi* microorganisms. For a_W of 0.75, the powder showed a different morphology, similar to merged spherical particles, forming irregular-shape agglomerates with sizes greater than 10 μm. At a_W of 0.856, the analyzed sample was no longer visualized as a powder but as a continuous mass.

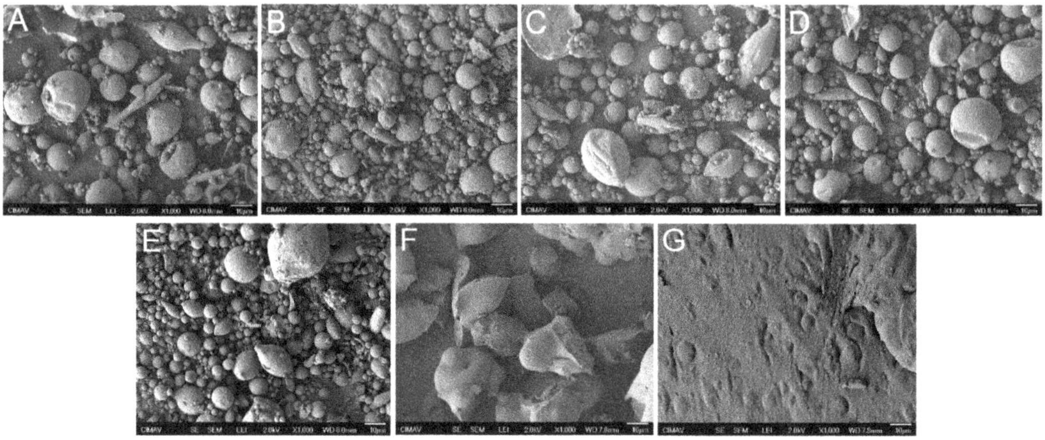

Figure 4. SEM micrographs of the functional food blend MX-IN:25-75 at the different a_w values: (**A**) 0.073, (**B**) 0.229, (**C**) 0.331, (**D**) 0.43, (**E**) 0.531, (**F**) 0.75, and (**G**) 0.856.

These observations evidenced the changes exerted by the adsorption of water at the microscopic level. The functional food retains powder characteristics up to a maximum a_w value of 0.531. The morphology of the functional food changed dramatically above this humidity level, and it behaved as a crystallized solid and sticky material. Clearly, this analysis is of great importance because it sets the water activity range where the morphology is preserved before the microstructural changes may affect the properties of the powder.

3.5. Thermal Analysis

From the MDSC measurements, the glass transition temperature (Tg) of blend MX-IN:25-75 was determined after storage at different humidity conditions. Figure 5 shows the reversible heat curves at each a_w. The determination of the Tg was performed graphically by identifying the onset of the reversible heat flow curve. The summary of Tg values is reported in Table 1. It is observed that the Tg tends to decrease with the increase of a_w, and the variation of the Tg was in the range of 32.53 to −3.05 °C. This is a common behavior in sugar-rich systems, where the adsorbed water molecules act as plasticizers, decreasing the temperature to the point where the amorphous solid passes from a glassy to a rubbery state. Thus, it is preferable to keep the functional food in its glassy state because the microstructure of the solid remains rigid, preventing active ingredient degradation due to exposure to ambient atmospheric conditions.

Table 1. Determined Tg values at different a_w. The values represent the average of two measurements, and the standard deviation is indicated in brackets.

a_w	Tg (°C)
0.0073	32.53 (1.3)
0.229	32.45 (0.9)
0.331	31.82 (1.1)
0.43	19.54 (0.6)
0.531	17.38 (1.8)
0.75	7.79 (0.4)
0.856	−3.05 (0.9)

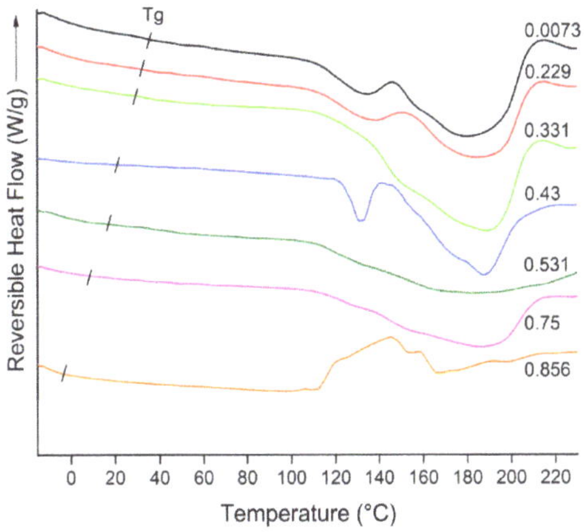

Figure 5. Reversible heat flow curves of the functional food blend MX-IN:25-75 at the different a_W.

Additionally, the Tg value indicates the maximum temperature at which the food can be stored before undergoing a microstructural change that may modify its composition. In this sense, samples with low moisture content (a_W of 0.0073–0.331) can be stored at temperatures close to 30 °C, while those with intermediate moisture content (a_W of 0.43 and 0.531) should be stored at temperatures below 20 °C, whereas samples with high moisture content (a_W of 0.75 and 0.856) should be kept at temperatures below 10 °C.

3.6. Functional Food Performance

The performance of blend MX-IN:25-75 as a functional food was tested by determining both the antioxidant activity of quercetin and the viability of *B. claussi* microorganisms. Figure 6 shows the AA and the viability of the blend MX-IN:25-75 subjected to different humidity conditions during storage at 35 °C.

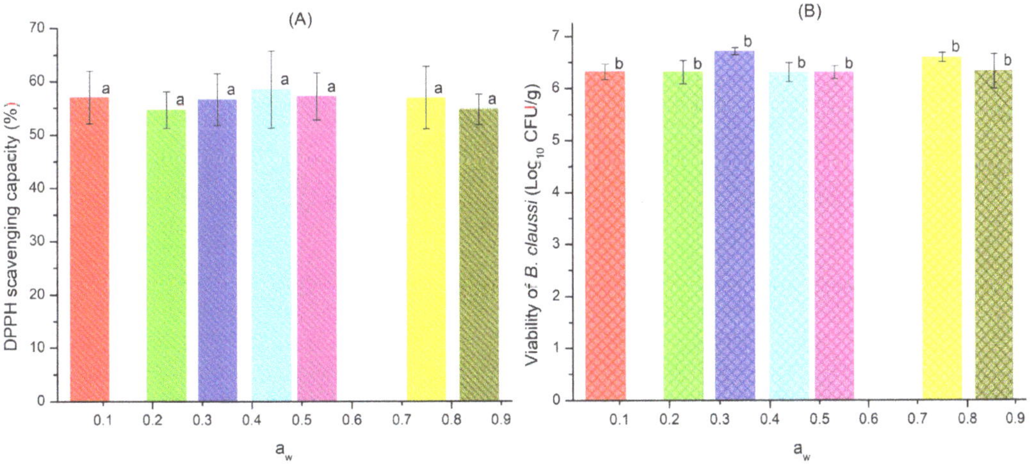

Figure 6. Functional performance of the blend MX-IN:25-75: (**A**) antioxidant activity of quercetin and (**B**) viability of *B. claussi* microorganisms. The lowercase letters indicate differences in the means at a significance value of 0.05.

Figure 6A shows that the powders presented antioxidant activity in the range of 54.6–58.5% of DPPH scavenging capacity. According to the ANOVA, the population means were not significantly different at the 0.05 level. By comparing the values reported herein against those reported in [10], the AA is the double of the former. They reported the AA for 1:1 ratio blends of MX-IN and compared them against samples prepared with pure MX or IN. They found that IN promotes the microencapsulation of quercetin, showing a synergistic effect in the blend. In the case of the blend reported here, it contains 75% of IN, which may favor the microencapsulation of quercetin and hence higher AA values.

Figure 6B shows the viability of *B. claussi* microorganisms at the different a_w. The variation in viability was narrow, showing values between 6.3–6.7 Log_{10} CFU/g. In consequence, the ANOVA showed that the population means were not significantly different at the 0.05 level. This suggested that the blend MX-IN:25-75 was able to microencapsulate the microorganisms and keep them alive after the drying process and during storage at 35 °C under the different humidity conditions. On the other hand, the values found here were relatively lower than those reported in [10]. This can be explained in terms of the MX concentration, since MX favors the microencapsulation of the microorganism and its protecting effect is more pronounced than that of IN. This may be caused by the difficulty that microorganisms may have in metabolizing the long carbohydrate chains of IN [40]. Vázquez-Maldonado et al. compared IN and lactose as co-microencapsulating agents of resveratrol and *B. claussi*. They found that IN showed higher capacity for the conservation of resveratrol, while both materials showed similar viability values for encapsulating the microorganism [41]. Although the viability values reported herein were not very high, they were slightly above the minimum value recommended by the FAO/WHO (2003) to cause therapeutic benefits in humans. Therefore, the tested blend MX-IN:25-75 can be considered as functional food with probiotic properties.

In terms of storage stability, it is evident that the results are very promising, since the blend MX-IN:25-75 presented functional characteristics such as antioxidant activity and viability in the complete interval of water activities. This indicates that although the morphology of the particles may collapse at high humidity values, the active ingredients remain protected and active to convert the carbohydrate polymers into a functional food.

4. Discussion

Equilibrium State Diagram of Functional Food Blend MX-IN:25-75

Commonly, two types of state diagrams are reported to estimate stability during storage. The first is based on determining the glass line, freezing curve, maximal freeze concentration, and solid mass fraction (Xs) [25]. For example, Wan et al. [42] determined for indica rice starch a Tg of −42.5 °C, an Xs of 0.71 g/g (wet basis), and a monolayer content of 7.43 g of water per 100 g of dry solids. Sablani et al. [43] determined for *basmati* rice a Tg of −11.6 °C, an Xs of 0.6, and a monolayer content value of 0.136 g of water per 100 g of dry solids. On the other hand, equilibrium state diagrams can be established by combining the water sorption isotherms and Tg for determining critical values for water content and water activity at a given storage temperature [44].

Figure 7 shows the equilibrium state diagram constructed for the functional food blend MX-IN:25-75 from the results of the physicochemical characterization. One of the main features in the state diagram is the moisture adsorption isotherm at 35 °C (the red solid curve). Here is indicated the water content at the monolayer level (M_0 of 2.79 g of water per 100 g of dry sample), corresponding to an a_w of 0.43. Moisture values greater than those of the monolayer will produce microstructural changes that can affect the behavior of the powder material. Thomsen et al. [45] constructed a state diagram for amorphous lactose to set the storage temperature and water activity at which lactose may remain in the glassy state. The depicted isotherm at 25 °C was employed as the borderline for separating the storage conditions for stable from unstable amorphous lactose. Fabra et al. [46] employed the concepts of a_w, Tg, and the critical water content (CWC) to set the storage conditions of grapefruit. Because of the high water content of the fruit, it must be stored at freezing temperatures.

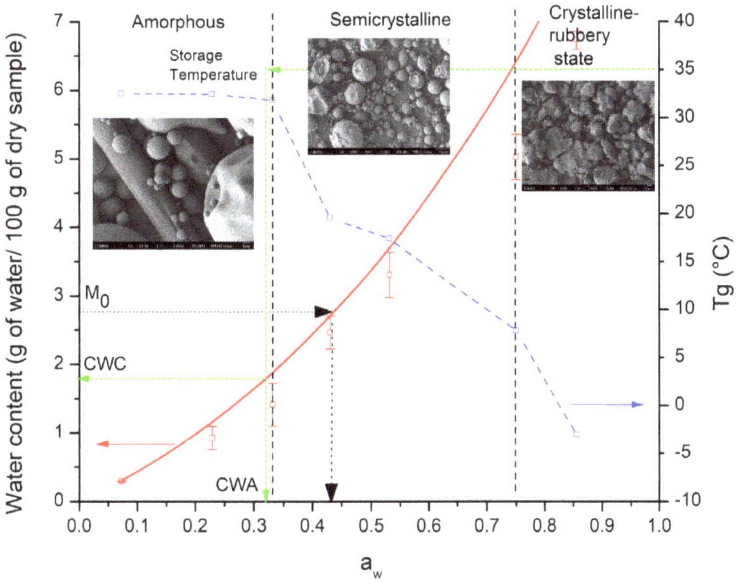

Figure 7. Equilibrium state diagram at 35 °C of functional food blend MX-IN:25-75.

The Tg (blue dashed curve) shows a tendency to decrease with moisture adsorption. Three regions with similar glass transition temperatures were observed: (i) For moisture values lower than the monolayer level, the Tg showed values slightly higher than 30 °C. (ii) In the region of intermediate moisture values, Tg values were close to 19 °C. (iii) At high moisture contents, the Tg drops to values below 10 °C. According to Sá and Sereno [47], if the storage temperature equals the Tg, degradation of fresh fruits and vegetables may be prevented. Employing spray drying, Roos [48] constructed a state diagram employing the Tg and the mass fraction of lactose to establish the drying and storage temperatures of non-sticking powders from lactose-containing dairy liquids. In accordance with this, the powder must be stored at a lower temperature than the Tg of the main component.

Additionally, from both the equilibrium state diagram and storage temperature (35 °C), the CWC and critical water activity (CWA) can be extrapolated. These values (indicated by the dashed green line) were a CWC of 1.77 g of water per 100 g of dry sample and a CWA of 0.32. Both values are lower than the corresponding values obtained from the moisture content of the monolayer. These results agree with those in the literature. For example, Shi et al. [49] reported lower CWC values for seafood meat (*Penaeus vannamei*) than those obtained from the monolayer. They also consider the CWC parameter to be of greater importance for the evaluation of the storage conditions of dehydrated food products. Vásquez et al. [50] emphasized the importance of CWC over M_0, and stated that lyophilized blueberries must be stored at a_w less than 0.1 in order to preserve the glassy state of the matrix and avoid deteriorative changes in the amorphous components. Conversely, Sablani et al. [51] found constant CWA values of about 0.56 and CWC values of 0.11, 0.09, and 0.06 g of water per g of dry solid for abalone samples stored at 23, 40, and 60 °C, respectively. These CWC values were relatively higher than those determined from the BET monolayer content and may disturb the stability of the food product. Shi and collaborators [52] depicted an equilibrium moisture content diagram to evaluate the stability settings for storage of freeze-dried edible fungi (*Agaricus bisporus*) and reported a monolayer moisture content of 0.062 g of water per g of *A. bisporus* and a Tg of −27.2 °C.

The addition of a carrying agent such as MX to freeze-dried seafood meat promoted an increment in the CWC from 0.151 to 0.258 and in a_w from 0.038 to 0.083 g of water per g of dry solids [49]. The joint use of concepts such as a_w, CWC, and Tg in an equilibrium state diagram is of great importance to explain variations in time-dependent mechanical and flow properties, such as crystallization of amorphous sugars, non-enzymatic browning, and enzymatic reactions [44].

The vertical dashed lines indicate the water activities at which the microstructural changes occur. From here, three regions were identified: amorphous, semi-crystalline, and crystalline-rubbery states. This means that the microstructural change of this mixture is not abrupt but gradual, since it presents a transition state such as the semi-crystalline region. The value of the monolayer approximately matches the region where the microstructure transforms from the amorphous state to the semi-crystalline state. Roos [44] established that the effect of Tg on the mechanical properties results in rapid changes in viscosity and modulus, which cause the crystallization of amorphous solids. He identified three regions that depend on storing temperature and water activity or content. These regions were identified as: (i) a stability zone, which corresponds to a glassy state; (ii) a critical zone, where the glass transition occurs; and (iii) a mobility zone, where the flow occurs. According to Rahman [53], the microbial stability in food is affected by the moisture content, while the oxidation phenomenon is caused by the crystallization at a temperature higher than the Tg.

A representative micrograph of the morphology of the powdered food particles is shown in each of the abovementioned regions. The behavior of the morphology corresponds with the microstructural fluctuations, resulting in spherical particles with smooth surfaces in the amorphous region. In the transition region, spherical particles with smooth surfaces were observed, as were particles with irregular shapes and contracted or collapsed surfaces. In the crystalline-rubbery state region, the morphology was completely irregular, with clusters of fused particles. According to Sahin et al. [53], the encapsulation efficiency can be affected by the presence of surface defects such as dents and heterogeneities since the active ingredients can escape through there. Al-Ghamdi et al. [54] used state diagrams to relate color change and browning with a_w and storage temperature of pumpkin powders. Both characteristics increased linearly with moisture content and at storage temperatures greater than 35 °C. Thanatuksorn, Kajiwara, and Suzuki [55] studied by SEM the structural alteration of fried wheat flour with various moisture contents. They observed that the porous structure reached its maximum value after 3 and 5 min of drying in samples containing 400 and 600 g of moisture per kg of sample, respectively.

In addition to the information on phase and state changes provided by the microstructural and thermal characterizations, the morphological analysis provides information about the heterogeneities in the microstructure of the food, which is required for understanding the properties and kinetic behavior of amorphous food systems [44]. According to Buera et al. [56], among the characterization methods for studying the crystallization of amorphous sugars are DSC, XRD, microscopy, infrared, and Raman spectroscopies, whereas proton nuclear magnetic resonance (^{1}H NMR) and dynamic mechanical analysis (DMA) have been employed as alternative techniques to determine Tg.

5. Conclusions

In this work, the use of a blend based on carbohydrate polymers as a matrix for a functional food with antioxidant and probiotic properties was proposed. The functional food was made up of 25% maltodextrin, 75% inulin, and the active ingredients quercetin and *Bacillus claussi*. The physicochemical characterization of the powders obtained by spray drying included moisture adsorption isotherms, X-ray diffraction, scanning electron microscopy, and thermal analysis. With the characterization results, an equilibrium state diagram for the storage stability of the functional food was constructed. The storage conditions were identified as a monolayer water content of 2.79 g of water per 100 g of dry sample, corresponding to an a_w of 0.43, and a maximum storage temperature slightly above

30 °C. The state diagram presented three microstructural regions identified as amorphous, semicrystalline, and crystalline-rubbery states. The morphology of particles varied in each of these regions, with spherical particles having smooth surfaces, a mixture of spherical and irregular particles having heterogeneous surfaces, and agglomerated and fused irregular-shaped particles being observed. The equilibrium state diagram was of great importance for identifying the storage conditions that promote the preservation of functional food properties and those where the collapse of the microstructure occurs.

The analysis of the functional properties of the blend showed antioxidant activities of about 50% of DPPH scavenging capacity and similar viability values of 6.5 Log_{10} CFU/g. These results demonstrated that the blend could be considered a functional food prepared by spray drying.

Supplementary Materials: The following supporting information can be downloaded at: https://www.mdpi.com/article/10.3390/polym15020367/s1, Figure S1. Evaluation of the different sorption isotherm models; Table S1. Estimated parameters for the evaluated sorption models.

Author Contributions: Conceptualization, C.L.-P. and Z.S.-L.; methodology, Z.S.-L., M.R.-A. and C.A.-Q.; software, F.G.-L.; validation, C.L.-P., Z.S.-L. and C.A.-Q.; formal analysis, C.L.-P.; investigation, C.L.-P. and F.G.-L.; resources, C.L.-P., Z.S.-L. and C.A.-Q.; data curation, A.R.C.-G. and A.I.G.-J.; writing—original draft preparation, C.L.-P.; writing—review and editing, C.L.-P., Z.S.-L., A.R.C.-G. and C.A.-Q.; visualization, F.G.-L.; supervision, C.L.-P. and Z.S.-L.; project administration, C.L.-P.; funding acquisition, C.L.-P., Z.S.-L. and C.A.-Q. All authors have read and agreed to the published version of the manuscript.

Funding: Centro de Investigación en Materiales Avanzados S.C. (CIMAV). Grant number PI-14-22.

Institutional Review Board Statement: Not applicable.

Data Availability Statement: The data presented in this study are available on request from the corresponding author.

Acknowledgments: The authors thank to Daniel Lardizabal for his support with the MDSC measurements, and the Laboratorio Nacional de Nanotecnología (Nanotech) for the use of the SEM. The financial support from CIMAV is also acknowledged.

Conflicts of Interest: The authors declare no conflict of interest. The funders had no role in the design of the study, in the collection, analysis, or interpretation of data, in the writing of the manuscript, or in the decision to publish the results.

References

1. Ferreira, A.R.; Alves, V.D.; Coelhoso, I.M. Polysaccharide-based membranes in food packaging applications. *Membranes* **2016**, *6*, 22. [CrossRef] [PubMed]
2. Cazón, P.; Velazquez, G.; Ramírez, J.A.; Vázquez, M. Polysaccharide-based films and coatings for food packaging: A review. *Food Hydrocoll.* **2017**, *68*, 136–148. [CrossRef]
3. Kontogiorgos, V.; Smith, A.M.; Morris, G.A. The parallel lives of polysaccharides in food and pharmaceutical formulations. *Curr. Opin. Food Sci.* **2015**, *4*, 13–18. [CrossRef]
4. Ribeiro, A.M.; Shahgol, M.; Estevinho, B.N.; Rocha, F. Microencapsulation of Vitamin A by spray-drying, using binary and ternary blends of gum arabic, starch and maltodextrin. *Food Hydrocoll.* **2020**, *108*, 106029. [CrossRef]
5. Meena, S.; Gote, S.; Prasad, W.; Khamrui, K. Storage stability of spray dried curcumin encapsulate prepared using a blend of whey protein, maltodextrin, and gum Arabic. *J. Food Process. Preserv.* **2021**, *45*, e15472. [CrossRef]
6. Fernandes, R.V.D.B.; Silva, E.K.; Borges, S.V.; de Oliveira, C.R.; Yoshida, M.I.; da Silva, Y.F.; do Carmo, E.L.; Azevedo, V.M.; Botrel, D.A. Proposing novel encapsulating matrices for spray-dried ginger essential oil from the whey protein isolate-inulin/maltodextrin blends. *Food Bioprocess Technol.* **2017**, *10*, 115–130. [CrossRef]
7. Klein, M.; Aserin, A.; Ishai, P.B.; Garti, N. Interactions between whey protein isolate and gum Arabic. *Colloids Surf. B Biointerfaces* **2010**, *79*, 377–383. [CrossRef] [PubMed]
8. Saxena, J.; Adhikari, B.; Brkljaca, R.; Huppertz, T.; Chandrapala, J.; Zisu, B. Inter-relationship between lactose crystallization and surface free fat during storage of infant formula. *Food Chem.* **2020**, *322*, 126636. [CrossRef]
9. Haładyn, K.; Tkacz, K.; Wojdyło, A.; Nowicka, P. The Types of Polysaccharide Coatings and Their Mixtures as a Factor Affecting the Stability of Bioactive Compounds and Health-Promoting Properties Expressed as the Ability to Inhibit the α-Amylase and α-Glucosidase of Chokeberry Extracts in the Microencapsulation Process. *Foods* **2021**, *10*, 1994.

10. Saavedra-Leos, M.Z.; Román-Aguirre, M.; Toxqui-Terán, A.; Espinosa-Solís, V.; Franco-Vega, A.; Leyva-Porras, C. Blends of Carbohydrate Polymers for the Co-Microencapsulation of Bacillus clausii and Quercetin as Active Ingredients of a Functional Food. *Polymers* **2022**, *14*, 236. [CrossRef]
11. Kaur, S.; Das, M. Functional foods: An overview. *Food Sci. Biotechnol.* **2011**, *20*, 861–875. [CrossRef]
12. Milner, J. Functional foods and health promotion. *J. Nutr.* **1999**, *129*, 1395S–1397S. [CrossRef] [PubMed]
13. Konstantinidi, M.; Koutelidakis, A.E. Functional foods and bioactive compounds: A review of its possible role on weight management and obesity's metabolic consequences. *Medicines* **2019**, *6*, 94. [CrossRef] [PubMed]
14. Cederholm, T.; Barazzoni, R.; Austin, P.; Ballmer, P.; Biolo, G.; Bischoff, S.C.; Compher, C.; Correia, I.; Higashiguchi, T.; Holst, M. ESPEN guidelines on definitions and terminology of clinical nutrition. *Clin. Nutr.* **2017**, *36*, 49–64. [CrossRef]
15. Favaro-Trindade, C.S.; Patel, B.; Silva, M.P.; Comunian, T.A.; Federici, E.; Jones, O.G.; Campanella, O.H. Microencapsulation as a tool to producing an extruded functional food. *LWT* **2020**, *128*, 109433. [CrossRef]
16. Pandey, K.R.; Naik, S.R.; Vakil, B.V. Probiotics, prebiotics and synbiotics-a review. *J. Food Sci. Technol.* **2015**, *52*, 7577–7587. [CrossRef]
17. Lee, L.; Choi, E.; Kim, C.; Sung, J.; Kim, Y.; Seo, D.; Choi, H.; Choi, Y.; Kum, J.; Park, J. Contribution of flavonoids to the antioxidant properties of common and tartary buckwheat. *J. Cereal Sci.* **2016**, *68*, 181–186. [CrossRef]
18. Masisi, K.; Beta, T.; Moghadasian, M.H. Antioxidant properties of diverse cereal grains: A review on in vitro and in vivo studies. *Food Chem.* **2016**, *196*, 90–97. [CrossRef]
19. Nunes, A.R.; Gonçalves, A.C.; Falcão, A.; Alves, G.; Silva, L.R. *Prunus avium* L. (Sweet Cherry) By-Products: A Source of Phenolic Compounds with Antioxidant and Anti-Hyperglycemic Properties—A Review. *Appl. Sci.* **2021**, *11*, 8516. [CrossRef]
20. Rostami, M.; Yousefi, M.; Khezerlou, A.; Mohammadi, M.A.; Jafari, S.M. Application of different biopolymers for nanoencapsulation of antioxidants via electrohydrodynamic processes. *Food Hydrocoll.* **2019**, *97*, 105170. [CrossRef]
21. Piñón-Balderrama, C.I.; Leyva-Porras, C.; Terán-Figueroa, Y.; Espinosa-Solís, V.; Álvarez-Salas, C.; Saavedra-Leos, M.Z. Encapsulation of active ingredients in food industry by spray-drying and nano spray-drying technologies. *Processes* **2020**, *8*, 889. [CrossRef]
22. Ye, Q.; Georges, N.; Selomulya, C. Microencapsulation of active ingredients in functional foods: From research stage to commercial food products. *Trends Food Sci. Technol.* **2018**, *78*, 167–179. [CrossRef]
23. Sablani, S.S.; Syamaladevi, R.M.; Swanson, B.G. A review of methods, data and applications of state diagrams of food systems. *Food Eng. Rev.* **2010**, *2*, 168–203. [CrossRef]
24. Leyva-Porras, C.; López-Pablos, A.L.; Alvarez-Salas, C.; Pérez-Urizar, J.; Saavedra-Leos, Z. Physical properties of inulin and technological applications. In *Polysaccharides*; Ramawat, K.M.J., Ed.; Springer: New York, NY, USA, 2015; pp. 959–984.
25. Rahman, M.S. State diagram of foods: Its potential use in food processing and product stability. *Trends Food Sci. Technol.* **2006**, *17*, 129–141. [CrossRef]
26. Icoz, D.Z.; Kokini, J.L. State diagrams of food materials. In *Food Materials Science*; Springer: New York, NY, USA, 2008; pp. 95–121.
27. Bhandari, B.; Roos, Y.H. *Introduction to Non-Equilibrium States and Glass Transitions-The Fundamentals Applied to Foods Systems*; Woodhead: Cambridge, MA, USA, 2017.
28. García-Coronado, P.; Flores-Ramírez, A.; Grajales-Lagunes, A.; Godínez-Hernández, C.; Abud-Archila, M.; González-García, R.; Ruiz-Cabrera, M.A. The Influence of Maltodextrin on the Thermal Transitions and State Diagrams of Fruit Juice Model Systems. *Polymers* **2020**, *12*, 2077. [CrossRef] [PubMed]
29. Leyva-Porras, C.; Saavedra-Leos, M.; López-Pablos, A.; Soto-Guerrero, J.; Toxqui-Terán, A.; Fozado-Quiroz, R. Chemical, thermal and physical characterization of inulin for its technological application based on the degree of polymerization. *J. Food Process Eng.* **2017**, *40*, e12333. [CrossRef]
30. Roos, Y. Solid and liquid states of lactose. In *Advanced Dairy Chemistry*; Springer: New York, NY, USA, 2009; pp. 17–33.
31. Higl, B.; Kurtmann, L.; Carlsen, C.U.; Ratjen, J.; Först, P.; Skibsted, L.H.; Kulozik, U.; Risbo, J. Impact of water activity, temperature, and physical state on the storage stability of Lactobacillus paracasei ssp. paracasei freeze-dried in a lactose matrix. *Biotechnol. Prog.* **2007**, *23*, 794–800. [CrossRef]
32. Lara-Mota, E.E.; Alvarez-Salas, C.; Leyva-Porras, C.; Saavedra-Leos, M.Z. Cutting-edge advances on the stability and state diagram of pure β-lactose. *Mater. Chem. Phys.* **2022**, *289*, 126477. [CrossRef]
33. Xiao, Z.; Xia, J.; Zhao, Q.; Niu, Y.; Zhao, D. Maltodextrin as wall material for microcapsules: A review. *Carbohydr. Polym.* **2022**, *298*, 120113. [CrossRef]
34. Labuza, T.; Kaanane, A.; Chen, J. Effect of temperature on the moisture sorption isotherms and water activity shift of two dehydrated foods. *J. Food Sci.* **1985**, *50*, 385–392. [CrossRef]
35. Platzer, M.; Kiese, S.; Herfellner, T.; Schweiggert-Weisz, U.; Eisner, P. How Does the Phenol Structure Influence the Results of the Folin-Ciocalteu Assay? *Antioxidants* **2021**, *10*, 811. [CrossRef] [PubMed]
36. Andrade, R.D.; Lemus, R.; Pérez, C.E. Models of sorption isotherms for food: Uses and limitations. *Vitae* **2011**, *18*, 325–334.
37. Stępień, A.; Witczak, M.; Witczak, T. Sorption properties, glass transition and state diagrams for pumpkin powders containing maltodextrins. *LWT* **2020**, *134*, 110192. [CrossRef]
38. Lewicki, P.P. The applicability of the GAB model to food water sorption isotherms. *Int. J. Food Sci. Tech.* **1997**, *32*, 553–557. [CrossRef]

39. Saavedra-Leos, Z.; Leyva-Porras, C.; Araujo-Díaz, S.B.; Toxqui-Terán, A.; Borrás-Enríquez, A.J. Technological application of maltodextrins according to the degree of polymerization. *Molecules* **2015**, *20*, 21067–21081. [CrossRef]
40. Pandey, P.; Mishra, H.N. Co-microencapsulation of γ-aminobutyric acid (GABA) and probiotic bacteria in thermostable and biocompatible exopolysaccharides matrix. *LWT* **2021**, *136*, 110293. [CrossRef]
41. Vázquez-Maldonado, D.; Espinosa-Solis, V.; Leyva-Porras, C.; Aguirre-Bañuelos, P.; Martinez-Gutierrez, F.; Román-Aguirre, M.; Saavedra-Leos, M.Z. Preparation of spray-dried functional food: Effect of adding *Bacillus clausii* bacteria as a co-microencapsulating agent on the conservation of resveratrol. *Processes* **2020**, *8*, 849. [CrossRef]
42. Wan, J.; Ding, Y.; Zhou, G.; Luo, S.; Liu, C.; Liu, F. Sorption isotherm and state diagram for indica rice starch with and without soluble dietary fiber. *J. Cereal Sci.* **2018**, *80*, 44–49. [CrossRef]
43. Sablani, S.S.; Bruno, L.; Kasapis, S.; Symaladevi, R.M. Thermal transitions of rice: Development of a state diagram. *J. Food Eng.* **2009**, *90*, 110–118. [CrossRef]
44. Roos, Y. Thermal analysis, state transitions and food quality. *J. Therm. Anal. Calorim.* **2003**, *71*, 197–203. [CrossRef]
45. Thomsen, M.K.; Jespersen, L.; Sjøstrøm, K.; Risbo, J.; Skibsted, L.H. Water activity-temperature state diagram of amorphous lactose. *J. Agric. Food Chem.* **2005**, *53*, 9182–9185. [CrossRef] [PubMed]
46. Fabra, M.; Talens, P.; Moraga, G.; Martínez-Navarrete, N. Sorption isotherm and state diagram of grapefruit as a tool to improve product processing and stability. *J. Food Eng.* **2009**, *93*, 52–58. [CrossRef]
47. Sá, M.; Sereno, A. Glass transitions and state diagrams for typical natural fruits and vegetables. *Thermochim. Acta* **1994**, *246*, 285–297. [CrossRef]
48. Roos, Y.H. State and Supplemented Phase Diagrams for the Characterization of Food. In *Water Activity in Foods*; John Wiley and Sons: Chicago, IL, USA, 2020; pp. 45–60.
49. Shi, Q.; Lin, W.; Zhao, Y.; Zhang, P. Thermal characteristics and state diagram of Penaeus vannamei meat with and without maltodextrin addition. *Thermochim. Acta* **2015**, *616*, 92–99. [CrossRef]
50. Vásquez, C.; Díaz-Calderón, P.; Enrione, J.; Matiacevich, S. State diagram, sorption isotherm and color of blueberries as a function of water content. *Thermochim. Acta* **2013**, *570*, 8–15. [CrossRef]
51. Sablani, S.; Kasapis, S.; Rahman, M.; Al-Jabri, A.; Al-Habsi, N. Sorption isotherms and the state diagram for evaluating stability criteria of abalone. *Food Res. Int.* **2004**, *37*, 915–924. [CrossRef]
52. Shi, Q.; Wang, X.; Zhao, Y.; Fang, Z. Glass transition and state diagram for freeze-dried Agaricus bisporus. *J. Food Eng.* **2012**, *111*, 667–674. [CrossRef]
53. Sahin, C.C.; Erbay, Z.; Koca, N. The physical, microstructural, chemical and sensorial properties of spray dried full-fat white cheese powders stored in different multilayer packages. *J. Food Eng.* **2018**, *229*, 57–64. [CrossRef]
54. Al-Ghamdi, S.; Hong, Y.; Qu, Z.; Sablani, S.S. State diagram, water sorption isotherms and color stability of pumpkin (Cucurbita pepo L.). *J. Food Eng.* **2020**, *273*, 109820. [CrossRef]
55. Thanatuksorn, P.; Kajiwara, K.; Suzuki, T. Characterization of deep-fat frying in a wheat flour-water mixture model using a state diagram. *J. Sci. Food Agric.* **2007**, *87*, 2648–2656. [CrossRef]
56. Buera, M.P.; Roos, Y.; Levine, H.; Slade, L.; Corti, H.R.; Reid, D.S.; Auffret, T.; Angell, C.A. State diagrams for improving processing and storage of foods, biological materials, and pharmaceuticals (IUPAC Technical Report). *Pure Appl. Chem.* **2011**, *83*, 1567–1617. [CrossRef]

polymers

MDPI

Article

Enhancement of Strawberry Shelf Life via a Multisystem Coating Based on *Lippia graveolens* Essential Oil Loaded in Polymeric Nanocapsules

Barbara Johana González-Moreno [1], Sergio Arturo Galindo-Rodríguez [2], Verónica Mayela Rivas-Galindo [1], Luis Alejandro Pérez-López [1], Graciela Granados-Guzmán [1] and Rocío Álvarez-Román [1,*]

[1] Departamento de Química Analítica, Facultad de Medicina, Universidad Autónoma de Nuevo León, Monterrey 64460, Nuevo León, Mexico; barbara.gonzalezmrn@uanl.edu.mx (B.J.G.-M.); veronica.rivasgl@uanl.edu.mx (V.M.R.-G.); luis.perezlp@uanl.edu.mx (L.A.P.-L.); graciela.granadosgu@uanl.edu.mx (G.G.-G.)

[2] Departamento de Química Analítica, Facultad de Ciencias Biológicas, Universidad Autónoma de Nuevo León, San Nicolás de los Garza 66455, Nuevo León, Mexico; sergio.galindord@uanl.edu.mx

* Correspondence: rocio.alvarezrm@uanl.edu.mx; Tel.: +52-(81)-8329-4185

Abstract: Strawberries (*Fragaria xannanasa*) are susceptible to mechanical, physical, and physiological damage, which increases their incidence of rot during storage. Therefore, a method of protection is necessary in order to minimize quality losses. One way to achieve this is by applying polymer coatings. In this study, multisystem coatings were created based on polymer nanocapsules loaded with *Lippia graveolens* essential oil, and it was found to have excellent optical, mechanical, and water vapor barrier properties compared to the control (coating formed with alginate and with nanoparticles without the essential oil). As for the strawberries coated with the multisystem formed from the polymer nanocapsules loaded with the essential oil of *Lippia graveolens*, these did not present microbial growth and only had a loss of firmness of 17.02% after 10 days of storage compared to their initial value. This study demonstrated that the multisystem coating formed from the polymer nanocapsules loaded with the essential oil of *Lippia graveolens* could be a viable alternative to preserve horticultural products for longer storage periods.

Keywords: polymeric coating; essential oil; *Lippia graveolens*; nanoparticles; strawberries

Citation: González-Moreno, B.J.; Galindo-Rodríguez, S.A.; Rivas-Galindo, V.M.; Pérez-López, L.A.; Granados-Guzmán, G.; Álvarez-Román, R. Enhancement of Strawberry Shelf Life via a Multisystem Coating Based on *Lippia graveolens* Essential Oil Loaded in Polymeric Nanocapsules. *Polymers* **2024**, *16*, 335. https://doi.org/10.3390/polym16030335

Academic Editor: Dan Cristian Vodnar

Received: 19 December 2023
Revised: 19 January 2024
Accepted: 22 January 2024
Published: 26 January 2024

1. Introduction

According to the Food and Agriculture Organization (FAO), food losses are defined as a decrease in the quantity or quality of food. An important part of food loss is waste, which is food initially intended for consumption and discarded or, in some cases, used alternatively, i.e., in a non-food way. Globally, between a quarter and a third of the food produced annually for human consumption is lost or wasted. This is equivalent to about 1.3 billion tons of food, including 40 to 50% of roots, fruits, vegetables, and oilseeds [1,2]. The highest losses occur in horticultural products, that is, fruits and vegetables [3]; this is because they are much less resistant and more perishable compared to seeds, tubers, or roots [4]. Strawberries (*Fragaria ananassa* L.) are highly nutritious but, at the same time, are extremely perishable due to their susceptibility to deterioration, mechanical injury, postharvest physiological disorders, and microbial decay [5]. Therefore, strawberries are considered a delicate fruit with a short shelf life: losses during postharvest storage are estimated to be as high as 40% [6]. The use of coatings is an innovative method of food preservation whose application would allow the following: (i) the generation of a physical barrier to protect the surface of the product; (ii) the control of the migration of solutes and humidity as well as, gas exchange and oxidation reactions to prolong shelf life; and (iii) a reduction in the risk of pathogen growth on the surfaces of fruit and vegetable products [7,8].

The principle of using coatings is very similar to modified atmosphere packaging, where an atmosphere consisting of high CO_2 and low O_2 concentration is created [9]. This environment can effectively slow down the respiration rate, conserve stored energy, slow microbial growth, and, therefore, extend the shelf life of the fruit [10]. The coating may contain ingredients such as antioxidant agents, additional nutrients, flavorings, preservatives, and antimicrobial compounds of natural origin, such as essential oils (EOs) [11]. Oregano is the common name applied to more than 40 species of the families *Verbenaceae*, *Lamiaceae*, *Compositae*, and *Leguminoseae*, of which the most important are Mediterranean or European oregano (*Origanum vulgare*) and Mexican oregano (*Lippia graveolens*). The antioxidant and antimicrobial properties of *Lippia graveolens* essential oil (EO-*Lg*) [12] make it a strong candidate as a natural preservative for foods, such as fruit and vegetable products. Regarding the chemical mechanism of peroxidation inhibition, the components of the EO act mainly as radical trapping agents that would lead to the formation of hydroperoxides, epoxides, and other oxygenated derivatives. Gutiérrez-Grijalva et al. [13] reported that EO-*Lg* may have significant potential as an auxiliary antioxidant and against enzymes involved in lipid and carbohydrate metabolism. On the other hand, previous studies have shown that EO-*Lg* is effective against bacteria, yeast, and fungi [14] and increased antioxidant activity without negative effects on sensory acceptability in tomatoes [15]. The antimicrobial property of EO-*Lg* is attributed to two of its main components, i.e., thymol and carvacrol [16], which have the ability to break the cell membrane of Gram-negative bacteria. This is thanks to their chemical structure since they contain an -OH group and have a non-polar character. The damage is reflected in the dissipation of the two components of the proton motive force, the pH gradient and the electrical potential, which causes greater permeability and the leakage of ions and other compounds that are necessary for the survival of the bacteria [17]. The antimicrobial effects of oregano EO have been previously reported against microorganisms such as *Candida albicans*, *Alternaria alternate*, *Escherichia coli*, *Pseudomonas aeruginosa*, and *Staphylococcus aureus*, among others [15,18,19]. However, the application of EOs in coatings is often problematic due to their physicochemical properties, and they can deteriorate due to environmental factors, such as light and oxygen, and have high volatility [20]. Furthermore, EOs have low solubility in water, which makes their incorporation into commercial products difficult. To overcome these effects, the nanoencapsulation of EOs in polymeric nanoparticles has become a possible solution [21]. Today, nanotechnology represents an area of opportunity for the development of vehicles that transport certain EOs, vitamins, and other plant extracts, such as polyphenols, with antimicrobial and antioxidant properties. Although the development of nanoparticles (NPs) was first reported in the pharmaceutical field for drug delivery systems less than two decades ago, these systems caught the interest of the food sector. NPs have great potential to guarantee the maintenance of color (by encapsulating dyes), flavor, and nutritional values (by encapsulating vitamins), thus increasing the shelf life of foods [22].

It is possible to produce coatings when antioxidants or antimicrobials are encapsulated in NPs. In the coating, the compound can be administered, prolonged, or controlled to create a specific microenvironment to improve the shelf life of fruit and vegetable products [23]. In addition, a plasticizer such as sodium alginate can be added to the coating; this is widely used because it is biodegradable, biocompatible, has a low price, and is non-toxic. This polysaccharide can form strong and homogeneous films at room temperature [24]. Previous studies have shown that the application of NPs with EOs allowed the color and firmness of the grapes to be preserved and delayed the presence of microbiological damage due to a prolonged storage time [25]. Therefore, the objective of this study was to evaluate the conservative effect of a coating obtained from nanocapsules of *Lippia graveolens* EO (NC-EO-*Lg*) and carry out its optical, mechanical, and permeability characterization for its application in strawberries.

2. Materials and Methods

2.1. Materials

Lippia graveolens was collected from Cuatro Ciénegas, Coahuila de Zaragoza (26°59′ N 102°03′59″ OE), México. The specimens were identified in the Herbarium of the School of Biological Sciences, Universidad Autónoma de Nuevo León, México. Methanol, acetone (Tedia, Fairfield, OH, USA), and isopropyl alcohol (Chromadex, Los Angeles, CA, USA) were of HPLC grade. Purified water was from a Milli-Q water-purification system (Veolia, Boston, MA, USA). The standard solution n-alkanes (C_8–C_{20}, C_{22}, and C_{24}), myrcene (≥99.5%), p-cymene (≥97%), carvacrol (≥98%), anethole (99%) GC grade, and sodium alginate were purchased from Sigma-Aldrich, St. Louis, MO, USA. Eudragit L100-55 polymer (1:1 methacrylic acid: ethyl acrylate) was purchased from Evonik Industries, Essen, Germany.

2.2. Extraction and Physicochemical Characterization of Essential Oil (EO)

2.2.1. Isolation of EO

For the extraction of EO from *Lippia graveolens* (EO-*Lg*) by hydrodistillation using a modified Clevenger-type apparatus, a quantity of 100 g of the aerial parts of fresh plant was accurately weighed and added to 800 mL of distilled water in a flask. Then, it was placed in a balloon heater attached to a refrigerator to ensure condensation of EO for 4 h. The EO was collected and stored in sealed vials in the dark, at 4 °C, until used. The yield percentage was calculated as the weight (g) of EO per 100 g of the plant [26].

2.2.2. Physical Characterization of EO

The relative density of the EO-*Lg* was determined at 25 °C according to the general method 0251 of the Pharmacopeia of the United Mexican States (FEUM) [27] with an Anton Paar Densimeter (DMA35, Ashland, VA, USA). The refractive index was measured at 25 °C on an Anton Paar Refractometer (Abbermat 300, Ashland, VA, USA) according to Method 0741 of the FEUM [27]. The specific rotation was determined on a polarimeter (Perkin Elmer, Waltham, MA, USA) according to Method 0771 of the FEUM [27]. The relative density and the refractive index test were performed in triplicate, while the specific rotation was performed in sextuplicated. An average value was calculated, and anethole was used as a control.

2.2.3. Chemical Composition of EO Using Gas Chromatography–Mass Spectrometry (GC-MS) and GC with Flame Ionization Detection (GC-FID)

The composition of volatile constituents of the EO-*Lg* was analyzed using a gas chromatograph (Agilent Technologies, 6890N, Santa Clara, CA, USA) equipped with a 5973 INERT mass selective spectrometer (ionization energy 70 eV) and an HP-5MS capillary column (5% phenylmethylpolysiloxane, 30 m × 0.2 mm, 0.25 µm, Agilent J and W, Santa Clara, CA, USA). The ionization-source temperature was 230 °C, the quadrupole temperature was 150 °C, and the injector temperature was 220 °C. Data acquisition was performed in the scan mode. The oven temperature was programmed as follows: 35 °C for 9 min, increased to 150 °C at 3 °C min^{-1} and held for 10 min, increased to 250 °C at 10 °C min^{-1}, and increased to 270 °C at 3 °C min^{-1} and held for 10 min. The flow rate of the helium carrier gas (99.999% purity) was 0.5 mL min^{-1}. The EO's components were identified by comparing retention indices relative to C_8–C_{20}, C_{22}, and C_{24} n-alkanes, and MS results were compared with the mass spectra from the US National Institute of Standards and Technology (NIST) library and reference data. To determine the proportion of each component, a quantitative analysis was performed with a GC-FID (Autosystem XL, Perkin Elmer, Boston, MA, USA) using the same HP-5MS column. The injector temperature was 270 °C; the oven temperature program was the same as the GC-MS analysis. The percentage composition of each component was calculated using the peak-normalization method.

2.3. Preparation and Characterization of the Nanocapsules

The EO-*Lg* nanocapsules (NC-EO-*Lg*) were prepared by the solvent displacement procedure developed by Fessi et al. [28]. The organic phase was prepared by dissolving 175 mg of Eudragit L100-55® and 100 mg of the EO-*Lg* in 15 mL of organic solvent mixture (acetone: isopropyl alcohol 50:50) under magnetic stirring at room temperature. The organic solution was added to 25 mL of the aqueous phase (Milli-Q water) under magnetic stirring (125 rpm). Finally, the organic solvent mixture was then evaporated under reduced pressure (Control Laborota 4003, Heidolph Instruments, Schwabach, Germany). The NP-BCO was prepared by the same method, omitting the EO-*Lg*. Subsequently, the NC physicochemical characterization was determined three times at 25 °C. The mean particle size and polydispersity index (PI) were measured at a 90-degree scattering angle using dynamic light scattering (DLS), while the zeta potential measurement was by laser Doppler microelectrophoresis (Zetasizer Nano-ZS90, Malvern Instruments, Worcestershire, UK). In addition, the size and morphology of the NC were investigated using a scanning electron microscope (SEM), JEOL brand JSM-7401F (Kyoto, Japan), with a field-emission gun (FEG) and an accelerating voltage of 4.0 kV. For this, the NC-EO-*Lg* samples were previously coated with gold-palladium for 15 s using the sputtering technique (Denton vacuum model Desk II equipment, Cherry Hill, NJ, USA). On the other hand, the encapsulation percentage (%E) and the encapsulation efficiency percentage (%EE) of the three main components of the EO-*Lg* (myrcene, p-cymene, and carvacrol) were determined by a previously validated GC-FID method. A calibration curve was made with the mixture of the three standard components of EO-*Lg*, and the retention times were 4.7, 5.4, and 12.2 min, respectively. The detection limits for myrcene, p-cymene, and carvacrol were 0.76, 0.35, and 0.83 µg/mL, respectively. The %E and %EE were calculated by the following formulas:

$$\%E = \frac{(\text{mg of main component encapsulated in the NC})}{(\text{mg of polymer} + \text{mg of main component in total EO})} \times 100 \quad (1)$$

$$\%EE = \frac{(\text{mg of main component encapsulated in the NC})}{(\text{mg of main component in total EO})} \times 100 \quad (2)$$

2.4. Preparation of the Coatings

For the formation of the coatings, the direct casting method [29,30] on Teflon was used, and sodium alginate (AL) powder was incorporated as film-coating into an aqueous suspension of NC-EO-*Lg* under magnetic stirring at 350 rpm. NC-EO-*Lg* coating (NC-EO-*Lg*-AL) was obtained with a sodium alginate concentration of 1% (*w*/*v*) and a drying time of 24 h. The NP-BCO-AL was prepared by the same method, omitting the EO-*Lg*. In addition, the coating's morphology was investigated using a scanning electron microscope (SEM), JEOL brand JSM-7401F (Kyoto, Japan), with a field-emission gun (FEG) and an accelerating voltage of 2.0 kV. For this, the coating samples were previously coated with gold–palladium for 25 s using the sputtering technique (Denton vacuum model Desk II equipment, Cherry Hill, NJ, USA).

2.5. Physical Characterization of the Coatings

2.5.1. Optical Evaluation

The opacity of the coatings was calculated as the ratio of absorbance (A) (in nm) to the coating thickness (in µm) using a UV–VIS spectrophotometer (Genesys, Thermo Scientific, Waltham, MA, USA). Rectangular coating strips (1 cm × 4 cm) were placed directly in glass cuvettes, using an empty glass cuvette as reference. The absorbance was measured at 600 nm, and the opacity of the coating was calculated by the following equation:

$$\text{Opacity} = \text{A } 600\text{ nm}/\delta \quad (3)$$

where O = opacity, A 600 nm = absorbance of the sample at 600 nm, and δ = Thickness of the coating in mm.

2.5.2. Mechanical Evaluation

Mechanical properties such as adhesion, tensile strength, and elongation at break for the coatings were measured by a texturometer (CT3 Texture Analyzer, Brookfield-Ametek, Middleborough, MA, USA). The coatings were cut into strips and gripped at each end with the necessary accessories, and then the jaws were moved at the controlled speed (2 mm/s) until the modulus was automatically recorded.

2.5.3. Fourier-Transform Infrared (FTIR) Spectroscopy Analysis

The chemical composition of the coating (the polymer, NC, and free EO) was analyzed using an FT-IR Optical Frontier Optical Spectrophotometer (PerkinElmer, Waltham, MA, USA). Each coating was scanned 30 times from 4000 to 400 cm^{-1} with a scanning interval of 4 cm^{-1}, and the air spectrum was used as a background correction.

2.5.4. Water Vapor Barrier Properties

The water vapor transmission rate (WVTR) and water vapor permeability (WVP) of the coatings were measured in accordance with ASTM E-96-95 [31]. The determination of the WVTR and WVP were carried out by gravimetric analysis. In a desiccator, coatings were placed on top of a vial containing a saturated solution with a relative humidity (RH) of 95%. Subsequently, the vial was weighed every 2 h for the first 8 h and then at 24 h, in quintuplicate. The water transpiration rate and water vapor permeability were determined with the following formulas [32]:

$$\mathrm{WVTR} = \frac{\mathrm{S}}{\mathrm{AT}} \tag{4}$$

where S is the weight gain of the test setup (g), A is the area of film exposed, and T is the time (h).

$$\mathrm{WVP} = \frac{(\mathrm{WVTR})\delta}{\mathrm{Pw}(\mathrm{RH1} - \mathrm{RH2})} \tag{5}$$

where δ is the average thickness of the films in mm, and Pw is the partial difference of water vapor between the 50% RH of the desiccator and the 0% RH of the saturated solution.

2.6. Application of the Coatings on Strawberries

Strawberries with homogeneous characteristics of color, size, and without mechanical damage were selected. They were washed with distilled water and dried. Subsequently, for the coating formation, the spraying method was used, which consisted of spraying the samples of the horticultural product with AL, NP-BCO-AL, or NC-EO-*Lg*-AL for 1 min and then drying them at room temperature [25]. After the application of the coating, the strawberries were stored for 15 days at 4 °C.

2.6.1. Weight Loss Percentage and Texture Analyses: Penetration Test

The strawberries from the two groups and the control (strawberries without coating) were individually weighed after the storage time at 4 °C. The weight loss was calculated as the difference between the initial and final weights of the fruit, and the values were reported on a percentage basis in accordance with the AOAC standard method. The penetration test was determined by a texturometer (CT3 Texture Analyzer, Brookfield-Ametek, Middleborough, MA, USA). The penetration test outlined a mechanical force–displacement using a 5 kg loading cell and a cylindrical flat head probe with a diameter of 5 mm (P/5) entering the fruit (placed on the plate with the receptacle cavity upright to the compression probe to assess its firmness). The data were acquired with the following instrumental settings: pretest speed, 10.00 mm/s; test speed, 5 mm/s; post-test speed, 10.00 mm/s; trigger force, 2.0 g. For each sample, ten replicates were used.

2.6.2. Visual Microbiological Damage

On the other hand, the presence of microbiological damage on the strawberries and the control were monitored for the storage time at 4 °C.

2.7. Statistical Analysis

The results obtained in this study are shown in the tables and figures as means ± SDs of different measurements. The statistical differences were evaluated using a one-way analysis of variance (ANOVA) with Tukey's post hoc test ($p < 0.05$), performed using SPSS software (Version 20.0, SPSS Inc., Chicago, IL, USA).

3. Results and Discussion

3.1. Isolation of Essential Oil of Lippia graveolens (EO-Lg)

Initially, the EO-*Lg* was obtained from *Lippia graveolens* by the hydrodistillation method in a Clevenger apparatus. This technique has been commonly used for the extraction of EOs since it avoids the degradation of plant material. The EO is extracted from the plant, together with water vapor, and is separated after condensation. A yield percentage of 5.32% was obtained in this study. Dilworth et al. [33] obtained a similar result, which was 4.29%. The extraction of EOs through hydrodistillation is variable and depends on certain parameters, such as the collection season, soil type, harvest location, and plant variety. On the other hand, lower extraction yields than those obtained in this work have been reported (0.92% to 4.41%) using another extraction technique: steam distillation [34].

3.2. Physical Properties of the EO-Lg

For the physical characterization of the EO-*Lg*, the physical parameters of density, refractive index, and optical rotation were evaluated based on the FEUM [27]. In the EO industry, the importance of evaluating the physical characteristics of EOs to guarantee their quality control on a routine basis has been established since variations in physical parameters allow for the possible adulteration or degradation of their components to be detected, therefore ensuring their biological activity [35]. In addition, the anethole standard was selected as a control for physical evaluation due to its physicochemical similarity with some of the components of EO-*Lg* (i.e., polarity and vapor pressure) and the fact that its physical characterization is already reported in the FEUM [27]. In Table 1, the values obtained from the physical characterization of anethole and EO-*Lg* are presented. In the case of anethole, it can be observed that the values obtained are within the intervals reported in the FEUM, which confirms the repeatability and reproducibility of the physical evaluations.

Table 1. Physical characterization of anethole control and essential oil of *Lippia graveolens*.

Parameter	Relative Density [1]	Refractive Index (°) [1]	Specific Rotation (g/mL) [2]
Anethole (FEUM)	0.983 – 0.988	1.557 – 1.561	−0.150 – 0.150
Anethole	0.987 ± 0.000	1.559 ± 0.000	0.050 ± 0.000
EO-*Lippia graveolens*	0.987 ± 0.000	1.503 ± 0.000	−0.200 ± 0.000
EO-*Lippia alba*	0.945 ± 0.005	1.462 ± 0.000	−0.117 ± 0.028

[1] ($\overline{x} \pm \sigma$, $n = 3$); [2] ($\overline{x} \pm \sigma$, $n = 6$).

In the case of EO-*Lg*, the relative density, refractive index, and specific rotation values obtained also coincide with those reported by Torrenegra-Alarcón et al. [36], presenting only small differences, which could be attributed to the geographical origin and the species of the plant. In addition, Domínguez [37] reported that a refractive index greater than 1.47 and a relative density greater than 0.9 indicates that the EO contains oxygenated aliphatic and/or aromatic compounds. These values coincide with what is expected according to Table 2, which shows that more than 99% of the EO-*Lg* compounds are monoterpenes (oxygenated and aliphatic) and aliphatic sesquiterpenes.

Table 2. Chemical composition of essential oil of *Lippia graveolens* by gas chromatography–mass spectrometry.

No. [1]	Composition	tR (min) [2]	Abundance	Component Type
1	α-thujene	16.51	0.19	Monoterpene hydrocarbon
2	α-pinene	17.83	0.08	Monoterpene hydrocarbon
3	myrcene	18.37	16.93	Monoterpene hydrocarbon
4	α-terpinene	18.92	0.08	Monoterpene hydrocarbon
5	p-cymene	20.25	7.56	Monoterpene hydrocarbon
6	1,8-sineole	22.88	0.04	Monoterpene hydrocarbon
7	γ-terpinene	26.18	0.08	Monoterpene hydrocarbon
8	linalool	26.63	0.12	Oxygenated monoterpene
9	terpinen-4-ol	29.29	0.10	Oxygenated monoterpene
10	thymol methyl ether	30.94	0.06	Oxygenated monoterpene
11	thymol	32.49	4.91	Oxygenated monoterpene
12	carvacrol	33.02	66.58	Oxygenated monoterpene
13	z-caryophyllene	36.96	2.99	Sesquiterpene hydrocarbon
14	α-humulene	38.33	0.11	Sesquiterpene hydrocarbon
15	butyl hydroxyanisole	41.49	0.07	Sesquiterpene oxygenated
16	caryophyllene oxide	43.52	0.03	Sesquiterpene oxygenated
	TOTAL		100.00	

[1] Elution order; [2] retention time.

3.3. *Chemical Composition of the EO Using Gas Chromatography–Mass Spectrometry (GC-MS) and GC with Flame Ionization Detection (GC-FID)*

The chemical characterization of EO-*Lg* was carried out using GC-MS and GC-FID. The chromatogram of EO-*Lg* obtained through GC-MS is shown in Figure 1. A total of 16 components were identified in the EO-*Lg* based on a comparison with the NIST (National Institute of Standards and Technology) library, the arithmetic index, and the Kovats index [38]. Next, the percentage abundance (%A) of the 16 components was determined based on their relative areas and is shown in Table 2.

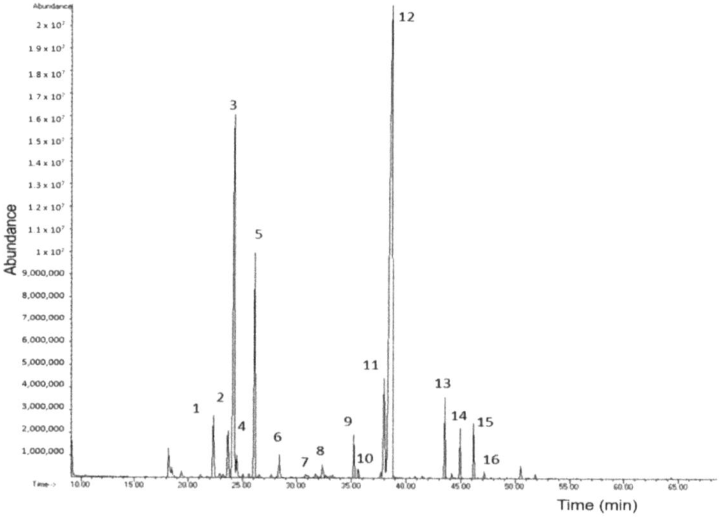

Figure 1. Chromatogram of essential oil of *Lippia graveolens* by gas chromatography–mass spectrometry.

As is shown in Figure 1, the EO-*Lg* was composed of 24.96% of monoterpenes hydrocarbons and 71.77% of oxygenated terpenes. The major components were myrcene (16.93%), p-cymene (7.56%), and carvacrol (66.58%). This coincides with what was reported by Chacón-Vargas et al. [39], who also identified carvacrol as the majority component, with an abundance percentage of 33.78%. Likewise, in another investigation, carvarol and p-cymene were found to be major components with abundance percentages of 25.19% and 13.78%, respectively [40], which also coincides with what was reported in this study.

However, various authors have reported other major components of EO-Lippia. Such is the case of Capatina et al. [41], who reported thymol (38.82%), p-cymene (20.28%), and γ-terpinene (19.58%) as the majority components. In another study carried out by Sharififard et al. [42], terpineol (22.85%) and α-terpinene (20.60%) were reported as major components. Other authors mention 1,8-cineol as the majority component [43]. The different abundances of the components reported in previous studies are due to variations in the collection season, the year of collection, and the region and/or climate in which the plant grows, as well as the location of the species analyzed and the methods of obtaining the EO such as steam distillation, extraction with organic solvents, ultrasound-assisted extraction, microwave-assisted extraction, and supercritical fluid extraction [44].

In order to monitor the components of EO-*Lg* in the NCs and coating, myrcene, p-cymene, and carvacrol were selected as the monitoring components based on their abundance and antibacterial activity.

3.4. Preparation and Characterization of the NC-EO-Lg

Subsequently, the formation of the NC was carried out using the nanoprecipitation technique. The nanoprecipitation technique (also called the solvent displacement technique) was described for the first time by Fessi et al. [28]. This technique has certain advantages over other encapsulation techniques, namely, (1) good reproducibility, (2) ability to obtain nanoparticle sizes with a narrow distribution, (3) simplicity, (4) ease of scaling, and (5) being environmentally friendly since the use of large amounts of toxic solvents is limited. Due to the above, nanoprecipitation has become an important strategy in the pharmaceutical, agricultural, food, and cosmetic industries [45].

Once the NCs were obtained, their physicochemical characterization was carried out. Their size, polydispersity index (PI), and zeta potential are shown in Table 3. A particle size of 287 ± 5.11 nm was obtained, which is close to the desired size of 200 nm. This nanometric size of NCs allows the increase in the ratio between the surface area and the contact area [46]. That is, the nanometric size would allow for the greater surface area of the NCs to be directly in contact with the surface of the fruit and vegetable product, from which the subsequent release of the active components of EO-*Lg* would occur. In previous studies, poly (lactic acid) nanocapsules with lemongrass EO, with a particle size of 96.4 nm, were applied to apples, and it was found that these had rot lesions three times smaller than those treated with non-encapsulated EO and the control [47]. Likewise, in another study, edible coatings for avocados were made from chitosan nanoparticles (355 ± 25.30 nm) loaded with *Schinus molle* EO, and it was found that this reduced weight and firmness losses [48].

Table 3. Content of myrcene, p-cymene, and carvacrol in nanocapsules of *Lippia graveolens* essential oil (NC-EO-*Lg*) purified through evaporation under reduced pressure.

	Mean Size (nm) [1]	PI [1]	Zeta Potential (mV) [1]	Component	%E [1]	%EE [1]
NC-EO-*Lg*	287 ± 5.11	0.10 ± 0.03	−50.90 ± 1.44	Myrcene	0.10 ± 0.04	0.24 ± 0.02
				p-cymene	0.17 ± 0.07	0.32 ± 0.01
				Carvacrol	27.91 ± 0.59	63.80 ± 1.47

[1] ($\bar{x} \pm \sigma$, $n = 3$).

In relation to the PI, this parameter is associated with the measurement of the degree of variability in the size of the NC. The PI values vary from 0 to 1. The highest value indicates a less homogeneous size distribution, while values close to zero indicate that the sample is monodispersed; that is, it presents minimal variability in population size [49]. A PI of 0.10 ± 0.03 was obtained (Table 3), which indicates a homogeneity of the NCs, and this is also observed in the size distribution curve in Figure 2A. The homogeneity in the size of the NC is a characteristic that would guarantee that individual interactions of the NCs with the surface of the horticultural products (i.e., adhesion or deposition) are also homogeneous.

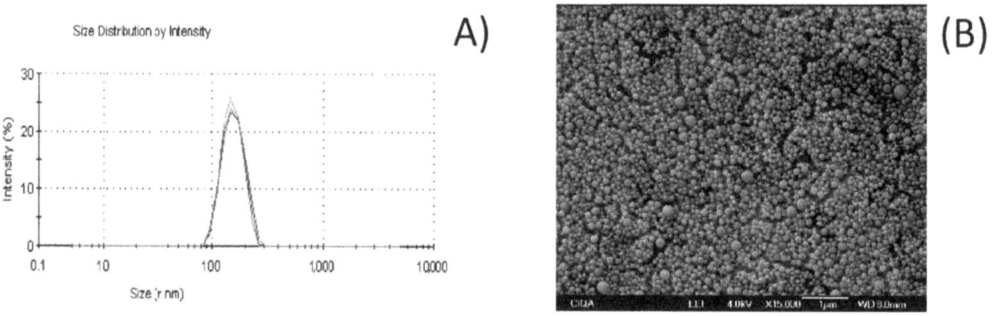

Figure 2. Size evaluation of essential oil *Lippia graveolens* nanocapsules obtained using the nanoprecipitation technique: (**A**) dynamic light scattering analysis and (**B**) scanning electron microscope image.

The SEM image presented in Figure 2B shows the presence of NCs with a spherical form. There are no significant particle aggregates, and the particle size value (in the 200 nm range) is close to that observed in the DLS analysis (287 ± 5.11 nm). This coincides with a previous study in which chitosan and cashew gum NPs loaded with EO from a species of *Lippia* with a spherical shape were obtained [50]. Likewise, in another study, polycaprolactone NCs loaded with EO from a species of *Lippia* were prepared, in which the morphology of the NCs was spherical, as in the current work [51].

Another physicochemical characteristic evaluated was the zeta potential. NCs with a zeta potential greater than +30 mV or less than −30 mV are considered strongly cationic or anionic, respectively, and typically have high degrees of stability [25]. Furthermore, the measurement of the zeta potential ensures that there will be greater separation distances between the NCs in the suspension, which is because the occasional aggregation caused by van der Waals interactions is reduced. These interactions are a consequence of the electrostatic repulsion forces between the NCs. The zeta potential also determines whether a charged active is encapsulated in the center of the NC or on the surface [52]. In this study, the zeta potential of the NCs was −50.90 ± 1.44 mV; this negative value indicates that negative charges are dominant on the surface of the NCs, and this can be attributed to Eudragit L100-55, which is an anionic copolymer. This negative potential would facilitate the interaction of NCs with the membranes of pathogenic bacteria present in fruit and vegetable products, such as strawberries. Subsequently, the direct interaction of the active components of EO-*Lg* with the bacteria would be favored, thus increasing its antimicrobial effectiveness.

On the other hand, as part of the physicochemical characterization of the NCs, the encapsulation percentage (%E) and the encapsulation efficiency percentage (%EE) of the three majority components were also determined: myrcene, p-cymene, and carvacrol. These parameters allow the establishment of the content of the EO components in the NCs in terms of their correct dosage in the fruit and vegetable product. To quantify the three majority components of EO-*Lg* in the NCs, *HS*-SPME validated via the GC-FID method was used. Table 3 shows the %E and %EE of the three components of EO-*Lg* in the NCs purified through evaporation under reduced pressure.

A %E of 0.10 was obtained for myrcene, 0.17 for p-cymene, and 27.91% for carvacrol, and the above represents a total %E of 28.18%. This means that approximately 30% of an NC is formed through these three components of EO-*Lg*, and the remaining 70% corresponds to the NC-forming polymer. It should be noted that Salas-Cedillo [53] used the same polymer as was used in this study to encapsulate the *Schinus molle* EO using the nanoprecipitation technique and obtained a total %E of 7.84% for the three majority components (myrcene, α-phellandrene, and limonene). This value is almost three times smaller than the one obtained in this study. This could be because the encapsulation depends on the physicochemical properties of the components (i.e., polarity and volatility) as well as the % abundance in the EO.

The %EEs obtained from the NCs were 0.24, 0.32, and 63.80% for myrcene, p-cymene, and carvacrol, respectively; thus, a total %EE of 64.36% was obtained. The above indicates that the three components were encapsulated by approximately 65% in comparison with their abundance in the initial EO. This is similar to what was reported by Pinto et al. [51], who obtained an %EE of 60% when they encapsulated EO from a species of *Lippia* in NPs with chitosan and 1% cashew gum. Likewise, this is similar to what was reported in other publications, which mention that the encapsulation efficiency percentages of non-polar active ingredients are around 80% when the nanoprecipitation technique is used [54]. This could be because the polar components have a greater tendency to diffuse from the organic phase to the aqueous phase, which could favor the encapsulation of non-polar compounds.

3.5. Preparation of the Coating

Once the NCs were obtained and characterized, the formation of the multisystem coating was carried out through the conventional casting method. This technique involves obtaining the dispersion of the coating components (NC and AL) and the evaporation of the solvent or water at a controlled temperature and humidity to form the multisystem coating [55]. For the formation of the multisystem coating, a drying time of 24 h and an AL concentration of 1% w/v were used. Macroscopic visual and microscopic evaluation of the multisystem coating were carried out. These physical properties of the multisystem coating are fundamental since they influence the appearance of the fruit and vegetable product once the coating is applied. Figure 3A shows that the multisystem coating obtained was homogeneous and completely formed.

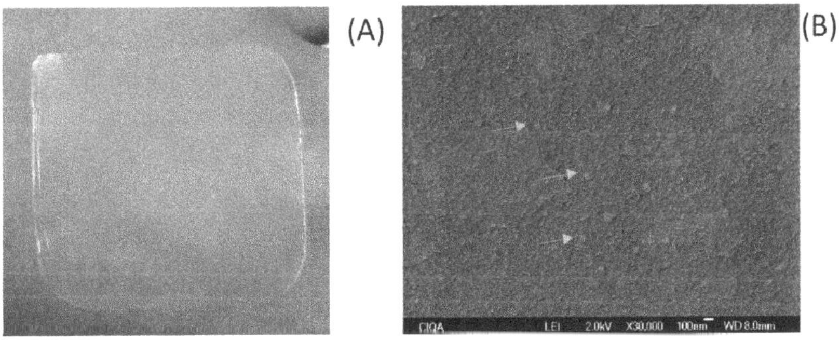

Figure 3. Multisystem coating based on NC-EO-*Lg*-AL formed through the casting method (**A**); NC-EO-*Lg*-AL multisystem coating scanning electron microscope image (**B**) Blue arrows indicate particles of approximately 200 nm homogeneously distributed in the coating.

SEM analysis is commonly used for the description of packaging surfaces, describing their homogeneity, integrity, smoothness, and the presence of cracks, which can influence the mechanical properties of multisystem coatings. Figure 3B shows that the NC-EO-*Lg*-AL exhibited a soft, uniform, and continuous surface, indicating homogeneity and structural integrity. This could be due to the fact that the presence of the phenolic components of EO-*Lg* in the NC of the multisystem coating acts as crosslinkers and, therefore, results in more uniform multisystem coating [56].

The SEM image shows a coating without cracks as well as particles of approximately 200 nm (indicated by blue arrows) homogeneously distributed. It has been reported that the presence of NC-EO favors the formation of homogeneous coatings [57]. However, previous studies also report that high amounts of EO in NCs generate thicker and more heterogeneous coatings [58]. It is worth mentioning that the amount of the EO-*Lg* encapsulated in polymeric NCs in this study favored the homogeneity of the coating.

3.6. Physical Characterization of the Coatings

3.6.1. Optical Evaluation

The optical properties of multisystem coatings are an essential sensory aspect for their acceptance by consumers. They are generally expected to be transparent, similar to polymeric packaging materials, or close in color to the food to which the coating will be applied [59]. In this analysis, higher values indicate less transparency and more opacity. Table 4 shows the results of the optical evaluation. It should be noted that all the multisystem coatings were clear enough to be used as see-through packaging. The AL multisystem coating showed lower opacity (higher transparency) than the NC-BCO-AL multisystem coating. This coincides with a previous study in which it was found that when adding a nanosystem (nanofibers), the opacity increased. This is possibly due to the scattering of light by nanometer-sized particles [60]. However, a decrease in opacity was observed due to the incorporation of the NC-EO-*Lg* compared with the multisystem coating from AL and NP-BCO-AL (nanoparticles without EO-*Lg*). It should be noted that the opacities of the AL and NP-BCO-AL coatings showed significant differences ($p < 0.05$) with respect to the multisystem coating's opacity. These results are similar to those obtained by Bathia et al. [56], who formed gelatin-based (porcine and bovine) edible films loaded with spearmint EO. They observed that gelatin-based films with the EO showed a decrease in opacity compared to control films that did not utilize the EO. However, it is necessary to mention that the mechanism by which EOs in NCs increase transparency is not entirely clear; in fact, there are various instances where EOs decrease transparency [58].

Table 4. Optical, mechanical, and water vapor barrier properties evaluation of the multisystem coating.

Coating	Opacity [1] (UA mm^{-1})	Adhesion [1] (Dynes cm^{-2})	Tensile Strength [1] (g cm^{-2})	Elongation at Break (%) [1]	WVP [1] (10^{-7} g mm cm^{-2} Pa^{-1} h^{-1})	WVTR [1] (10^{-3} g cm^{-2} h^{-1})
AL	0.71 ± 0.10	4690.86 ± 2.00	977.56 ± 3.94	90.76 ± 0.71	2.79 ± 0.04	4.68 ± 0.09
NP-BCO-AL	0.96 ± 0.26	4989.22 ± 1.03	575.54 ± 5.27	78.28 ± 0.76	1.97 ± 0.03	3.72 ± 0.08
NC-EO-*Lg*-AL	0.60 ± 0.10	5802.36 ± 2.15	1555.26 ± 4.31	182.27 ± 2.14	1.01 ± 0.03	2.03 ± 0.06

[1] ($\bar{x} \pm \sigma$, $n = 3$).

3.6.2. Mechanical Evaluation

The evaluation of mechanical parameters such as adhesion, tensile strength, and elongation at break will allow the verification of the characteristics of the related coatings from the preparation stage to their application. A coating must possess the following: (i) a good percentage of elongation in order to avoid breaking during packaging or handling; (ii) adequate adhesion to the surfaces of the fruit and vegetable products onto which it will be applied; (iii) good resistance to breaking to ensure integrity; and (iv) a soft, smooth, and transparent texture to not compromise physical appearance.

The adhesion indicates the magnitude of the force that has to be applied to the multisystem coating to be detached from the surface where it was placed. The force that must be applied to the multisystem coating-based NC-EO-*Lg*-AL for detachment was 5802.36 ± 2.15 dynes cm^{-2} (Table 4). This value showed a significant difference ($p < 0.05$) to those of the NP-BCO-AL and AL coatings. That is, the NC-EO-*Lg* favors the adhesion of the multisystem coating on the applied surface. Likewise, this effect was observed in the study carried out by González-Moreno et al. [24], where the adhesion of multisystem coatings based on NCs loaded with *Thymus vulgaris* EO was evaluated. A value of 4768.27 ± 2.63 dynes cm^{-2} was obtained. The adhesion parameter is indicative of the permanence of the coating on the fruit and vegetable products. The adhesion properties of the coatings are determined by the intrinsic properties of the NC-forming polymer and the environment in which it is placed. Among the intrinsic properties are the molecular weight, the concentration of the polymer, the flexibility of the polymer chains, and the chemical groups that are able to form electrostatic or secondary bonds such as hydrogen bonds and van der Waals forces.

On the other hand, the tensile strength was also influenced by the presence of the NC-EO-*Lg*. This parameter indicates the weight that the coating can withstand before breaking. In Table 4, it can be observed that the multisystem coating formed from NC-EO-*Lg*-AL can resist 1555.26 $\pm$ 4.31 g cm^{-2} before breaking. This value showed a significant difference ($p < 0.05$) to those of NP-BCO-AL and AL coating. That is, EO-*Lg* incorporated into NCs increases the resistance of multisystem coating when a force is applied. The increase in breaking strength by the EOs was also observed in the study developed by Jancy et al. [61], who utilized a multisystem coating-based NP-EO derived from fennel seed. It was found that the presence of the NP-EO increased the breaking strength by up to seven times. This can be explained by the results of other studies, which have shown that an NC-EO is able to increase the tensile strength of multisystem coatings, probably due to cross-linking processes. In this sense, EOs with phenolic compounds, such as the EO-*Lg*, can act as cross-linkers in multisystem coatings [62].

The percentage of elongation (%) indicates the maximum percentage change in the length of the multisystem coating before breaking [63]. Table 4 shows that the NC-EO-*Lg*-AL multisystem coating can be stretched or elongated by 182.27 $\pm$ 2.14% of its initial size. This value showed a significant difference ($p < 0.05$) to those of NP-BCO and AL coatings. This characteristic of increasing stretch is attributed to the presence of NC-EO-*Lg* in the multisystem coating. The effect of the addition of the NC-EO on the mechanical properties of multisystem coatings is quite complex, and the improvement in stretching or elongation characteristics has been reported in the literature. For instance, in the research carried out by Liang et al. [64], multisystem films based on chitosan and NCs loaded with epigallocatechin gallate were developed. The addition of these NCs to chitosan multisystem films increased the percentage of elongation in comparison to the edible films formed only with chitosan. Likewise, similar findings were reported by Noronha et al. [65], who obtained a multisystem coating from methylcellulose with NC-α-tocopherol. Significant changes in the percentage of elongation were reported with the incorporation of NCs into multisystem coatings compared to the control coating. The increase in the elongation percentage values indicated that the incorporation of NCs provides greater flexibility in the coatings. This could indicate that the addition of NCs would also modify the interactions with coating-forming agents such as AL. It has been proposed that a homogeneous dispersion of hydrophobic NC in an AL polymer network increases the spacing between the chains of macromolecules, which reduces ionic and hydrogen bonding between the polymer chains and induces the development of structural discontinuities in the chains of the multisystem coatings, which would increase the percentage of elongation [66].

It should be noted that the mechanical evaluation of the multisystem coating will depend on the type of material used in the formulation of the solution to make the multisystem coating, the conditions of formation, the type of plasticizer, the nature of the solvent, the evaporation of the solvent, and the thickness. It is worth highlighting that depending on use and application, the desired physical and mechanical properties will change in order to meet the objectives for which multisystem coatings were created.

3.6.3. Fourier-Transform Infrared (FTIR) Spectroscopy Analysis

An analysis via infrared spectroscopy was carried out in order to establish the interactions between the components of the coatings. Each of the components was analyzed separately: (A) the Eudragit L100-55 polymer, (B) the NP-BCO, (C) the NC-EO-*Lg*, (D) the NC-EO-*Lg*-AL, and (E) the free EO-*Lg*. The spectra are shown below in Figure 4.

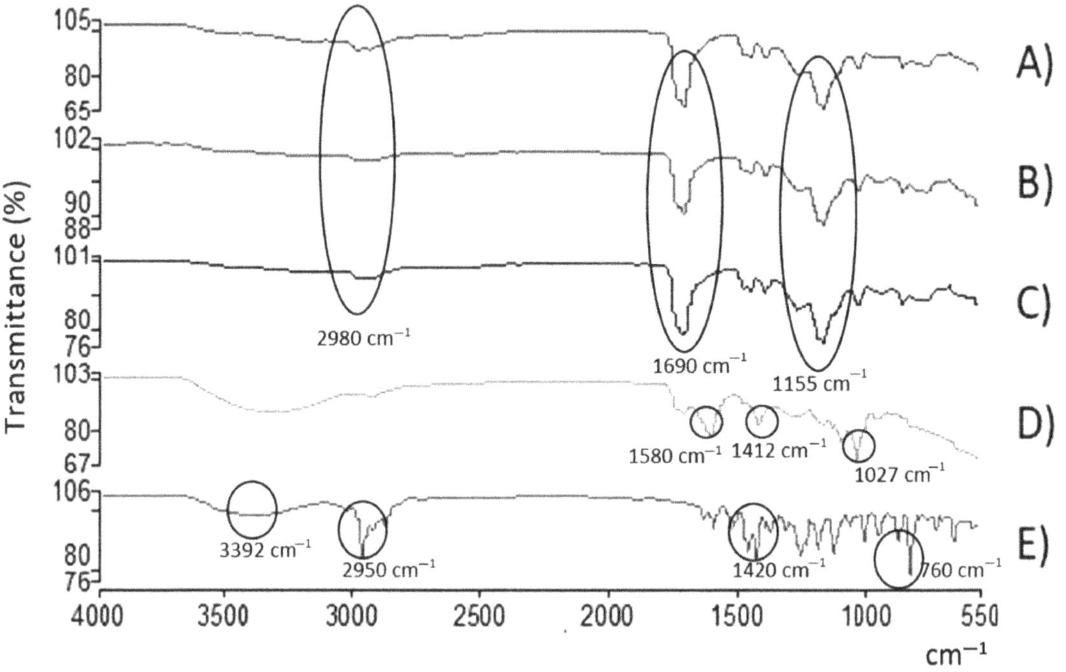

Figure 4. FT-IR spectrum of (**A**) polymer, (**B**) NP-BCO, (**C**) NC-EO-*Lg*, (**D**) NC-EO-*Lg*-AL, and (**E**) EO-*Lg*. The characteristic bands of the functional groups are indicated.

FT-IR spectroscopy is used to identify functional groups and chemical interactions between formulation components. The middle region of the infrared spectrum (400–4000 cm^{-1}) is widely used to study fundamental vibrations and the rotational–vibrational structure associated with different chemical bonds. Figure 4A shows the spectrum of the Eudragit L100-55 polymer, in which four important signals of the main chemical groups present in the polymer unit of this copolymer can be identified. In the region between 2500 and 3500 cm^{-1}, the stretching of the hydroxyl group (O–H) belonging to the carboxylic acid present in the side chain of the polymer unit can be observed, and in the region of 2980 cm^{-1}, the C–H signals can be observed. In the region of 1690 cm^{-1}, stretching, which corresponds to the carbonyl groups (C=O) present in the side chains of the carboxylic acid and the ester, can be observed. Finally, in the region of 1155 cm^{-1}, the stretching corresponding to the C–O–C group present in the side chain of the ester can be observed. The aforementioned signals coincide with what was reported in the technical sheet of the Eudragit L100-55 polymer. Likewise, in Figure 4B,C, the IR spectra obtained from the NP-BCO and the NC-EO are observed, respectively. In both spectra, it can be seen that the signals obtained correspond mainly to the signals obtained in the polymer spectrum (Figure 4A). Therefore, no signals are attributed to the spectrum of the EO-*Lg*. Regarding the NC-EO-*Lg*, it can be inferred that the oil components are encapsulated in the framework of the NC and are not in a free or a superficial form. It is important to highlight that the signals with greater intensity correspond to the majority components of the EO-*Lg*. On the other hand, Figure 4D shows the spectrum obtained for the NC-EO-*Lg*-AL. It can be seen that the signals obtained correspond mainly to the AL, which could indicate that at the time of adding AL to the NC suspension, the chains of this polysaccharide were organized on the polymeric wall of the NC. And, since no signals attributed to the spectrum of the EO-*Lg* were observed, it is inferred that, as in the NC-EO-*Lg*, the oil components are encapsulated in the NC and are not in free form. These results showed that EO-*Lg* was encapsulated in the NC without changing its structure or function, which coincides with what was reported by Ma et al. [67]. Finally, in Figure 4E, the IR spectrum obtained from the EO-*Lg* is shown, where a band of 3392 cm^{-1} can be seen that belongs to the O–H bond vibrations characteristics of the phenolic components present in EO, such as carvacrol and thymol. Another observable band is that of 2950 cm^{-1}, which appears due to the presence of the CH_3 or CH_2 bonds of the methyl and methylene groups that are frequently found in the structures of organic compounds. Within the region of 1101–1209 cm^{-1}, characteristic bands of the C–OH bonds of primary and secondary alcohols are present in molecules such as terpinen-4-ol and 1,8-cineole, which agrees with what was reported by Matadamas-Ortiz et al. [14]. In the region of 850–1080 cm^{-1}, bands belonging to the C–O groups are observable.

3.6.4. Water Vapor Barrier Properties

An important property of coatings is their ability to prevent moisture exchange between the medium and the interior of the food. WVTR is the measure of the moisture that moves through a unit space of material per unit time [68]. On the other hand, WVP is a vital property used to evaluate the ability of films to act as a moisture barrier. The results obtained for WVTR and WVP are presented in Table 4. The NC-EO-*Lg*-AL multisystem coating performed better as a moisture barrier compared to the AL and NP-BCO-AL coatings because it presented lower WVTR and WVP values with significant differences ($p < 0.05$). This decrease in the WVTR and WVP values could be due to the fact that the presence of nanoencapsulated EO-*Lg* modifies the hydrophilic/hydrophobic ratio of the coating, thus affecting water transmission through the hydrophilic part of this coating. Lower WVTR and WVP values are preferred since this reduces the moisture loss of wrapped food products [69]. This agrees with the results reported by Al-Harrasi et al. [70], who reported that multisystem gelatin–sodium alginate films loaded with ginger EO decrease WVP values.

Therefore, the EO-*Lg* incorporated into the NCs acted as a plasticizing agent in the coating, favoring the accommodation of the NCs in the epicarp and strawberry seeds, thus reducing water loss. Thus, the hydrophobic bioactive substances that increase the hydrophobicity of the multisystem coatings reduce water vapor migration.

3.7. Application of Multisystem Coatings in Strawberries

3.7.1. Weight Loss Percentage and Texture Analyses: Penetration Test

To determine the effectiveness of the multisystem antimicrobial coating in improving the quality of fresh strawberries, the percentage of weight loss was utilized as a quality marker during storage time and compared with strawberries without a coating (control). The results show that the percentage of weight loss increased during storage time in all strawberries (Figure 5). The weight loss in the untreated strawberries (control) was much greater than in the strawberries coated with a multisystem coating.

Overall, based on Figure 5, the percentage of weight loss in the control strawberries on the 10th observation day was 22.35%, while in the strawberries coated with NC-EO-*Lg*-AL, the weight loss was 7.12%. The weight loss percentage value showed significant differences ($p < 0.05$) in strawberries with the multisystem coating (NC-EO-*Lg*-AL) in comparison to strawberries with the NP-BCO-AL and AL coatings. These results coincide with those reported by Martínez et al. [71], who developed and applied multisystem chitosan-based coatings with *Thymus capitatus* EO to strawberries. In this study, less weight loss was observed in strawberries that had the chitosan coating and EO in comparison to strawberries that did not have a coating and those that had the chitosan-based coating without the EO. The results of this study support the advantages of applying multisystem coating to strawberries since it has been shown to prevent water loss from freshly cut strawberries.

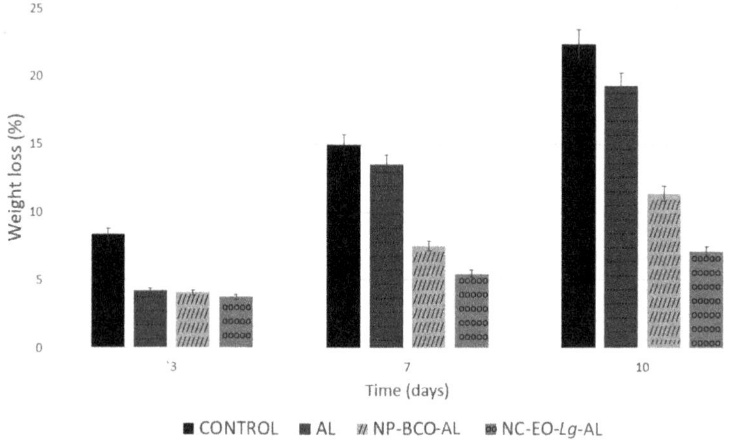

Figure 5. Weight loss percentages of strawberry fruits coated with alginate, NP-BCO-AL, or NC-EO-*Lg*-AL and the control (without multisystem coating) during storage at 4 °C (mean ± SD, $n = 7$).

Weight loss is commonly attributed entirely to water loss [72]. Strawberries are highly perishable, so they are easily susceptible to water loss, which results in the weakening of the fruit tissue due to their very thin epicarp. This negatively affects the appearance of the fruit, causing softening, the development of pathogens, and wrinkling, as well as a change in color and aroma, which accelerates senescence and subsequently causes serious economic losses. Multisystem coatings with NC-EO-*Lg*-AL provide a barrier function, protecting the fruit from the outside, in addition to also limiting transpiration, which delays dehydration [73]. On the other hand, the fruit firmness of the coated strawberries during the storage time was analyzed because this represents one of the essential parameters for determining fruit quality. Strawberries have a very active metabolism and are subject to mechanical damage, attacks by microorganisms, and loss of quality during storage. The decrease in firmness is due to the greater loss of water vapor from the surface of the fruit to the outside atmosphere, which can encourage the growth of fungi and bacteria. These microorganisms cause structural damage to tissues and allow them to soften. Furthermore, the decrease in firmness is related to the increase in moisture loss. It should be noted that the decrease in firmness could be related to the degradation of the cell wall, specifically the cortical parenchyma, as a consequence of the enzymatic degradation processes and the loss of moisture during the storage period.

Overall, based on Figure 6, at time 0, the firmness of the fruit was 3.11 N, while after ten storage days, decreased firmness was observed, both in the uncoated and coated samples, but the lowest value was found in the control strawberries (1.92 N). In percentage terms, after 10 days of cold storage, the control group lost 43.64% of firmness, while the AL and NP-BCO-AL coating treatment only lost 37.16 and 32.8%, respectively. On the other hand, the highest firmness values were recorded in strawberries with an NC-EO-*Lg*-AL multisystem coating (2.78 N), which lost only 17.02% firmness. It should be noted that the firmness value of the multisystem coating showed significant differences ($p < 0.05$) in comparison to the AL and NP-BCO-AL coatings. The results of this study coincide with those reported by De-Bruno et al. [74], who applied edible coating enriched with bergamot EO to strawberries and obtained a lower loss of firmness compared to the control group. Furthermore, it can be inferred that in this study, the NC-EO-*Lg*-AL multisystem coating had the ability to act as a barrier that interferes with gas exchange, which leads to a reduction in the respiration rate of the fruit and prevents water loss, as observed in Table 4; furthermore, it is possible that the antioxidant activity of the EO-*Lg* components decreased the enzymatic activity of the fruit, resulting in the slower ripening of the strawberry.

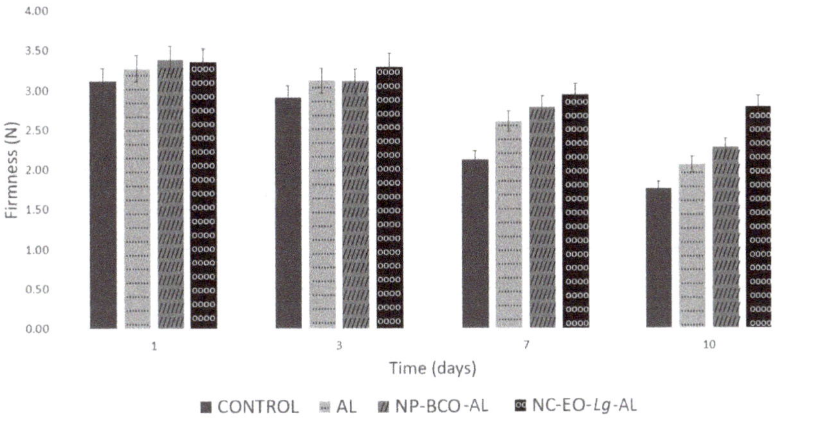

Figure 6. Firmness of strawberries coated with alginate, NP-BCO-AL, or NC-EO-*Lg*-AL and the control (without multisystem coating) after storage for 10 days at 4 °C (mean ± SD, $n = 7$).

3.7.2. Visual Microbiological Damage

Regarding visual microbiological damage, fungal growth on the surface of the strawberry began to be observed rapidly in the control samples after 7 days of storage at 4 °C. After 15 days, they already showed considerable microbiological damage (Figure 7A). Likewise, the strawberries coated with the AL (Figure 7B) and NP-BCO-AL coating (Figure 7C) also showed microbiological damage after storage time. On the other hand, the strawberries coated with NC-EO-*Lg*-AL multisystem coating did not present microbiological damage (Figure 7D).

Figure 7. Visual microbiological damage of strawberries after 15 days of storage at 4 °C: uncoated (control) (**A**), AL coating (**B**), NP-BCO coating (**C**), and NC-EO-*Lg* multisystem coating (**D**).

It should be noted that strawberries with the NC-EO-*Lg*-AL multisystem coating were the most efficient in reducing microbial decomposition and could be considered an effective treatment to reduce the deterioration of fresh strawberries during their storage period. The growth of pathogens such as fungi and bacteria on the surface of strawberries in refrigerated conditions is a serious problem that reduces the number of purchases by consumers, so the multisystem coatings used in this study are a good alternative to increase shelf life in refrigerated conditions. Different studies have been carried out using EO coatings in order to control the decomposition of fungi and bacteria and extend the shelf life of fruits and vegetables [75].

4. Conclusions

Smart biodegradable and multisystem coatings may be good alternative candidates for the postharvest preservation of fruits, especially strawberries, which suffer a rapid decrease in quality due to their high metabolism and rapid development of pathogens on their surface. Our observations in this study demonstrated that multisystem coating fruit with NC-EO-*Lg* decreased the percentage of weight loss and loss of firmness of strawberries, which is why it is advisable to use this to maintain the quality of strawberries during storage life and marketing periods. Thus, it is considered to be a safe and effective method as well as an environmentally friendly technology, and it also has the advantage of being a method that can be reproduced on an industrial scale.

Author Contributions: Methodology, B.J.G.-M. and G.G.-G.; validation, R.Á.-R.; formal analysis, B.J.G.-M., S.A.G.-R., L.A.P.-L. and R.Á.-R.; investigation, B.J.G.-M., V.M.R.-G., L.A.P.-L. and G.G.-G.; writing—original draft preparation, B.J.G.-M.; writing—review and editing, S.A.G.-R., V.M.R.-G., G.G.-G. and R.Á.-R.; visualization, L.A.P.-L.; supervision, S.A.G.-R. and V.M.R.-G.; project administration, R.Á.-R. All authors have read and agreed to the published version of the manuscript.

Funding: This research was funded by UANL/ProACTI, grant number 40-BQ-2023.

Institutional Review Board Statement: Not applicable.

Data Availability Statement: Data are contained within the article.

Acknowledgments: B.J.G.-M. is thankful to the CONACYT México Program of Scholarships for Postgraduate Study 725423.

Conflicts of Interest: The authors declare no conflicts of interest.

References

1. FAO. El Estado de la Alimentación y la Agricultura. In *Avanzar en la Reducción de la Pérdida y el Desperdicio de Alimentos*; FAO: Rome, Italy, 2019.
2. Curiel, J.A. Application of biotechnological techniques aimed to obtain bioactive compounds from food industry by-products. *Biomolecules* **2021**, *12*, 88. [CrossRef]
3. Pace, B.; Cefola, M. Innovative preservation technology for the fresh fruit and vegetables. *Foods* **2021**, *29*, 719. [CrossRef] [PubMed]
4. Torres-Sánchez, R.; Martínez-Zafra, M.T.; Castillejo, N.; Guillamón-Frutos, A.; Artés-Hernández, F. Real-time monitoring system for shelf life estimation of fruit and vegetables. *Sensors* **2020**, *20*, 1860. [CrossRef]
5. Mendoza-Barboza, C.R.; Borges, C.D.; Kringel, A.L.; da-Silveira, R.P.; da-Silva, F.A.; Schulz, G.A.S. Application of microemulsions as coating in fresh cut strawberries. *J. Food Sci. Technol.* **2020**, *57*, 2764–2770.
6. Trinetta, V.; McDaniel, A.; Batziakas, K.; Yucel, U.; Nwadike, L.; Pliakoni, E. Antifungal packaging film to maintain quality and control postharvest diseases in strawberries. *Antibiotics* **2020**, *9*, 618. [CrossRef] [PubMed]
7. Salsabiela, S.; Sukma-Sekarina, A.; Bagus, H.; Audiensi, A.; Azizah, F.; Heristika, W.; Susanto, E.; Munawaroh, H.S.H.; Show, P.L.; Ningrum, A. Development of edible coating from gelatin composites with the addition of black tea extract (*Camellia sinensis*) on minimally processed watermelon (*Citrullus lanatus*). *Polymers* **2022**, *28*, 2628. [CrossRef] [PubMed]
8. Paolucci, M.; Di-Stasio, M.; Sorrentino, A.; La-Cara, F.; Volpe, M.G. Active edible polysaccharide-based coating for preservation of fresh figs (*Ficus carica* L.). *Foods* **2020**, *3*, 1793. [CrossRef]
9. Nor, S.; Ding, P. Trends and advances in edible biopolymer coating for tropical fruit: A review. *Food Res. Int.* **2020**, *134*, 109208. [CrossRef]

10. Salas-Méndez, E.; Pinheiro, A.C.; Ballesteros, L.F.; Silva, P.; Rodríguez-García, R.; Jasso-de-Rodríguez, D. Application of edible nanolaminate coatings with antimicrobial extract of *Flourensia cernua* to extend the shelf-life of tomato (*Solanum lycopersicum* L.) fruit. *Postharvest Biol. Technol.* **2019**, *150*, 19–27. [CrossRef]
11. Yin, C.; Huang, C.; Wang, J.; Liu, Y.; Lu, P.; Huang, L. Effect of chitosan and alginate-based coatings enriched with cinnamon essential oil microcapsules to improve the postharvest quality of mangoes. *Materials* **2019**, *12*, 2039. [CrossRef]
12. Tavares, L.S.; de-Souza, V.C.; Schmitz-Nunes, V.; Nascimento-Silva, O.; de-Souza, G.T.; Farinazzo-Marques, L.; Capriles-Goliatt, P.V.Z.; Facio-Viccini, L.; Franco, O.L.; de-Oliveira-Santos, M. Antimicrobial peptide selection from *Lippia* spp leaf transcriptomes. *Peptides* **2020**, *129*, 170317. [CrossRef] [PubMed]
13. Gutiérrez-Grijalva, E.P.; Antunes-Ricardo, M.; Acosta-Estrada, B.A.; Gutiérrez-Uribe, J.A.; Heredia, J.B. Cellular antioxidant activity and in vitro inhibition of α-glucosidase, α- amylase and pancreatic lipase of oregano polyphenols under simulated gastrointestinal digestion. *Int. Food Res. J.* **2019**, *116*, 676–686. [CrossRef] [PubMed]
14. Matadamas-Ortiz, A.; Hernández-Hernández, E.; Castaño-Tostado, E.; Amaro-Reyes, A.; García-Almendárez, B.E.; Velazquez, G.; Regalado-González, C. Long-term refrigerated storage of beef using an active edible film reinforced with mesoporous silica nanoparticles containing oregano essential oil (*Lippia graveolens* Kunth). *Int. J. Mol. Sci.* **2022**, *24*, 92. [CrossRef]
15. Rodríguez-García, I.; Cruz-Valenzuela, M.R.; Silva-Espinoza, B.A.; Gonzalez-Aguilar, G.A.; Moctezuma, E.; Gutierrez-Pacheco, M.M.; Tapia-Rodriguez, M.R.; Ortega-Ramirez, L.A.; Ayala-Zavala, J.F. Oregano (*Lippia graveolens*) essential oil added within pectin edible coatings prevents fungal decay and increases the antioxidant capacity of treated tomatoes. *J. Sci. Food Agric.* **2016**, *11*, 3772–3778. [CrossRef] [PubMed]
16. Zhang, J.; Ma, S.; Du, S.; Chen, S.; Sun, H. Antifungal activity of thymol and carvacrol against postharvest pathogens *Botrytis cinerea*. *J. Food Sci. Technol.* **2019**, *56*, 2611–2620. [CrossRef] [PubMed]
17. Leyva-López, N.; Gutiérrez-Grijalva, E.P.; Vazquez-Olivo, G.; Heredia, J.B. Essential oils of oregano: Biological activity beyond their antimicrobial properties. *Molecules* **2017**, *14*, 989. [CrossRef]
18. Herrera-Rodríguez, S.E.; López-Rivera, R.J.; García-Márquez, E.; Estarrón-Espinosa, M.; Espinosa-Andrews, H. Mexican oregano (*Lippia graveolens*) essential oil-in-water emulsions: Impact of emulsifier type on the antifungal activity of *Candida albicans*. *Food Sci. Biotechnol.* **2019**, *28*, 441–448. [CrossRef]
19. Bhargava, K.; Conti, D.S.; da-Rocha, S.R.P.; Zhang, Y. Application of an oregano oil nanoemulsion to the control of foodborne bacteria on fresh lettuce. *Food Microbiol.* **2015**, *47*, 69–73. [CrossRef]
20. Asbahani, A.E.; Miladi, K.; Badri, W.; Sala, M.; Addi, E.H.A.; Casabianca, H.; Elaissari, A. Essential oils: From extraction to encapsulation. *Int. J. Pharm.* **2015**, *483*, 220–243. [CrossRef]
21. Rai, M.; Paralikar, P.; Jogee, P.; Agarkar, G.; Ingle, A.P.; Derita, M.; Zacchino, S. Synergistic antimicrobial potential of essential oils in combination with nanoparticles: Emerging trends and future perspectives. *Int. J. Pharm.* **2017**, *15*, 67–78. [CrossRef]
22. Odetayo, T.; Tesfay, S.; Ngobese, N.Z. Nanotechnology-enhanced edible coating application on climacteric fruits. *Food Sci. Nutr.* **2022**, *20*, 2149–2167. [CrossRef]
23. Zhang, M.; Luo, W.; Yang, K.; Li, C. Effects of sodium alginate edible coating with cinnamon essential oil nanocapsules and nisin on quality and shelf life of beef slices during refrigeration. *J. Food Prot.* **2022**, *1*, 896–905. [CrossRef] [PubMed]
24. González-Moreno, B.J. Cubierta Biopolimérica de Nanoingredientes a Base de *Thymus vulgaris* con Potencial Aplicación en Productos Hortofrutícolas. Master's Thesis, Universidad Autónoma de Nuevo León, Monterrey Nuevo León, Mexico, 2020.
25. Piña Barrera, A.M.; Álvarez-Román, R.; Báez-González, J.G.; Amaya-Guerra, C.A.; Rivas-Morales, C.; Gallardo-Rivera, C.T.; Galindo-Rodríguez, S.A. Application of a multisystem coating based on polymeric nanocapsules containing essential oil of *Thymus vulgaris* L. to increase the shelf life of table grapes (*Vitis vinifera* L.). *IEEE Trans. Nanotechnol.* **2019**, *18*, 549–557. [CrossRef] [PubMed]
26. Samba, N.; Aitfella-Lahlou, R.; Nelo, M.; Silva, L.; Coca, R.; Rocha, P.; López-Rodilla, J.M. Chemical composition and antibacterial activity of *Lippia multiflora* moldenke essential oil from different regions of Angola. *Molecules* **2020**, *26*, 155. [CrossRef] [PubMed]
27. Permanent Commission of Pharmacopoeia of the United Mexican States. *Farmacopea de los Estados Unidos Mexicanos*, 10th ed.; Secretaría de Salud: Mexico City, Mexico, 2011.
28. Fessi, H.; Puisieux, F.; Devissaguet, J.P.; Ammoury, N.; Benita, S. Nanocapsule formation by interfacial polymer deposition following solvent displacement. *Int. J. Pharm.* **1989**, *55*, R1–R4. [CrossRef]
29. Ye, J.; Ma, D.; Qin, W.; Liu, Y. Physical and antibacterial properties of sodium alginate-sodium carboxymethylcellulose films containing *Lactococcus lactis*. *Molecules* **2018**, *23*, 2645. [CrossRef] [PubMed]
30. Bhatia, S.; Al-Harrasi, A.; Al-Azri, M.S.; Ullah, S.; Bekhit, A.E.A.; Pratap-Singh, A.; Chatli, M.K.; Anwer, M.K.; Aldawsari, M.F. Preparation and physiochemical characterization of bitter orange oil loaded sodium alginate and casein based edible films. *Polymers* **2022**, *15*, 3855. [CrossRef]
31. *ASTM E-96-95*; Standard Test Method for Water Vapor Transmission of Materials. ASTM International: West, Conshohocken, PA, USA, 1995.
32. Siracusa, V.; Romani, S.; Gigli, M.; Mannozzi, C.; Cecchini, J.P.; Tylewicz, U.; Lotti, N. Characterization of active edible films based on citral essential oil, alginate and pectin. *Materials* **2018**, *11*, 1980. [CrossRef]
33. Dilworth, L.; Riley, C.K.; Stennett, D. Plant Constituents: Carbohydrates, oils, resins, balsams, and plant hormones. In *Pharmacognosy Fundamentals, Applications and Strategie*; Academic Press: Cambridge, MA, USA, 2017; pp. 61–80.

34. Hernández-Hernández, E.; Regalado-González, C.; Vázquez-Landaverde, P.; Guerrero-Legarreta, I.; García-Almendárez, B.E. Microencapsulation, chemical characterization, and antimicrobial activity of Mexican (*Lippia graveolens* H.B.K.) and European (*Origanum vulgare* L.) oregano essential oils. *Sci. World J.* **2014**, *64*, 1814.
35. Lugo-Estrada, L.; Galindo-Rodríguez, S.A.; Pérez-López, L.A.; Waksman-deTorres, N.; Álvarez-Román, R. Headspace–solid-phase microextraction gas chromatography method to quantify *Thymus vulgaris* essential oil in polymeric nanoparticles. *Pharmacogn. Mag.* **2019**, *15*, 473–478.
36. Torrenegra-Alarcón, M.E.; Matiz-Melo, G.E.; González, J.G.; León-Méndez, G. In vitro antibacterial activity of the essential oil against microorganisms involved in acné. *Rev. Cuba. Farm* **2015**, *49*, 512–523.
37. Domínguez, X. *Métodos de Investigación Fitoquímica*; Limusa S.A.: Ciudad de México, Mexico, 1985.
38. Adams, R.P. *Identification of Essential Oil Components by Gas Chromatography/Mass Spectrometry*; Allured Publishing Corporation: Carol Stream, IL, USA, 2007; pp. 1–804.
39. Chacón-Vargas, K.F.; Sánchez-Torres, L.E.; Chávez-González, M.L.; Adame-Gallegos, J.R.; Nevárez-Moorillón, G.V. Mexican oregano (*Lippia berlandieri* Schauer and *Poliomintha longiflora* Gray) essential oils induce cell death by apoptosis in *Leishmania* (leishmania) *mexicana* promastigotes. *Molecules* **2022**, *15*, 5183. [CrossRef]
40. Medina-Romero, Y.M.; Rodriguez-Canales, M.; Rodriguez-Monroy, M.A.; Hernandez-Hernandez, A.B.; Delgado-Buenrostro, N.L.; Chirino, Y.I.; Cruz-Sanchez, T.; Garcia-Tova, C.G.; Canales-Martinez, M.M. Effect of the essential oils of *Bursera morelensis* and *Lippia graveolens* and five pure compounds on the mycelium, spore production, and germination of species of fusarium. *J. Fungi* **2022**, *9*, 617. [CrossRef] [PubMed]
41. Capatina, L.; Napoli, E.M.; Ruberto, G.; Hritcu, L. *Origanum vulgare* ssp. hirtum (Lamiaceae) essential oil prevents behavioral and oxidative stress changes in the scopolamine zebrafish model. *Molecules* **2021**, *23*, 7085. [CrossRef]
42. Sharififard, M.; Alizadeh, I.; Jahanifard, E.; Wang, C.; Azemi, M.E. Chemical composition and repellency of *Origanum vulgare* essential oil against cimex lectularius under laboratory conditions. *J. Arthropod. Borne Dis.* **2018**, *25*, 387–397. [CrossRef]
43. Ribeiro, F.P.; Santana-de-Oliveira, M.; de-Oliveira-Feitosa, A.; Santana-Barbosa- Marinho, P.; Moacir-do-Rosario-Marinho, A.; de-Aguiar-Andrade, E.H.; Favacho-Ribeiro, A. Chemical composition and antibacterial activity of the *Lippia origanoides* Kunth essential oil from the Carajás National Forest, Brazil. *Evid. Based Complement. Alternat. Med.* **2021**, *19*, 9930336. [CrossRef]
44. Amiri, H. Essential oils composition and antioxidant properties of three thymus species. *Evid.-Based Complement. Altern. Med.* **2012**, *2012*, 728065. [CrossRef] [PubMed]
45. Miladi, K.; Sfar, S.; Fessi, H.; Elaissari, A. Encapsulation of alendronate sodium by nanoprecipitation and double emulsion: From preparation to in vitro studies. *Ind. Crops. Prod.* **2015**, *72*, 24–33. [CrossRef]
46. Sotelo-Boyás, M.E.; Correa-Pacheco, Z.N.; Bautista-Baños, S.; Corona-Rangel, M.L. Physicochemical characterization of chitosan nanoparticles and nanocapsules incorporated with lime essential oil and their antibacterial activity against food-borne pathogens. *Food Sci. Technol.* **2017**, *77*, 15–20. [CrossRef]
47. Antonioli, G.; Fontanella, G.; Echeverrigaray, S.; Longaray-Delamare, A.P.; Fernandes-Pauletti, G.; Barcellos, T. Poly(lactic acid) nanocapsules containing lemongrass essential oil for postharvest decay control: In vitro and in vivo evaluation against phytopathogenic fungi. *Food Chem.* **2020**, *1*, 326. [CrossRef]
48. Chávez-Magdaleno, M.E.; González-Estrada, R.R.; Ramos-Guerrero, A.; Plascencia-Jatomea, M.; Gutiérrez-Martínez, P. Effect of pepper tree (*Schinus molle*) essential oil-loaded chitosan bio-nanocomposites on postharvest control of *Colletotrichum gloeosporioides* and quality evaluations in avocado (*Persea americana*) cv. Hass. *Food Sci. Biotechnol.* **2018**, *12*, 1871–1875. [CrossRef] [PubMed]
49. Galindo-Rodríguez, S.A.; Allemann, E.; Fessi, H.; Doelker, E. Physicochemical parameters associated with nanoparticle formation in the salting-out, emulsification-difussion, and nanoprecipitation methods. *Pharm. Res.* **2004**, *21*, 1428–1439. [CrossRef] [PubMed]
50. Abreu, F.O.; Oliveira, E.F.; Paula, H.C.; de-Paula, R.C. Chitosan/cashew gum nanogels for essential oil encapsulation. *Carbohydr. Polym.* **2012**, *1*, 1277–1282. [CrossRef] [PubMed]
51. Pinto, N.; Rodrigues, T.H.S.; Pereira, R.; Silva, L.M.A.; Cáceres, C.A.; Azeredo, H.M.; de-Canuto, K.M. Production and physicochemical characterization of nanocapsules of the essential oil from *Lippia sidoides* Cham. *Ind Crops Prod.* **2016**, *86*, 279–288. [CrossRef]
52. Fraj, A.; Jaâfar, F.; Marti, M.; Coderch, L.; Ladhari, N. A comparative study of oregano (*Origanum vulgare* L.) essential oil-based polycaprolactone nanocapsules/microspheres: Preparation, physicochemical characterization, and storage stability. *Ind. Crops Prod.* **2019**, *140*, 111669. [CrossRef]
53. Salas-Cedillo, H.I. Desarrollo de un Potencial Insecticida Nanoparticulado de *Schinus molle* Para el Control de *Aedes aegypti*. Master's Thesis, Universidad Autónoma de Nuevo León, Monterrey Nuevo León, Mexico, 2016.
54. Cartaxo, A.L.; Costa-Pinto, A.R.; Martins, A.; Faria, S.; Gonçalves, V.M.F.; Tiritan, M.E.; Ferreira, H.; Neves, N.M. Influence of PDLA nanoparticles size on drug release and interaction with cells. *J. Biomed. Mater. Res. A.* **2019**, *107*, 482–493. [CrossRef] [PubMed]
55. Solano-Doblado, L.G.; Alamilla-Beltrán, L.; Jiménez-Martínez, C. Películas y recubrimientos comestibles funcionalizados. *Rev. Espec. Cienc. Químico-Biológicas* **2018**, *21*, 30–42. [CrossRef]
56. Bhatia, S.; Al-Harrasi, A.; Shah, Y.A.; Jawad, M.; Al-Azri, M.S.; Ullah, S.; Anwer, M.K.; Aldawsari, M.F.; Koca, E.; Aydemir, L.Y. Physicochemical characterization and antioxidant properties of chitosan and sodium alginate based films incorporated with ficus extract. *Polymers* **2023**, *28*, 1215. [CrossRef]

57. Yadav, R.K.; Nandy, B.C.; Maity, S.; Sarkar, S.; Saha, S. Phytochemistry, pharmacology, toxicology, and clinical trial of *Ficus racemosa*. *Pharmacogn. Rev.* **2015**, *9*, 73–80.
58. Ma, Y.; Chen, S.; Liu, P.; He, Y.; Chen, F.; Cai, Y.; Yang, X. Gelatin improves the performance of oregano essential oil nanoparticle composite films-application to the preservation of mullet. *Foods* **2023**, *29*, 2542. [CrossRef]
59. Galus, S.; Kadzińska, J. Moisture sensitivity, optical, mechanical and structural properties of whey protein-based edible films incorporated with rapeseed oil. *Food Technol. Biotechnol.* **2016**, *54*, 78–89. [CrossRef] [PubMed]
60. Medina-Jaramillo, C.; Quintero-Pimiento, C.; Gómez-Hoyos, C.; Zuluaga-Gallego, R.; López-Córdoba, A. Alginate-edible coatings for application on wild andean blueberries (*Vaccinium meridionale* Swartz): Effect of the addition of nanofibrils isolated from cocoa by-products. *Polymers* **2020**, *5*, 824. [CrossRef]
61. Jancy, S.; Shruthy, R.; Preetha, R. Fabrication of packaging film reinforced with cellulose nanoparticles synthesised from jack fruit non-edible part using response surface methodology. *Int. J. Biol. Macromol.* **2020**, *142*, 63–72. [CrossRef] [PubMed]
62. Xue, Y.; Lofland, S.; Hu, X. Thermal conductivity of protein-based materials: A review. *Polymers* **2019**, *11*, 456. [CrossRef] [PubMed]
63. Rangel-Marrón, M.; Mani-López, E.; Palou, E.; López-Malo, A. Effects of alginate-glycerol-citric acid concentrations on selected physical, mechanical, and barrier properties of papaya puree-based edible films and coatings, as evaluated by response surface methodology. *LWT* **2018**, *101*, 83–91. [CrossRef]
64. Liang, J.; Yan, H.; Wang, X.; Zhou, Y.; Gao, X.; Puligundla, P.; Wan, X. Encapsulation of epigallocatechin gallate in zein/chitosan nanoparticles for controlled applications in food systems. *Food Chem.* **2017**, *231*, 19–24. [CrossRef] [PubMed]
65. Noronha, C.M.; de-Carvalho, S.M.; Lino, R.C.; Barreto, P.L.M. Characterization of antioxidant methylcellulose film incorporated with α-tocopherol nanocapsules. *Food Chem.* **2014**, *159*, 529–535. [CrossRef] [PubMed]
66. Sánchez-González, L.; Vargas, M.; González-Martínez, C.; Chiralt, A.; Cháfer, M. Characterization of edible films based on hydroxypropylmethylcellulose and tea tree essential oil. *Food Hydrocoll.* **2009**, *23*, 2102–2109. [CrossRef]
67. Ma, Y.; Liu, P.; Ye, K.; He, Y.; Chen, S.; Yuan, A.; Chen, F.; Yang, W. Preparation, characterization, in vitro release, and antibacterial activity of oregano essential oil chitosan nanoparticles. *Foods* **2022**, *22*, 3756. [CrossRef]
68. González-Sandoval, D.C.; Luna-Sosa, B.; Martínez-Ávila, G.C.G.; Rodríguez-Fuentes, H.; Avendaño-Abarca, V.H.; Rojas, R. Formulation and characterization of edible films based on organic mucilage from mexican opuntia ficus-indica. *Coatings* **2019**, *9*, 506. [CrossRef]
69. Li, Y.; Yao, M.; Liang, C.; Zhao, H.; Liu, Y.; Zong, Y. Hemicellulose and nano/microfibrils improving the pliability and hydrophobic properties of cellulose film by interstitial filling and forming micro/nanostructure. *Polymers* **2022**, *23*, 1297. [CrossRef] [PubMed]
70. Al-Harrasi, A.; Bhatia, S.; Al-Azri, M.S.; Ullah, S.; Najmi, A.; Albratty, M.; Meraya, A.M.; Mohan, S.; Aldawsari, M.F. Effect of drying temperature on physical, chemical, and antioxidant properties of ginger oil loaded gelatin-sodium alginate edible films. *Membranes* **2022**, *6*, 862. [CrossRef] [PubMed]
71. Martínez, K.; Ortiz, M.; Albis, A.; Gilma-Gutiérrez-Castañeda, C.; Valencia, M.E.; Grande-Tovar, C.D. The Effect of edible chitosan coatings incorporated with *Thymus capitatus* essential oil on the shelf-life of strawberry (*Fragaria ananassa*) during cold storage. *Biomolecules* **2018**, *21*, 155. [CrossRef] [PubMed]
72. Heristika, W.; Ningrum, A.; Supriyadi-Munawaroh, H.S.H.; Show, P.L. Development of composite edible coating from gelatin-pectin incorporated garlic essential oil on physicochemical characteristics of red chili (*Capsicum annnum* L.). *Gels* **2023**, *6*, 49. [CrossRef] [PubMed]
73. Shlekh, K.A.; Ngiwngam, K.; Tongdeesoontorn, W. Polysaccharide-Based Active coatings incorporated with bioactive compounds for reducing postharvest losses of fresh fruits. *Coatings* **2022**, *12*, 8. [CrossRef]
74. De Bruno, A.; Gattuso, A.; Ritorto, D.; Piscopo, A.; Poiana, M. Effect of edible coating enriched with natural antioxidant extract and bergamot essential oil on the shelf life of strawberries. *Foods* **2023**, *20*, 488. [CrossRef]
75. Pirozzi, A.; Del-Grosso, V.; Ferrari, G.; Donsì, F. Edible coatings containing oregano essential oil nanoemulsion for improving postharvest quality and shelf life of tomatoes. *Foods* **2020**, *4*, 1605. [CrossRef]

polymers

MDPI

Article

Development and Characterization of a Natural Antioxidant Additive in Powder Based on Polyphenols Extracted from Agro-Industrial Wastes (Walnut Green Husk): Effect of Chickpea Protein Concentration as an Encapsulating Agent during Storage

Daniela Soto-Madrid [1], Florencia Arrau [1], Rommy N. Zúñiga [2], Marlén Gutiérrez-Cutiño [3,4] and Silvia Matiacevich [1,*]

1 Food Properties Research Group (INPROAL), Department of Food Science and Technology, Technological Faculty, Universidad de Santiago de Chile, Obispo Umaña 050, Estación Central, Santiago 9170022, Chile; daniela.sotom@usach.cl (D.S.-M.); florencia.arrau@usach.cl (F.A.)
2 Department of Biotechnology, Universidad Tecnológica Metropolitana, Las Palmeras 3360, Ñuñoa, Santiago 8330381, Chile; rommy.zuniga@utem.cl
3 Molecular Magnetism & Molecular Materials Laboratory (LM4), Department of Chemistry of Materials, Chemistry and Biology Faculty, Universidad de Santiago de Chile, Avenida Libertador Bernardo O'Higgins 3363, Estación Central, Santiago 9170022, Chile; marlen.gutierrez@usach.cl
4 Center for the Development of Nanoscience and Nanotechnology (CEDENNA), Santiago 9170022, Chile
* Correspondence: silvia.matiacevich@usach.cl

Citation: Soto-Madrid, D.; Arrau, F.; Zúñiga, R.N.; Gutiérrez-Cutiño, M.; Matiacevich, S. Development and Characterization of a Natural Antioxidant Additive in Powder Based on Polyphenols Extracted from Agro-Industrial Wastes (Walnut Green Husk): Effect of Chickpea Protein Concentration as an Encapsulating Agent during Storage. *Polymers* **2024**, *16*, 777. https://doi.org/10.3390/polym16060777

Academic Editors: Zenaida Saavedra-Leos and César Leyva-Porras

Received: 6 February 2024
Revised: 29 February 2024
Accepted: 5 March 2024
Published: 12 March 2024

Abstract: Developing a powder-form natural antioxidant additive involves utilizing polyphenols extracted from agro-industrial wastes (walnut green husk). This research explores chickpea proteins (CPP) as an emergent encapsulating agent to enhance the stability and shelf life of the antioxidant additive. This study aims to develop a natural antioxidant powder additive based on polyphenols obtained from walnut green husks encapsulated by chickpea protein (5%, 7.5%, and 10% *w*/*v*) to evaluate their effect under storage at relative humidities (33 and 75% RH). The physicochemical and structural properties analysis indicated that better results were obtained by increasing the protein concentration. This demonstrates the protective effect of CPP on the phenolic compounds and that it is potentially non-toxic. The results suggest that the optimal conditions for storing the antioxidant powder, focusing on antioxidant activity and powder color, involve low relative humidities (33%) and high protein concentration (10%). This research will contribute to demonstrating chickpea protein as an emerging encapsulating agent and the importance of the cytotoxic analysis of extracts obtained from agroindustrial wastes.

Keywords: storage stability; vegetal protein; by-products; encapsulation; physicochemical properties

1. Introduction

There is constant concern about managing agro-industrial wastes. In Chile, the horticultural sector generates approximately 42.5 million tons of waste per year, of which about a million correspond to fruit peels and husks [1] One solution to this problem is to apply processes that reduce waste generation and/or minimize it through its use in the production of secondary goods, known as the circular economy [2].

In addition, consumers worldwide tend to prefer food products containing natural additives, driving the industry to develop these products and ingredients [3]. In this sense, it has been reported that the main active compounds present in various agro-industrial wastes (such as fruit husks or peels) are polyphenols. They have interesting biological and active properties, such as antioxidant and antimicrobial properties [4,5]. Hence, polyphenols could be used as a basis for the development of natural additives that replace synthetic ones. Moreover, according to the market research report from Grand View Research, Inc., the global market size for polyphenols was 1.28 billion USD in 2018, and the annual growth rate for the polyphenol market is expected to reach 7.2% from 2019 to 2025 [6].

Chile occupies the second place among the leading walnut exporters and generates a large amount of waste during the harvest that corresponds mainly to the green husk (exocarp and mesocarp), equivalent to approximately 20% of the total production [7]. This husk has a few uses as a fertilizer due to its organic matter, wood dye, and a replacement for Chinese ink [8,9]. Several studies have shown that green walnut husks are rich in active compounds, mainly polyphenolic compounds with antimicrobial and antioxidant capacities [8–11]. Therefore, the walnut green husks represent a natural source of active compounds (polyphenols) and a material for extracting these compounds for their potential use as natural additives.

However, the stability of polyphenols is affected by oxygen, light, heat, and water, challenging their incorporation into foods [12]. A possible solution to the instability of polyphenols is the use of an encapsulation technique since it protects, masks, and retains the properties of the active compound [13]. Several materials are used as encapsulating agents, including proteins of animal origin, vegetable proteins, and carbohydrates [14]. A material used as an encapsulating agent should meet several characteristics: low viscosity and hygroscopicity, high solubility in water, absence of odor and taste, ability to form films, and low cost [15]. At the same time, proteins from plant sources showed other advantages, such as biocompatibility, biodegradability, and good amphiphilic and technofunctional properties (such as water solubility, emulsifying, and foaming capacity) [16,17]. Various studies of proteins from plant sources as the encapsulating agent material of active compounds have been reported. For example, soy protein, wheat proteins, zein or prolamin from corn, barley proteins, and other vegetable proteins with high nutritional value from legumes such as lentils, peas, rice, beans, sunflower, and chickpea [16,18–20].

Chickpea proteins are emerging biopolymers to be used as an encapsulating agent of drug carriers [20]. They are the third most abundant legume crop globally, with a high protein content (14.9–24.6%) [21] and higher bioavailability than other legumes [22]. Therefore, they could be an alternative to replace animal proteins and protect polyphenolic compounds to develop a plant-based food additive.

It has been reported that phenolic compounds can interact with proteins through non-covalent bonds (hydrophobic interactions and hydrogen bonds), which are generated spontaneously in most food systems. It has been described that the interaction positively influences the sensory, functional, and antioxidant properties of food products [23,24]. For that, chickpea proteins could interact with polyphenols, which can help to protect the antioxidant capacity of phenolic compounds extracted from the walnut green husk and be used to develop new natural food additives. However, food additives must be stable during storage at different relative humidities, which could affect their effectiveness in food matrices. Hence, the objective of this work was to develop a natural antioxidant powder additive based on polyphenols obtained from the walnut green husk (waste from the Chilean agroindustry) and encapsulated using chickpea protein to evaluate their effect under storage at two relative humidities. Moreover, the polyphenolic extract's identification and cytotoxicity was also evaluated because the walnut green husk was obtained from a traditional Chilean agricultural crop.

2. Materials Methods

2.1. Samples

The green walnut open husks were obtained from a walnut tree cultivation (*Juglans regia* L.), Chandler variety, in April 2021 in Cuncumen, Province of San Antonio, V Region, Chile. Random sampling was carried out following the methodology of Soto-Madrid et al. [9]. Once the walnut green husks were collected, they were dried in a forced air oven (Zenithlab, DHG-9053 A, Changzhou, China) at 40 °C for 48 h. Then, dried husks were ground in Thermomix equipment

(Vorkwerk, Wuppertal, Germany) and stored at room temperature in glass bottles covered with aluminum foil to protect the samples from light.

The chickpea protein (CPP) used as encapsulating material was extracted from commercial chickpea flour (Extrumol, Santiago, Chile) according to the methodology of Soto-Madrid et al. [25]. Briefly, the chickpea protein fraction was obtained by dispersing the defatted flour at alkaline pH (pH = 11.5) and via subsequent isoelectric precipitation (pH = 4.5). Subsequently, the protein was washed with purified water and neutralized to pH 7. Finally, the protein obtained was freeze-dried (Virtis SP Scientific, Benchtop Pro 9L ES-55, Warminster, PA, USA).

2.2. *Walnut Green Husk Characterization*

Proximal analysis of the walnut green husk and their extracts was performed. Moisture content, proteins, lipids, carbohydrates, ashes, crude fiber, and non-nitrogen extracts were determined according to methods of the Official Association of Analytical Chemistry [26].

2.3. *Extraction of Phenolic Compounds from the Walnut Green Husk*

The extraction of phenolic compounds was carried out through ultrasound-assisted extraction (UAE) (Sonics Materials, VCX 500, Newtown, CT, USA) using ethanol–purified water mixture (75:25) as a solvent, with a solid–solvent ratio of 1:25 (*w*:*v*) according to the methodology described by Soto-Madrid et al. [9]. Subsequently, the extract was filtered using a vacuum pump (Rocker, model 300 C, Kaohsiung, Taiwan) and the Whatman paper (N°1). The ethanol was evaporated in a rotary evaporator (Buchi R-100, Flawil, Switzerland) at a temperature of 40 °C. Finally, the extract obtained was stored in a 200 mL amber bottle and refrigerated until further analysis.

2.3.1. Quantification of Total Polyphenol Content and Antioxidant Capacity

Total phenolic content (TPC) was determined using the Folin–Ciocalteu method with some modifications [27]. Briefly, 0.1 mL of sample was added to a 10 mL volumetric flask with 4.9 mL of distilled water and 0.5 mL of Folin–Ciocalteu reagent (Merck, Darmstadt, Germany), followed by 1.7 mL of Na_2CO_3 (20% *w*/*v*, Merck, Darmstadt, Germany) addition. Then, distilled water was added until it reached 10 mL. The reactive mixture was allowed to stand in darkness for 2 h as an indicator of TPC, and the formation of a blue color was quantified at 740 nm using a spectrophotometer (Shimadzu UVmini-1240, Kyoto, Japan). Gallic acid (Merck, Darmstadt, Germany) was used to construct the standard curve (0.1 to 0.8 mg/mL). Results were expressed as milligrams of gallic acid equivalents/g sample dry weight (mgGAE/g dw). All assays were performed in triplicate.

The antioxidant capacity was determined by scavenging 2,2-diphenyl-1 picrylhydrazyl (DPPH), according to the method reported by Brand-Williams et al. [28], with some modifications. Briefly, 50 µL of diluted concentrations of the walnut green husk extract and the powders developed (reconstituted at 1% *w*/*v*) were mixed with 2950 µL of a methanolic solution containing the DPPH radical (concentration 80 mg/L) (Sigma-Aldrich, St. Louis, MI, USA). The mixture was stirred and left in the dark for 30 min, and subsequently, its absorbance at 517 nm was measured using a spectrophotometer (Shimadzu, UVmini-1240, Kyoto, Japan). The standard curve was constructed using Trolox (Sigma-Aldrich, St. Louis, MI, USA) (0 to 1600 µM), and the results were expressed as mg Trolox/g sample dry weight (mg Trolox/g dw). All assays were performed in triplicate.

2.3.2. Identification of Phenolic Compounds

The identification of the phenolic compounds present in the walnut green husk extract was carried out in a performance liquid chromatography system coupled to mass spectrometry (UPLC-QTOF-ESI-MS, Waters Xevo G2-XS QTof/Tof, Waters, Milford, MA, USA). Chromatographic separation was conducted on an ACQUITY UPLC® Hss T3 column (2.1 × 100 mm, 1.8 µm particle size). The mobile phase gradient followed the following sequence: at time 0 (97% A and 3% B), at 30 min (3% A and 97% B), and from 35 to 40 min (97% A and 3% B).

The mass spectrometer was operated in both positive and negative ion modes. The identification method was carried out using the Progenesis QI MetaScope v2.3 software (Waters, Milford, MA, USA), and the search parameters were established through the HMDB library. The precursor ion and the fragments used an error tolerance of 10 ppm. All compounds found with adduct formation in M-H and M+H modes are reported as results. A mass range of 100 to 1000 Da was taken into account.

2.4. *Cell Viability Assay Using Vero Cells*

Resazurin was used to quantify the viability of Vero cells. This compound fluoresces when cells metabolize resazurin and reduce it to resofurin. The amount of resofurin produced is proportional to the cell metabolic activity and can therefore be used to evaluate cell viability, where high fluorescence indicates high cell viability. Briefly, 5000 Vero cells were seeded per well in 96-well microplates. Then, 20 µL of 0.5 mg/mL resazurin solution in PBS (Phosphate Buffered Saline) was added per well and incubated at 37 °C for 4 h in a humidified environment with a concentration of 5% CO_2. The wells were then analyzed by fluorescence (Tecan, Infinite 200 Pro reader, Mennedorf, Switzerland), with an excitation wavelength of 560 nm and an emission wavelength of 590 nm. The data were analyzed with the statistical software GraphPad Prism 9.5.1 to determine the cytotoxic concentration of each extract that reduces viability by 50% (CC_{50}).

The Vero cell line was obtained from Dr. Cesar Echeverria from the University of Antofagasta, Chile.

2.5. *Development of the Active Antioxidant Additive*

For the assays, chickpea protein was mixed at three concentrations (5, 7.5, and 10% *w*/*v* in 40 mL) with the walnut green husk extract (20 mL) at room temperature for 3 h with stirring (490 rpm) and under dark conditions until the complete dispersion of the protein. This methodology favored complete protein dispersion and the non-covalent interaction between protein and polyphenols [29]. Afterward, the samples were freeze-dried at −18 °C and placed in the freeze-dryer chamber collector at −60 °C with a shelf at 30 °C under a pressure of 0.05 bar for 72 h (Ilshin FD5508, Siheung-si, Republic of Korea). Three samples were obtained: FDP 5%, corresponding to the additive that contains CPP at 5% *w*/*v*; FDP 7.5%, corresponding to the additive with 7.5% *w*/*v* of CPP; and FDP 10%, corresponding to the additive that contains 10% *w*/*v* CPP.

2.6. *Physicochemical Characterization of Additives in Powders*

2.6.1. Encapsulation Efficiency (E.E. %)

The encapsulation efficiency was calculated by the total polyphenol content of the powder additive and the extract using the following equation:

$$\text{E.E. } (\%) = \frac{TPCE - TPCP}{TPCP} * 100\%, \quad (1)$$

where TPCE corresponds to the total polyphenol content of the sample before freeze-drying, and TPCP corresponds to the polyphenol content of the powder additive.

2.6.2. Drying Process Yield (DY %)

The total solids of the extract with the encapsulating agent (SST, soluble solids/100 mL of extract) were measured via refractometry (RHB-32 ATC, YHEquipment, Shenzen, China) before drying. Once the sample was dried, the powders were weighed on an analytical balance (HR-120, A&D Co., Tokyo, Japan). With the weights, the extraction yield (DY%)

was calculated using the following Equation (2) according to the conditions described by Fenoglio et al. [30]:

$$DY(\%) = \frac{Total\ solid\ after\ freeze - drying}{Initial\ total\ solids} * 100\%. \quad (2)$$

2.6.3. Moisture Content and Water Activity

The moisture content of the dry powders was determined gravimetrically by the difference in mass before and after drying the samples in an oven (Shel Lab 1410-2E, Capovani Brothers, New York, NY, USA) at 105 °C until constant weight (AOAC, 1998). Results were expressed as dry basis percentage (% db; g water/100 g solids). The water activity (a_W) of walnut green husk dried and additive powders was determined using a chilled-mirror dew point device (Aqualab, Series 3 TE, Decagon, Washington, DC, USA) at 25 °C (AOAC, 1998).

2.6.4. Color Analysis

The color of the additives was determined through image analysis using a computer vision system (previously calibrated). It consists of a black box with four natural lights D65 (18 W, Phillips, Amsterdam, The Netherlands) and a digital camera (Canon, EOS Rebel XS, Tokyo, Japan) at a distance of 22.5 cm from the sample (camera lens angle and lights at 45°) [31]. Samples were measured as pellets by pressing powders with a Quick Press hand press (Perkin-Elmer, Waltham, MA, USA). The digital color parameters were obtained in the RGB space using the software Adobe Photoshop v7.0 (Adobe Systems Incorporated, 2007), which was subsequently converted to the CIELAB space, in which L* indicates lightness, a* the red-green axis, and b* the blue-yellow axis.

2.6.5. Structural Characterization: SEM and FTIR

Scanning electron microscopy (SEM) was employed to analyze the microstructure of the additives using a field emission scanning electron microscope (Zeiss, model EVO MA10, Jena, Germany). The samples were affixed to stubs using double-sided adhesive tape, coated with a gold layer, and images were captured with an acceleration voltage of 20 kV.

The chemical groups and bonding arrangement of components present in the samples were determined by Fourier transform infrared–attenuated total reflectance (FTIR-ATR) using an infrared spectrophotometer equipped with an ATR PRO ONE (Jasco FTIR-4600, Easton, MD, USA). Measurements were performed in a spectral range from 4000 to 400 cm^{-1}, with a resolution of 4 cm^{-1} and 32 scans per sample.

2.6.6. Isoelectric Point (IEP)

The isoelectric points (IEP) were determined through Zeta Potential (pZ) measurements (Zetasizer Nano Series, NanoZS90, Malvern Instruments, Malvern, UK). Dilute suspensions of the powder additive (approximately 0.05 g/L) were prepared in 10^{-3} mol/L KCl, and the pH was adjusted using 10^{-2} mol/L HCl or KOH. The IEP was identified as the pH value where pZ equals zero.

2.7. Stability at Different Relative Humidities (RH)

The developed additives were evaluated in hermetically sealed desiccators at different relative humidities for two weeks. Saturated solutions of $MgCl_2$ and NaCl were used to obtain 33% and 75% RH, respectively [32]. The polyphenol total contents, antioxidant activity, water activity, and color parameters (L*, a*, and b*) were measured at the beginning and end of the analysis.

2.8. Statistical Analysis

All experiments were run in triplicate. Data were reported as means with their corresponding standard deviation. ANOVA test was performed at a confidence level of 95% to determine statistical differences using Statgraphics Centurion XVI® software (StatPoint

Technologies Inc., Warrenton, VA, USA, Version XVI). Differences between samples were evaluated using multiple range tests, using the least significant differences (LSD) multiple comparison method. The significance of the differences was determined at a 95% confidence level ($p < 0.05$). The linear dependency between two independent variables was obtained by the r-Pearson coefficient using Microsoft Excel v10.

3. Results and Discussion

3.1. *Characterization of the Walnut Green Husk and Extracts*

3.1.1. Proximal Analysis

It is essential to characterize the walnut green husk and the extract obtained from it through proximal analysis to standardize the extraction process (Table 1). The raw material presented an a_W value lower than 0.6, ensuring their stability against microbial growth [33] before each extraction.

Table 1. Proximal analysis of the walnut green husk and liquid extract corresponding to the 2021 harvest.

Analysis	Walnut Green Husk (g/100 g Dry Base)	Extract Liquid (g/100 g Wet Base)
Moisture	7.5 ± 0.01	93.9 ± 0.01
Proteins (%Nx5.3 [1])	6.98 ± 0.23	* ND
Lipids	1.94 ± 0.11	** ND
Ash	12.6 ± 0.05	1.0 ± 0.06
Crude fiber	20.34 ± 1.19	*** ND
Non-nitrogen extract (N.N.E.)	50.66 ± 1.13	5.0 ± 0.08
Energy (Kcal)	248 ± 4.41	19.9 ± 0.31

* ND: detection limit ≤ 0.39 g/100 g; ** ND: detection limit ≤ 0.52 g/100 g; ***: detection limit ≤ 0.59 g/100 g. [1] Conversion factor for nuts [34].

As expected, due to the ultrasound-assisted extraction process, the extracts' proximal analyses (Table 1) showed non-detection (ND) of compounds such as lipids, proteins, and fibers, indicating no contamination during the extraction process and no possible interactions with proteins of these compounds that affect the polyphenol protection. Moreover, the extraction efficiency in the extract obtained was 100% from the walnut green husk, so the extraction methodology is validated to obtain non-nitrogenous extracts (N.N.E.). Besides, 83.3% of the total dry sample of N.N.E. was obtained, which comprises soluble compounds such as polyphenols, phenolic acids, and flavonoids since they do not have a group functional based on nitrogen in their structure [13]. However, the ash content (16.7% of the total dry sample) was attributed to soluble minerals of the raw material, which can act as electrolytes and could negatively affect the stability of the protein–polyphenols interaction [35]. Increasing chickpea protein concentrations must be studied to avoid this potential effect.

Independently of this, it is important to note that ultrasound-assisted extraction is a simple, efficient, and sustainable technique [36] that allows for better penetration of solvents, a shorter extraction time, and higher extraction yield of polyphenols, even at lower temperatures compared to other extraction methods of phenolic compounds from plant matrices [37].

In parallel, the polyphenol content and its antioxidant capacity were determined via the DPPH method to (i) confirm that the compounds present in the extract (N.N.E.) are polyphenols and (ii) if they maintain their antioxidant activity after the extraction process. The walnut green husk sample harvest in 2021 presented a 36% higher value for TPC and a similar value for antioxidant capacity (202 ± 1.2 mg GAE/g dry sample) compared to the harvest in 2019, previously reported by Soto-Madrid et al. [9]. It could be attributed to differences in the polyphenol type and quantity extracted but with the same activity. However, few compounds were reported with which to compare it. Moreover, the differences could also be due to raw material differences, which may vary according to ripeness stage, environmental factors, and the mode of collection and storage [38]. For that, it is crucial to identify the compounds in the polyphenolic extract and evaluate the efficacy of the extraction process.

3.1.2. Identification of Compounds in the Walnut Green Husk Extract

The compounds identified by UPLC-QTOF-ESI-MS in negative and positive modes, where the mass/charge (m/z) values were also compared to those reported by Sheng et al. [39], are shown in Table 2. Briefly, 64 compounds were identified, including hydrolyzable tannins, flavonoids, phenolic acids, phenolic glycosides, and quinones. Unexpectedly, herbicides and fungicides were also identified due to the traditional agricultural fields that use these pesticides as a common practice.

Table 2. Compounds identified in the walnut green husk extract by UPLC-QTOF-ESI-MS.

N°	RT (min)	Formula	Measured Mass (m/z)	Error (ppm)	Ion Mode	Compound Identification	Reference	Classification
1	1.30	$C_{11}H_{14}NO_7^+$	271.0687	−4.03	M-H	pyridine N-oxide glucuronide	[a]	Other (aromatic compound)
2	1.55	$C_4H_6O_6$	133.0138	−6.03	M-H	malic acid	[a]	organic acid
3	2.89	$C_{13}H_{16}O_{10}$	331.0669	−0.41	M-H	3-glycogallic acid	[a]	Phenolic glycosides
4	3.14	$C_7H_6O_5$	169.0140	−1.37	M-H	gallic acid	[a]; [b]	Phenolic acid
5	3.64	$C_{14}H_{16}N_2O_8$	321.0729	0.08	M-H	glutamic acid-betaxanthin	[a]	Vegetal pigment
6	3.86	$C_{14}H_{18}O_{10}$	345.0835	2.27	M-H	methyl 6-O-galloyl-beta-D-glucopyranoside	[a,b]	Hydrolyzable tannin
7	3.86	$C_8H_8O_5$	183.0303	2.35	M-H	methyl gallate	[a]	Phenolic compound
8	4.13	$C_{15}H_{20}O_{10}$	359.1002	5.13	M-H	3-methoxy-4-hydroxyphenylglycol-glucuronide	[a]	Phenolic glycosides
9	4.20	$C_9H_{10}O_4$	181.0505	−0.61	M-H	syringaldehyde	[a]; [b]	Aromatic aldehyde
10	4.21	$C_8H_8O_3$	151.0399	−1.58	M-H	vanillin	[a]; [b]	Phenolic aldehyde
11	4.47	$C_{13}H_{16}O_8$	299.0773	0.29	M-H	4-methylcatechol 1-glucuronide	[a]	Phenolic glycosides
12	4.52	$C_7H_6O_4$	153.0189	−2.89	M-H	protocatechuic acid	[a]; [b]	Phenolic acid
13	4.62	$C_9H_8O_4$	179.0356	3.67	M-H	caffeic acid	[a]; [b]	Phenolic acid
14	4.62	$C_7H_{12}O_6$	191.0561	0.02	M-H	quinic acid	[a]; [b]	Phenolic acid
15	4.62	$C_{16}H_{18}O_9$	353.0879	0.31	M-H	crypto chlorogenic acid	[a]	Phenolic acid
16	4.69	$C_{14}H_{18}O_9$	329.0885	2.13	M-H	vanillyl glucose	[a]	Hydrolyzable tannin
17	4.93	$C_{15}H_{20}O_{10}$	359.0985	0.29	M-H	glucosyringic acid	[a]	Phenolic glycosides
18	5.00	$C_{30}H_{26}O_{12}$	577.1342	−1.56	M-H	procyanidin B8	[a]; [b]	Flavonoid
19	5.02	$C_8H_8O_4$	169.0497	1.06	M+H	isovanilic acid	[a]; [b]	Phenolic acid
20	5.15	$C_{15}H_{18}O_9$	341.0878	0.06	M-H	glucocaffeic acid	[a]	Phenolic glycosides
21	5.21	$C_{21}H_{22}O_{11}$	449.1099	2.18	M-H	astilbin	[a]; [b]	Flavonoid
22	5.30	$C_{21}H_{20}O_{12}$	463.0888	1.38	M-H	myricitrin	[a]; [b]	Flavonoid
23	5.47	$C_{16}H_{20}O_9$	337.0930	0.37	M-H	gentiopicroside	[a]	Other
24	5.47	$C_9H_8O_3$	163.0398	−1.76	M-H	coumaric acid	[a]; [b]	Phenolic acid
25	5.59	$C_{15}H_{10}O_6$	287.0549	−0.56	M+H	kaempferol	[a]; [b]	Flavonoid
26	5.76	$C_{16}H_{18}O_9$	353.0877	−0.42	M-H	chlorogenic acid	[a]; [b]	Phenolic acid
27	5.81	$C_9H_{10}O_3$	165.0555	−1.30	M-H	4-hydroxyphenyl-2-propionic acid	[a]; [b]	Phenolic acid
28	5.81	$C_9H_{10}O_3$	167.0702	−1.68	M+H	ethylparaben	[a]; [b]	p-hydroxybenzoic acid ethyl ester
29	5.88	$C_7H_6O_3$	137.0241	−2.14	M-H	3-hydroxybenzoic acid	[a]; [b]	Phenolic acid
30	5.90	$C_{10}H_{12}O_4$	177.0562	2.26	M-H	xanthoxylin	[a]	Phenolic ketone
31	6.10	$C_{15}H_{18}O_8$	325.0929	−0.01	M-H	coumaric acid 2-glucoside isomer	[a]; [b]	Phenolic glycosides
32	6.23	$C_{16}H_{20}O_9$	355.1035	0.20	M-H	ferulic acid 4-glucoside isomer	[a]	Phenolic glycosides

Table 2. *Cont.*

N°	RT (min)	Formula	Measured Mass (*m/z*)	Error (ppm)	Ion Mode	Compound Identification	Reference	Classification
33	6.25	$C_{28}H_{28}N_4O_6S$	547.1661	0.72	M-H	1-((2-methoxy-4-(((phenylsulfonyl)amino)carbonyl) phenyl)methyl)-1H-indazol-6-yl)carbamic	[a]	Herbicide
34	6.55	$C_{15}H_{14}O_6$	289.0723	1.76	M-H	catechin	[a]; [b]	Flavonoid
35	6.55	$C_{13}H_{16}O_9$	315.0737	4.99	M-H	protocatechuic acid 4-glucoside	[a]	Phenolic glycosides
36	6.69	$C_{16}H_{18}O_8$	337.0929	0.02	M-H	3-p-coumaroylquinic acid	[a]; [b]	Phenolic acid
37	6.69	$C_9H_{10}O_5$	197.0458	1.25	M-H	syringic acid	[a]; [b]	Phenolic acid
38	6.76	$C_{15}H_{22}O_5$	281.1398	1.32	M-H	dihydrophasic acid	[a]; [b]	Other
39	6.93	$C_{41}H_{28}O_{26}$	935.0794	0.99	M-H	casuarinin	[a]	Hydrolyzable tannin
40	7.06	$C_{10}H_{10}O_3$	177.0557	−2.84	M-H	(S)-Isosclerone	[a]	Other
41	7.06	$C_{10}H_8O_3$	175.0397	−1.85	M-H	7-hydroxy-methyl coumarin	[a]; [b]	Phenolic acid
42	7.18	$C_{21}H_{20}O_{13}$	479.0824	−1.43	M-H	myricetin-3-glucoside	[a]	Phenolic glycosides
43	7.44	$C_{20}H_{20}O_{11}$	435.0928	−1.10	M-H	taxifolin 3-arabinoside	[a]	Flavonoid
44	7.61	$C_9H_6O_3$	163.0388	−1.07	M+H	3-hydroxycoumarin	[a]; [b]	Other
45	7.83	$C_{10}H_6O_3$	173.0251	3.79	M-H	juglone	[a]	Quinone
46	7.92	$C_{23}H_{22}O_{12}$	489.1054	3.13	M-H	quercetin 3-O-acetyl-rhamnoside	[a]; [b]	Flavonoid
47	7.95	$C_{21}H_{24}O_{11}$	433.1148	1.69	M-H	catechin 3-glucoside	[a]	Phenolic glycosides
48	8.02	$C_{21}H_{24}O_{24}$	435.1297	0.04	M-H	florizin	[a]; [b]	Glycoside
49	8.06	$C_{21}H_{20}O_{12}$	463.0880	−0.47	M-H	quercetin 3-galactoside	[a]	Flavonoid
50	8.06	$C_{14}H_6O_8$	300.9991	0.35	M-H	ellagic acid	[a]; [b]	Phenolic acid
51	8.46	$C_{21}H_{22}O_{11}$	449.1092	0.49	M-H	astilbin	[a]; [b]	Flavonoid
52	8.46	$C_{10}H_{10}O_4$	193.0516	4.91	M-H	cis-ferulic acid	[a]; [b]	Phenolic acid
53	8.51	$C_{20}H_{18}O_{11}$	433.0773	−0.84	M-H	quercetin 3-xyloside	[a]	Flavonoid
54	8.64	$C_{10}H_{12}O$	149.0961	−0.15	M+H	cuminaldehyde	[a]; [b]	Aldehído aromático
55	8.66	$C_{21}H_{20}O_{11}$	447.0934	0.20	M-H	quercitrin	[a]; [b]	Flavonoid
56	9.68	$C_9H_{16}O_4$	187.0978	0.89	M-H	azelaic acid	[a]; [b]	Other
57	9.95	$C_{11}H_{12}O_5$	225.0766	4.81	M+H	sinapic acid	[a]; [b]	Phenolic acid
58	11.58	$C_{15}H_{10}O_7$	301.0356	0.75	M-H	quercetin	[a]; [b]	Flavonoid
59	13.77	$C_{14}H_{10}O_8$	287.0207	3.06	M-H	2-(3,4-dihydroxybenzoyloxy)-4,6-dihydroxybenzoate	[a]	Phenolic compounds
60	16.69	$C_{15}H_{10}O_6$	285.0416	3.97	M-H	luteolin	[a]; [b]	Flavonoid
61	18.83	$C_{18}H_{12}Cl_2N_2O$	341.0261	2.19	M-H	boscalida	[a]	Fungicide
62	28.23	$C_{10}H_8O_2$	161.0599	0.91	M+H	naphthalen diol isomer	[a]; [b]	Quinone
63	31.59	$C_{20}H_{26}NO_3^+$	309.1744	3.10	M-H	8-O-Methyloblongin	[a]	Isoquinoline
64	31.86	$C_{21}H_{22}O_{12}$	465.1037	−0.28	M-H	(-)-epicatechin 3′-O-glucuronide	[a]	Flavonoid

[a] Data basis: Progenesis QI v2.3 software; [b] Sheng et al. [38].

Of the total identified compounds (64), 29% corresponded to phenolic acids such as gallic acid, protocatechuic acid, and ferulic acid, which had also been identified by Soto-Madrid et al. [9] (2021) via HPLC-RP. Then, 22% corresponded to flavonoids such as quercetin, quercitrin, catechin, kaempferol, and others; 15% to phenolic glycosides; 4.5% of hydrolyzable tannins; and 3% of quinones. This 73.5% phenolic compounds demonstrated antioxidant activity [13]. However, 3% was attributed to pesticides, which could negatively affect health.

Therefore, to use the walnut green husk extract as a base to develop a natural food additive, it is required to evaluate its cytotoxicity through in vitro studies with cell cultures.

3.1.3. Cytotoxicity Evaluation of Walnut Green Husk Extract

The literature has reported that an extract can be considered very toxic with a CC_{50} < 10 μg/mL, moderately toxic with CC_{50} = 11–30 μg/mL, slightly toxic at CC_{50} = 31–50 μg/mL, and potentially non-toxic at CC_{50} > 50 μg/mL [40]. The cell viability assay of this work's walnut green husk extracts was CC_{50} = 90 ± 9 μg/mL, demonstrating that it is potentially non-toxic and could be used to develop a natural antioxidant additive based on agroindustrial waste. However, it is essential to consider the traces of these compounds for another possible adverse effect, such as potential allergenicity, which was not evaluated and must be labeled and regulated. Considering this result, it is necessary to realize a cell viability study in extracts obtained from wastes due to the growing tendency of waste revalorization in agroindustry, where the use of pesticides is a common agricultural practice.

3.2. Development of the Natural Antioxidant Additive

3.2.1. Physicochemical Characterization of the Natural Antioxidant Powder Additive

Different concentrations of chickpea protein (5, 7.5, and 10% *w*/*v*) were studied as encapsulating material of the phenolic compounds to develop the additive. As expected, the freeze-drying showed high values of drying process yield (DY%), close to ~98.6% for all samples, confirming that this process generates low losses in terms of solids recovery. This is a positive aspect since, for example, in the spray drying process, the DY% is low (60–90%) due to losses of solids occurring by their adhesion to the drying chamber [41]. Also, high wall material concentrations are required to protect the active compound at high temperatures [16].

Table 3 shows the results of the physicochemical characterization of the powder obtained by freeze-drying. The highest E.E. (%) was obtained for the FDP 7.5% (60 ± 6%) sample. Compared with the literature, spray-drying is a better process by which to obtain a high E.E.% (65–92%) of polyphenol compounds using proteins as encapsulating agents [41]. The encapsulation efficiency differences reported in the literature could be related to the nature of the polyphenolic compounds (i.e., charge, type of compound, chemical structure) and the structure of the wall material, positively or negatively conditioning the polyphenol–polymer interaction since they are the most critical variables to consider for the encapsulation of polyphenols [42]. It considered that the drying technique and the material used as protection affected the retention capacity of compounds within the matrix. For that, selecting the wall material and the drying technique is crucial to balance high drying process yield and encapsulation efficiency to maximize the incorporation and retention of the functional compounds within the encapsulation matrix.

The water activity (a_W) and moisture content are critical physical parameters of powdered additives since they strongly influence their storage stability and safety. In this sense, all the powders analyzed presented a_W lower than 0.2, demonstrating safety and low biochemical kinetic reactions [33]. The FDP 5% sample showed the highest value moisture content (7.800 ± 0.003% db), and it was statistically different ($p < 0.05$) when compared with the FDP 7.5% and 10% samples (approx. 6% db) (Table 3). It could be attributed to the lower concentration of protein to protect the phenolic extract. The ice crystals' sublimation during freeze-drying generated many small porous and less compact structures, resisting mass transfer and acting as a barrier against sublimation [43]. It results in greater moisture retention and, consequently, higher moisture in the final product.

Table 3. Physicochemical parameters of the powder additive obtained via freeze-drying with different concentrations of chickpea protein as the encapsulating agent.

Analysis	Chickpea Protein Concentration (% *w/v*)		
	5	**7.5**	**10**
Encapsulation efficiency (%)	44 ± 3^{a}	60 ± 8^{b}	42 ± 5^{a}
a_W	0.17 ± 0.04^{a}	0.200 ± 0.001^{a}	0.197 ± 0.001^{a}
Moisture (% dry basis)	7.80 ± 0.003^{a}	6.49 ± 0.004^{b}	6.03 ± 0.003^{b}
Parameter L*	76.82 ± 3.42^{a}	78.12 ± 0.07^{a}	77.12 ± 2.74^{a}
Parameter a*	1.7 ± 0.1^{a}	5.9 ± 0.5^{b}	2.78 ± 0.4^{a}
Parameter b*	5.34 ± 0.5^{a}	6.04 ± 5.1^{a}	10.6 ± 0.6^{c}
Images	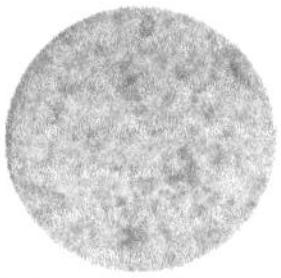		

Different letters (a, b, c) indicate significant differences ($p < 0.05$) between samples.

Table 3 also shows the color parameters considered in the CIELAB L*a*b* space. The lightness parameter (L*~77) showed no statistically significant differences ($p > 0.05$) in the three samples analyzed. This high L* value (scale 0–100) indicates light powders. However, for the a* parameter, the 7.5% sample presented the highest value (5.9 ± 0.5). It was significantly different ($p < 0.05$) compared to the other two samples, which shows a little tendency towards the red color. On the other hand, for the b* parameter, the FDP 10% sample presented the highest value (10.6 ± 0.6). It was statistically different ($p < 0.05$) from the other samples, indicating the tendency towards the yellow color. The changes in a* and b* parameters are attributed to a higher protein concentration, generating greater protection for the brown phenolic extract, which begins to turn yellow, the characteristic color of chickpea protein. Independent of the significant differences in each parameter observed, the visual color and chromaticity diagram (xy scale) indicated light powders with a little tendency to yellow at higher protein concentrations. Furthermore, in the images (Table 3), it can be observed that at high protein concentrations, the compaction of the powder additive increased. Therefore, all the samples analyzed had low water activity, this is favorable for their storage and shelf life and, combined with their light color, would allow for their addition to food matrices.

3.2.2. Structural Characterization of Powders

Scanning electron microscopy (SEM) was performed to evaluate the morphology of the powder additives based on the walnut green husk extract. Micrographs correspond to the CPP control (Figure 1A) used as wall material and to the powder additives with different percentages of CPP (Figure 1B–D). Figure 1A shows an irregular, brittle, and flake-like structure, a common structural characteristic (at 20 μm) of freeze-dried proteins [44,45]. A porous structure with irregular shapes and sizes is evident in the FDP 5% sample (Figure 1B). This sample porosity may result from ice formation in the material during the freeze-drying process. However, as the CPP concentration increases in the development of the powder additive, a more defined, almost spherical morphology is observed with the formation of larger capsules and a decrease in porosity (Figure 1C,D). This is evidence of the encapsulation of phenolic compounds from the walnut green husk extract using concentrations of 7.5 and 10% CPP.

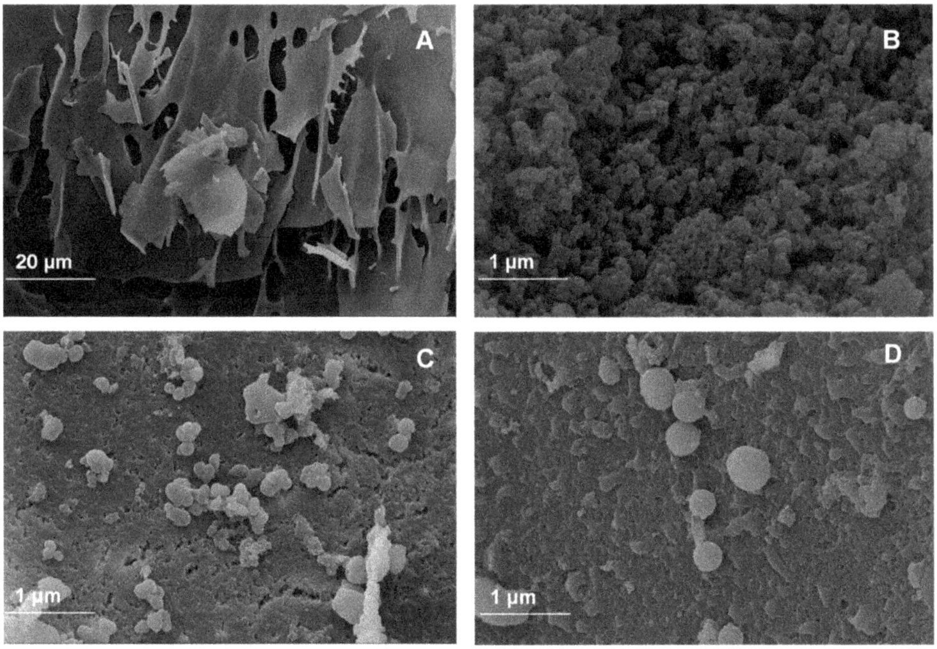

Figure 1. Images of antioxidant powder additives: (**A**) control protein; (**B**) freeze-dried powder additive with 5% chickpea protein; (**C**) freeze-dried powder additive with 7.5% chickpea protein; (**D**) freeze-dried powder additive with 10% chickpea protein.

The protective effect of CPP on the active compounds from the walnut green husk extract can be evidenced by a decrease or displacement of the typical signals of the bands measured by FTIR [20]. Figure 2 illustrates the FTIR spectra of the freeze-drying additives and the control sample corresponding to CPP. All FTIR spectra showed a typical absorption band at a wavelength of ~3278 cm^{-1}, characteristic of the water's hydroxyl group (–OH). As a result of encapsulation, the C=O stretching of the amide I band of CPP at ~1633 cm^{-1} has shifted slightly to ~1636 cm^{-1} in the spectrum of the FDP 7.5% sample. In parallel, amide II: N–H bending and C–N stretching of proteins at 1530 cm^{-1} has shifted to 1533 and 1538 cm^{-1} in the FDP 5% and FDP 7.5% samples. In addition, it also identified a band around ~1235 cm^{-1} corresponding to the amide III region (CN stretching, NH bending) [46], which has shifted slightly to 1233 cm^{-1} in the FDP 10% sample. Although the band displacements are between 2–3 cm^{-1}, the literature using chickpea protein reported these little changes as component interactions [20]. Moreover, CPP peaks are weakened in intensity due to the encapsulation of phenolic compounds [20].

3.2.3. Zeta Potential (pZ)

Zeta potential is an important and valuable indicator of particle surface charge, which can be used to predict and control the stability of suspensions [47,48]. Figure 3 shows the control protein's zeta potential versus pH curve and FDP 5, 7.5, and 10% samples. At pH ~3, all samples present values between |20–25| mV, independent of the protein concentration used. This indicates that they are outside of the flocculation region (|5–15| mV) and near the optimal region |30| mV, evidencing the stability of the powder additives at this pH. So, this indicates potential applications in acid food matrices. Moreover, at pH 6–7, there were differences in the values obtained, where the FDP 5% sample presented the highest value (|30| mV) compared to the other samples (|~20| mV). It can be attributed to the acid-base properties of different radicals charges or functional groups due to the structural characteristics of each flavonoid (present in the walnut green husk extract), which showed a negative charge at pH ~7 [49], contributing to the total surface charge and evidencing that they were not protected (free) in the FDP 5% sample.

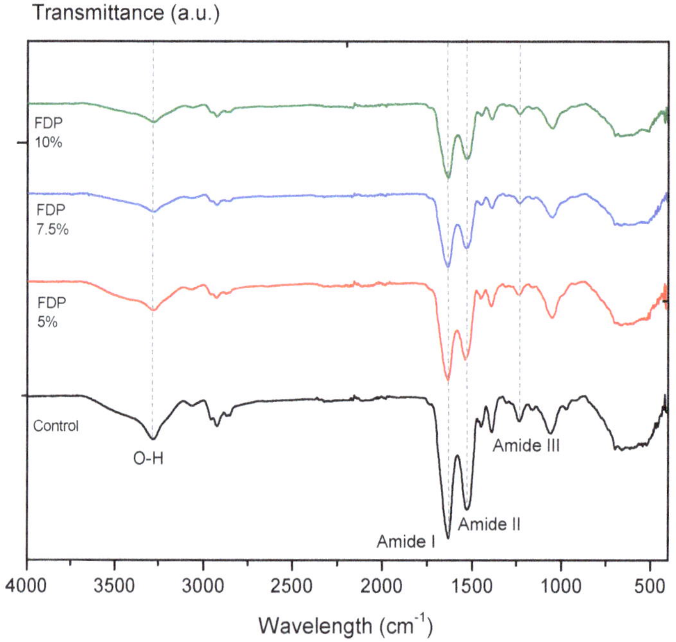

Figure 2. Infrared spectra by Fourier transform of additives in powders: Control: freeze-dried chick-pea protein; FDP 5%: freeze-dried powder additive with 7.5% chickpea protein; FDP 7.5%: freeze-dried powder additive with 7.5% chickpea protein; FDP 10%: freeze-dried powder additive with 10% chickpea protein.

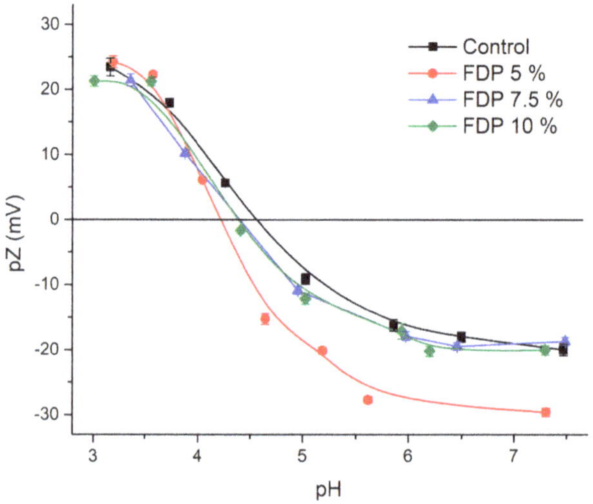

Figure 3. Zeta potential (pZ) as a function of pH for control protein (black), FDP 5% (red), FDP 7.5% (blue), and 10% FDP (green) samples. FDP = freeze-drying powder at different chickpea protein concentrations in % *w*/*v*.

As expected, the control exhibits an isoelectric point (IEP) value of 4.55, similar to the 4.5 reported by Soto-Madrid et al. [25], while FDP 7.55% and 10% display consistent IEP at 4.45 with similar behavior. However, FDP 5% exhibits a slightly lower IEP of 4.3. All values agree with Vani and Zayas. [50], Boye et al. [51], and Ma et al. [52]; the authors indicate that most plant proteins have an IP between 4.0 and 5.0. It is important to note that proteins could adsorb charges on their surfaces at their IEP due to the presence of other compounds, such as polyphenols [53]. In this case, the diminution of IEP in the FDP 5% sample can be attributed to negative charges from free phenolic compounds, independent of the total neutral charge at IEP.

3.3. Stability of the Antioxidant Additive at Different Relative Humidities (RH)

The stability of antioxidant powder additives depends mainly on their water activity since water can act as a reagent or solvent in different degradation reactions or contribute to microbial growth. Furthermore, the water content of a freeze-drying product depends on the residual moisture left in the product after drying and the water that it can adsorb from the surrounding atmosphere during storage [54].

The effect of two relative humidities (33 and 75% RH) on antioxidant capacity via DPPH and total polyphenol content, TPC, was analyzed in powder additives. The results are shown in Figure 4A (TPC) and Figure 4B (antioxidant capacity). Relative humidity affected TPC, decreasing significantly at 75% RH, dependent on protein concentration (Figure 4A). For the FDP 5% sample, a 60% diminution of TPC was observed; meanwhile, at 7.5%, it was 17%. However, relative humidity did not significantly affect ($p > 0.05$) the TPC when the protein concentration was 10%. This demonstrated the importance of protein concentration in protecting the antioxidant compounds from humidity during storage. The same behavior was observed for antioxidant capacity (Figure 4B).

Interestingly, TPC and antioxidant capacity diminished as protein concentration increased at 33% RH. It can be attributed to higher protection and protein–polyphenols interaction when increasing the protein concentration, as shown in Figure 2. Moreover, the results above indicated the presence of free polyphenols in the FDP 5% sample, which are corroborated in Figure 3 with higher TPC and antioxidant activity, independent of relative humidity.

Furthermore, for the stability of the powders during storage, it is essential to maintain the activity of antioxidant compounds and the powder color. Figure 5 shows the stability of the additive powders for the lightness parameter (L*) at the different relative humidities and the visual changes. The effect of relative humidity in lightness was insignificant at 33% RH, independent of protein concentration. However, at 75% RH, the effect was also dependent on protein concentration, with the lowest loss at 10% of protein concentration (FDP 10%). As expected, considering the above results, the stability of samples at 75% RH was lower, showing a dark powder after storage. For that, a 33% RH confirms the stability of the antioxidant powder additive during storage. The FDP 10% sample maintained the lightness at 33% RH and exhibited the lowest change in color when stored at higher relative humidity.

It is important to note that the higher encapsulation efficiency obtained at 7.5% *w*/*v* of the encapsulating agent is not correlated to a higher TPC and antioxidant capacity during storage conditions, as expected. In this case, it is attributed to oxidized phenolic compounds in the powder surface, which is correlated (r-Pearson = 0.9707) to the a* parameter (Table 3).

Nevertheless, considering the cost of the freeze-dried process, the study highlights the need for further investigations to bolster these findings compared to widely used encapsulation technologies like spray drying. Such comparative analyses will provide a more comprehensive understanding of the relative effectiveness and feasibility of the developed antioxidant powder additives. The ongoing pursuit of knowledge in this area will contribute valuable insights to the field and facilitate informed decision-making for industrial applications.

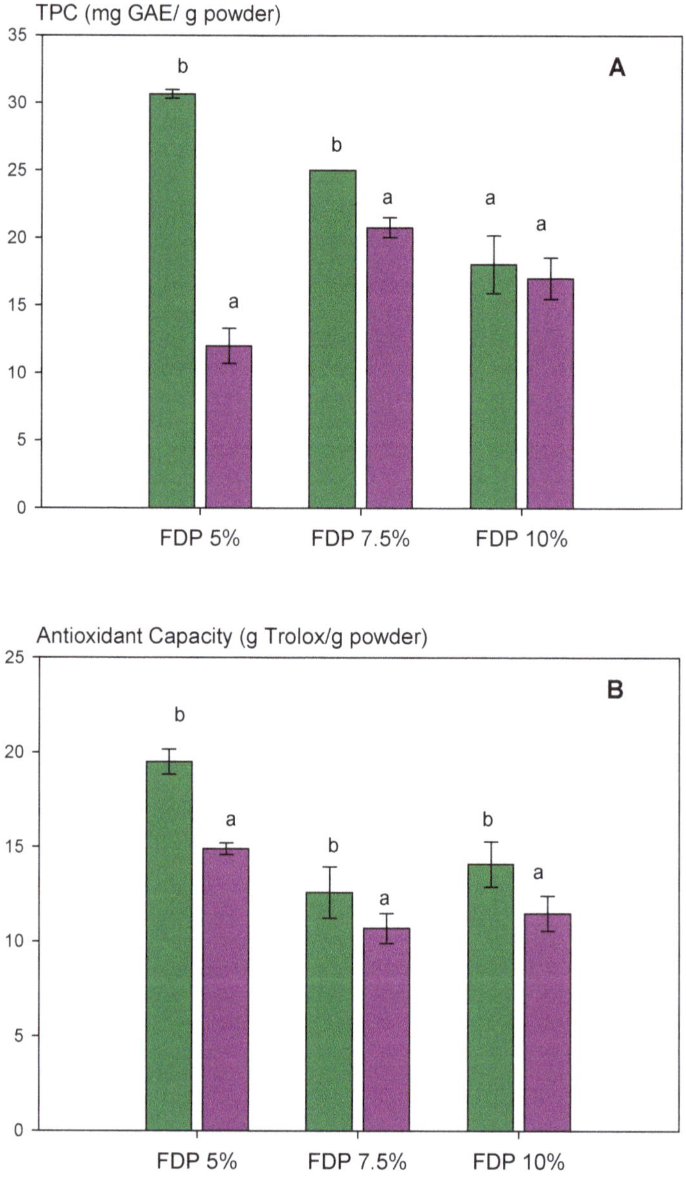

Figure 4. Analysis of the stability of the total polyphenol content and antioxidant activity of powder additives at different relative humidities (RH): (**A**) total polyphenol content (TPC); (**B**) antioxidant capacity of the powders quantified by DPPH radical scavenging activity. Green bars indicate relative humidity 33%, and violet bars indicate a relative humidity 75%. FDP = freeze-drying powder at different chickpea protein concentrations in % *w*/*v*. GAE = gallic acid equivalent. Different letters (a,b) indicate significant differences ($p < 0.05$) between samples.

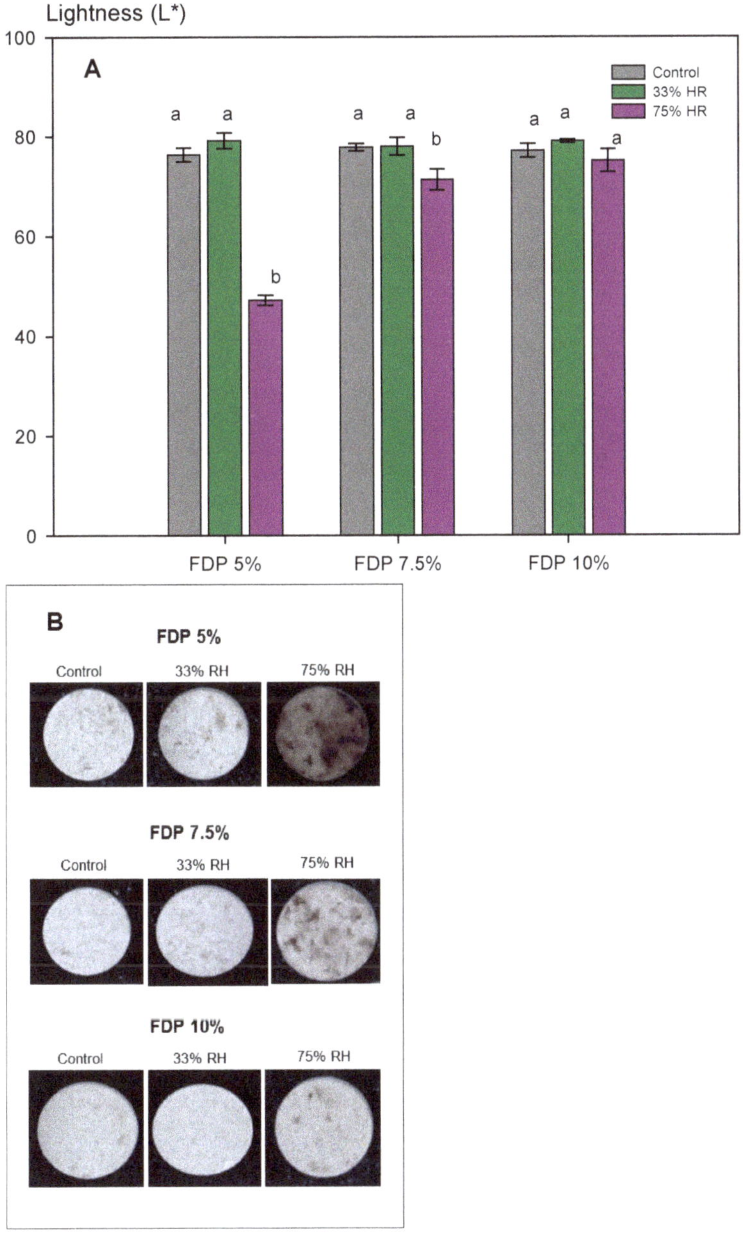

Figure 5. Stability of the additive powders for the lightness parameter (L*) at different relative humidities (RH) and protein concentrations (%). (**A**) Graph of the analysis of the L* parameter for powder additives. (**B**) Images of the powder additives: control, 33, and 75% RH. Green bars indicate relative humidity 33%, and violet bars indicate a relative humidity 75%. FDP = freeze-drying powder at different chickpea protein concentrations in % *w*/*v*. Different letters (a,b) indicate significant differences ($p < 0.05$) between samples.

4. Conclusions

The extracted compounds obtained from walnut green husk showed the presence of phenolic acids, flavonoids, hydrolyzable tannins, and quinones, which are responsible for the antioxidant capacity of the extract. Moreover, herbicides and pesticides were also identified. Still, the extract is potentially non-toxic and can be used as a matrix for phenolic extraction to develop natural additives. Chickpea proteins are shown in this study as emerging polymers for encapsulating the phenolic extract from walnut green husk. The FDP 10% sample presented the best values in the physicochemical and structural characterization, demonstrating the protective effect of chickpea protein on the active compounds. Considering only the antioxidant activity and powder color of additives developed at high protein concentrations (10%), the best storage condition for these powders is a low relative humidity (33%) to maintain the antioxidant compounds' stability. This study demonstrated the importance of storage stability studies for powder-form natural additives. Further studies will require applying this additive to different food matrices and studying its behavior as an antioxidant additive through concentration and sensorial analyses.

Author Contributions: Conceptualization, D.S.-M. and S.M.; Methodology, F.A., M.G.-C. and D.S.-M.; Software, D.S.-M. and M.G.-C.; Validation, D.S.-M., S.M. and R.N.Z.; Formal Analysis, D.S.-M., S.M. and R.N.Z.; Investigation, F.A., D.S.-M., M.G.-C., S.M. and R.N.Z.; Resources, S.M., D.S.-M. and R.N.Z.; Writing—Original Draft Preparation, D.S.-M.; Writing—Review and Editing, D.S.-M., M.G.-C., S.M. and R.N.Z.; Visualization, D.S.-M. and S.M.; Supervision, S.M. and R.N.Z.; Project Administration, S.M.; Funding Acquisition, S.M. and M.G.-C. All authors have read and agreed to the published version of the manuscript.

Funding: This research was funded by Fondecyt N° 1231198 from ANID-Agencia Nacional de Investigación y Desarrollo, Project AYUDANTE_DICYT, Código 082371MS_AYUDANTE, Vicerrectoría de Investigación, Innovación y Creación, University of Santiago de Chile, and Basal AFB220001 (CEDENNA).

Institutional Review Board Statement: The study was conducted in accordance with the Declaration of Helsinki, and approved by the Ethics Committee of University of Santiago de Chile (protocol code 310/2023 of 23 May 2023) for studies involving cell lines.

Data Availability Statement: Data are contained within the article.

Acknowledgments: Authors acknowledge the project FONDEQUIP (ANID) EQM180076 for UPLC-QTOF-ESI-MS measurements.

Conflicts of Interest: The authors declare no conflict of interest.

References

1. Gonzalez, C. El Proyecto Aconcagüino Que Convierte las Cáscaras de Nuez en Paneles para Construcción. País Circular. Chile. 2020. Available online: https://www.paiscircular.cl/industria/el-proyecto-aconcaguino-que-convierte-las-cascaras-de-nuez-en-paneles-para-construccion/ (accessed on 16 February 2023).
2. Matiacevich, S.; Soto Madrid, D.; Gutiérrez Cutiño, M. Circular Economy: Obtaining and encapsulating polyphenolic compounds from agroindustrial waste. *RIVAR Rev. Iberoam. Vitic. Agroind. Rural.* **2023**, *10*, 77–100. [CrossRef]
3. Procomer. Prefieren Productos Saludables Con Etiqueta Limpia. 2019. Available online: https://www.procomer.com/alertas_comerciales/exportador-alerta/prefierenproductos-saludables-con-etiqueta-limpia/ (accessed on 10 December 2023).
4. Benelli, P.; Riehl, C.; Smânia, A.; Smânia, E.; Ferreira, S. Bioactive extracts of orange (*Citrus sinensis* L. Osbeck) pomace obtained by SFE and low-pressure techniques: Mathematical modeling and extract composition. *J. Supercrit. Fluids* **2010**, *55*, 132–141. [CrossRef]
5. Di Donato, P.; Taurisano, V.; Tommonaro, G.; Pasquale, V.; Jiménez, J.; de Pascual-Teresa, S.; Poli, A.; Nicolaus, B. Biological properties of polyphenols extracts from agro industry's wastes. *Waste Biomass Valorization* **2018**, *9*, 1567–1578. [CrossRef]
6. Grand View Research, Inc. Polyphenols Market Size, Share & Trends Analysis Report by Product (Grape Seed, Green Tea, Cocoa), by Application (Beverages, Food, Feed, Dietary Supplements, Cosmetics), and Segment Forecasts. 2019, pp. 2019–2025. Available online: https://www.grandviewresearch.com/industry-analysis/polyphenols-market-analysis (accessed on 27 December 2023).
7. Chilenut. Asociación de Productores y Exportadores de Nueces. 2017. Available online: http://www.chilenut.cl/index.php?seccion=nuez-de-nogal (accessed on 15 December 2023).

8. Oliveira, I.; Sousa, A.; Ferreira, I.; Bento, A.; Estevinho, L.; Pereira, J. Total phenols, antioxidant potential, and antimicrobial activity of walnut (*Juglans regia* L.) green husks. *Food Chem. Toxicol.* **2008**, *46*, 2326–2331. [CrossRef] [PubMed]
9. Soto-Madrid, D.; Gutiérrez-Cutiño, M.; Pozo-Martínez, J.; Zúñiga-López, M.C.; Olea-Azar, C.; Matiacevich, S. Dependence of the ripeness stage on the antioxidant and antimicrobial properties of walnut (*Juglans regia* L.) green husk extracts from industrial by-products. *Molecules* **2021**, *26*, 2878. [CrossRef] [PubMed]
10. Carvalho, M.; Ferreira, P.; Mendes, V.; Silva, R.; Pereira, J.; Jerónimo, C.; Silva, B. Human cancer cell antiproliferative and antioxidant activities of *Juglans regia* L. *Food Chem. Toxicol.* **2010**, *48*, 441–447. [CrossRef]
11. Zhou, Y.; Yang, B.; Jiang, Y.; Liu, Z.; Liu, Y.; Wang, X.; Kuang, H. Studies on cytotoxic activity against HepG-2 cells of naphthoquinones from green walnut husks of Juglans mandshurica Maxim. *Molecules* **2015**, *20*, 15572–15588. [CrossRef]
12. Đorđević, V.; Balanč, B.; Belščak-Cvitanović, A.; Lević, S.; Trifković, K.; Kalušević, A.; Nedović, V. Trends in encapsulation technologies for delivery of food bioactive compounds. *Food Eng. Rev.* **2015**, *7*, 452–490. [CrossRef]
13. Galanakis, C. *Polyphenol Properties, Recovery, and Applications*; Woodhead Publishing, An Imprint of Elsevier: Cambridge, UK, 2018.
14. Labuschagne, P. Impact of wall material physicochemical characteristics on the stability of encapsulated phytochemicals: A Review. *Food Res. Int.* **2018**, *107*, 227–247. [CrossRef]
15. El Asbahani, A.; Miladi, K.; Badri, W.; Sala, M.; Aït Addi, E.; Casabianca, H.; El Mousadik, A.; Hartmann, D.; Jilale, A.; Renaud, F.; et al. Essential oils: From extraction to encapsulation. *Int. J. Pharm.* **2015**, *483*, 220–243. [CrossRef]
16. Akbarbaglu, Z.; Peighambardoust, S.; Sarabandi, K.; Jafari, S. Spray drying encapsulation of bioactive compounds within protein-based carriers; different options and applications. *Food Chem.* **2021**, *359*, 129965. [CrossRef]
17. Day, L.; Cakebread, J.; Loveday, S. Food proteins from animals and plants: Differences in the nutritional and functional properties. *Trends Food Sci. Technol.* **2022**, *119*, 428–442. [CrossRef]
18. Liu, L.; Jiang, L.; Xu, G.; Ma, C.; Yang, X.; Yao, J. Potential of alginate fibers incorporated with drug-loaded nanocapsules as drug delivery systems. *J. Mater. Chem. B* **2014**, *2*, 7596–7604. [CrossRef]
19. Coimbra, P.; Cardoso, F.; Goncalves, E. Spray-drying wall materials: Relationship with bioactive compounds. *Crit. Rev. Food Sci. Nutr.* **2021**, *61*, 2809–2826. [CrossRef]
20. Shakoor, I.; Pamunuwa, G.; Karunaratne, D. Efficacy of alginate and chickpea protein polymeric matrices in encapsulating curcumin for improved stability, sustained release, and bioaccessibility. *Food Hydrocoll. Health* **2023**, *3*, 100119. [CrossRef]
21. Glusac, J.; Isaschar-Ovdat, S.; Fishman, A. Transglutaminase modifies the physical stability and digestibility of chickpea protein-stabilized oil-in-water emulsions. *Food Chem.* **2020**, *315*, 126301. [CrossRef]
22. Jukanti, A.; Gaur, P.; Gowda, C.; Chibbar, R. Nutritional quality and health benefits of chickpea (*Cicer arietinum* L.): A review. *Br. J. Nutr.* **2012**, *108*, S11–S26. [CrossRef]
23. Quan, T.; Benjakul, S.; Sae-leaw, T.; Balange, A.; Maqsood, S. Protein–polyphenol conjugates: Antioxidant property, functionalities, and their applications. *Trends Food Sci. Technol.* **2019**, *91*, 507–517. [CrossRef]
24. Yan, X.; Zeng, Z.; McClements, D.; Gong, X.; Yu, P.; Xia, J.; Gong, D. A review of the structure, function, and application of plant-based protein–phenolic conjugates and complexes. *Compr. Rev. Food Sci. Food Saf.* **2023**, *22*, 1312–1336. [CrossRef] [PubMed]
25. Soto-Madrid, D.; Pérez, N.; Gutiérrez-Cutiño, M.; Matiacevich, S.; Zúñiga, R.N. Structural and physicochemical characterization of extracted proteins fractions from chickpea (*Cicer arietinum* L.) as a potential food ingredient to replace ovalbumin in foams and emulsions. *Polymers* **2023**, *15*, 110. [CrossRef] [PubMed]
26. *AOAC—Official Methods of Analysis*, 16th ed.; Association of Official Analytical Chemists: Washington, DC, USA, 1998.
27. Singleton, V.; Orthofer, R.; Lamuela-Raventós, R.; Lester, P. Analysis of total phenols and other oxidation substrates and antioxidants by means of Folin-Ciocalteu reagent. *Methods Enzym.* **1999**, *299*, 152–178. [CrossRef]
28. Brand-Williams, W.; Cuvelier, M.; Berset, C. Use of a free radical method to evaluate antioxidant activity. *LWT-Food Sci. Technol.* **1995**, *28*, 25–30. [CrossRef]
29. Jiang, J.; Zhang, Z.; Zhao, J.; Liu, Y. The effect of non-covalent interaction of chlorogenic acid with whey protein and casein on physicochemical and radical scavenging activity of in vitro protein digests. *Food Chem.* **2018**, *268*, 334–341. [CrossRef]
30. Fenoglio, D.; Soto Madrid, D.; Alarcón Moyano, J.; Ferrario, M.; Guerrero, S.; Matiacevich, S. Active food additive based on encapsulated yerba mate (*Ilex paraguariensis*) extract: Effect of drying methods on the oxidative stability of a real food matrix (mayonnaise). *J. Food Sci. Technol.* **2021**, *58*, 1574–1584. [CrossRef] [PubMed]
31. Matiacevich, S.; Mery, D.; Pedreschi, F. Prediction of mechanical properties of corn and tortillas chips using computer vision. *Food Bioprocess Technol.* **2012**, *5*, 2025–2030. [CrossRef]
32. Greenspan, L. Humidity fixed points of binary saturated aqueous solutions. *J. Res. Natl. Bur. Stand. A Phys. Chem.* **1977**, *81A*, 89–96. [CrossRef]
33. Tang, J.; Yang, T. Dehydrated Vegetables: Principles and Systems. In *Handbook of Vegetable Preservation and Processing*; Hui, Y.H., Ghazala, S., Graham, D.M., Murrell, K.D., Nip, W.-K., Eds.; Marcel Dekker: New York, NY, USA, 2004.
34. Tacon, A. *Nutricion y Alimentacion de Peces y Camarones Cultivados*; Manual de Capacitación (No. F009. 002); Food and Agricultural Organization (FAO): Geneva, Switzerland, 1989.
35. Bandyopadhyay, P.; Ghosh, A.K.; Ghosh, C. Recent developments on polyphenol–protein interactions: Effects on tea and coffee taste, antioxidant properties and the digestive system. *Food Funct.* **2012**, *3*, 592–605. [CrossRef] [PubMed]
36. Picó, Y. Ultrasound-assisted extraction for food and environmental samples. *Trends Anal. Chem.* **2013**, *43*, 84–99. [CrossRef]

37. Deng, J.; Xu, Z.; Xiang, C.; Liu, J.; Zhou, L.; Li, T.; Yang, Z.; Ding, Z. Comparative evaluation of maceration and ultrasonic-assisted extraction of phenolic compounds from fresh olives. *Ultrason. Sonochem.* **2017**, *37*, 328–334. [CrossRef]
38. do Nascimento Nunes, M.C. Impact of environmental conditions on fruit and vegetable quality. *Stewart Postharvest Rev.* **2008**, *4*, 1–14. [CrossRef]
39. Sheng, F.; Hu, B.; Jin, Q.; Wang, J.; Wu, C.; Luo, Z. The analysis of phenolic compounds in walnut husk and pellicle by UPLC-Q-Orbitrap HRMS and HPLC. *Molecules* **2021**, *26*, 3013. [CrossRef]
40. García-Huertas, P.; Pabón, A.; Arias, C.; Blair, S. Evaluación del efecto citotóxico y del daño genético de extractos estandarizados de *Solanum nudum* con actividad antiplasmodial. *Veter Parasitol. Reg. Stud. Rep.* **2012**, *33*, 78–87. [CrossRef]
41. Anandharamakrishnan, C.; Ishwary, S. *Spray Drying Techniques for Food Ingredient Encapsulation*; Wiley-Blackwell: Chichester, UK, 2015.
42. Robert, P.; Gorena, T.; Romero, N.; Sepulveda, E.; Chavez, J.; Saenz, C. Encapsulation of polyphenols and anthocyanins from pomegranate (*Punica granatum*) by spray drying. *Int. J. Food Sci. Technol.* **2010**, *45*, 1386–1394. [CrossRef]
43. Pikal, M.J.; Rambhatla, S.; Ramot, R. The impact of freezing stage in lyophilization: Effect of the ice nucleation temperature on process design and product quality. *Am. Pharm. Rev.* **2002**, *5*, 48–53.
44. Ezhilarasi, P.N.; Indrani, D.; Jena, B.S.; Anandharamakrishnan, C. Freeze drying technique for microencapsulation of Garcinia fruit extract and its effect on bread quality. *J. Food Eng.* **2013**, *117*, 513–520. [CrossRef]
45. Khazaei, K.M.; Jafari, S.M.; Ghorbani, M.; Kakhki, A.H. Application of maltodextrin and gum Arabic in microencapsulation of saffron petal's anthocyanins and evaluating their storage stability and color. *Carbohydr. Polym.* **2014**, *105*, 57–62. [CrossRef] [PubMed]
46. Kong, J.; Yu, S. Fourier transform infrared spectroscopic analysis of protein secondary structures. *Acta Biochim. Biophys. Sin.* **2007**, *39*, 549–559. [CrossRef] [PubMed]
47. Heurtault, B.; Saulnier, P.; Pech, B.; Proust, J.E.; Benoit, J.P. Physico-chemical stability of colloidal lipid particles. *Biomaterials* **2003**, *24*, 4283–4300. [CrossRef] [PubMed]
48. Salvia-Trujillo, L.; Rojas-Graü, M.A.; Soliva-Fortuny, R.; Martín-Belloso, O. Effect of processing parameters on physicochemical characteristics of microfluidized lemongrass essential oil-alginate nanoemulsions. *Food Hydrocoll.* **2013**, *30*, 401–407. [CrossRef]
49. Martínez-Flórez, S.; González-Gallego, J.; Culebras, J.M.; Tuñón, M.J. Los flavonoides: Propiedades y acciones antioxidantes. *Nutr. Hosp.* **2002**, *17*, 271–278.
50. Vani, B.; Zayas, J.F. Wheat germ protein flour solubility and water retention. *J. Food Sci.* **1995**, *60*, 845–848. [CrossRef]
51. Boye, J.; Zare, F.; Pletch, A. Pulse proteins: Processing, characterization, functional properties and applications in food and feed. *Food Res. Int.* **2010**, *43*, 414–431. [CrossRef]
52. Ma, K.K.; Greis, M.; Lu, J.; Nolden, A.A.; McClements, D.J.; Kinchla, A.J. Functional performance of plant proteins. *Foods* **2022**, *11*, 594. [CrossRef] [PubMed]
53. Lunkad, R.; Barroso da Silva, F.L.; Košovan, P. Both charge-regulation and charge-patch distribution can drive adsorption on the wrong side of the isoelectric point. *J. Am. Chem. Soc.* **2022**, *144*, 1813–1825. [CrossRef] [PubMed]
54. Moreira, T.; Delgado, H.; Urra, C.; Fando, R. Efecto de un aditivo antihumectante y de otro antioxidante sobre las características higroscópicas y sobre la viabilidad de dos formulaciones vacunales liofilizadas de Vibrio cholerae. *Rev. CENIC. Cienc. Biol.* **2010**, *41*, 1–10.

polymers

Article

Preliminary Assessment of Tara Gum as a Wall Material: Physicochemical, Structural, Thermal, and Rheological Analyses of Different Drying Methods

Elibet Moscoso-Moscoso [1,2,*], Carlos A. Ligarda-Samanez [1,2,*], David Choque-Quispe [1,2], Mary L. Huamán-Carrión [1], José C. Arévalo-Quijano [1,3], Germán De la Cruz [1,4], Rober Luciano-Alipio [1,5], Wilber Cesar Calsina Ponce [1,6], Reynaldo Sucari-León [1,7], Uriel R. Quispe-Quezada [1,8] and Dante Fermín Calderón Huamaní [1,9]

1 Nutraceuticals and Biomaterials Research Group, Universidad Nacional José María Arguedas, Andahuaylas 03701, Peru; dchoque@unajma.edu.pe (D.C.-Q.); huamancarrionmary@gmail.com (M.L.H.-C.); jcarevalo@unajma.edu.pe (J.C.A.-Q.); german.delacruz@unsch.edu.pe (G.D.l.C.); rluciano@unaat.edu.pe (R.L.-A.); wcalsina@unap.edu.pe (W.C.C.P.); rsucari@unah.edu.pe (R.S.-L.); uquispe@unah.edu.pe (U.R.Q.-Q.); dante.calderon@unica.edu.pe (D.F.C.H.)

2 Research Group in the Development of Advanced Materials for Water and Food Treatment, Universidad Nacional José María Arguedas, Andahuaylas 03701, Peru

3 Department of Education and Humanities, Universidad Nacional José María Arguedas, Andahuaylas 03701, Peru

4 Agricultural Science Faculty, Universidad Nacional de San Cristobal de Huamanga, Ayacucho 05000, Peru

5 Administrative Sciences Faculty, Universidad Nacional Autónoma Altoandina de Tarma, Junín 12731, Peru

6 Social Sciences Faculty, Universidad Nacional del Altiplano, Puno 21001, Peru

7 Engineering and Management Faculty, Universidad Nacional Autónoma de Huanta, Ayacucho 05000, Peru

8 Agricultural and Forestry Business Engineering, Universidad Nacional Autónoma de Huanta, Ayacucho 05000, Peru

9 Ambiental Engineering, Universidad Nacional San Luis Gonzaga, Ica 11001, Peru

* Correspondence: elibetmm22@gmail.com (E.M.-M.); caligarda@unajma.edu.pe (C.A.L.-S.)

check for updates

Citation: Moscoso-Moscoso, E.; Ligarda-Samanez, C.A.; Choque-Quispe, D.; Huamán-Carrión, M.L.; Arévalo-Quijano, J.C.; De la Cruz, G.; Luciano-Alipio, R.; Calsina Ponce, W.C.; Sucari-León, R.; Quispe-Quezada, U.R.; et al. Preliminary Assessment of Tara Gum as a Wall Material: Physicochemical, Structural, Thermal, and Rheological Analyses of Different Drying Methods. *Polymers* **2024**, *16*, 838. https://doi.org/10.3390/polym16060838

Academic Editors: Zenaida Saavedra-Leos and Cesar Leyva-Porras

Received: 31 January 2024
Revised: 12 March 2024
Accepted: 14 March 2024
Published: 19 March 2024

Abstract: Tara gum, a natural biopolymer extracted from *Caesalpinia spinosa* seeds, was investigated in this study. Wall materials were produced using spray drying, forced convection, and vacuum oven drying. In addition, a commercial sample obtained through mechanical methods and direct milling was used as a reference. The gums exhibited low moisture content (8.63% to 12.55%), water activity (0.37 to 0.41), bulk density (0.43 to 0.76 g/mL), and hygroscopicity (10.51% to 11.42%). This allows adequate physical and microbiological stability during storage. Polydisperse particles were obtained, ranging in size from 3.46 μm to 139.60 μm. Fourier transform infrared spectroscopy characterisation confirmed the polysaccharide nature of tara gum, primarily composed of galactomannans. Among the drying methods, spray drying produced the gum with the best physicochemical characteristics, including higher lightness, moderate stability, smaller particle size, and high glass transition temperature (141.69 °C). Regarding rheological properties, it demonstrated a non-Newtonian pseudoplastic behaviour that the power law could accurately describe. The apparent viscosity of the aqueous dispersions of the gum decreased with increasing temperature. In summary, the results establish the potential of tara gum as a wall material applicable in the food and pharmaceutical industries.

Keywords: tara gum; wall material; encapsulation; physicochemical properties; thermal properties; structural properties; rheological properties

1. Introduction

The quest for natural polymers has intensified interest in unconventional plant resources [1–3]. Among these, tara gum extracted from the seeds of *Caesalpinia spinosa* stands out as a promising polymer. Tara gum has become a valuable source of polysaccharides, boasting advantageous physicochemical and functional properties, making it a material of growing interest in polymer science and food applications [4–6]. Its potential as an encapsulating agent has been acknowledged in diverse industry applications [5,7,8], attributed to its capacity for enhancing active compounds' stability and controlled release. The increasing significance is underscored by the abundant gum production in Peru, accounting for 80% of the world's total [6], driving the need for an exhaustive characterisation to take full advantage of this resource.

Only a few natural biopolymers act as encapsulation matrices [9,10]; the most common are gum arabic, maltodextrin, xanthan gum, and emulsifying starches, which are more widely used as encapsulating agents for food applications [11,12]. Therefore, it is necessary to evaluate the characteristics of tara gum since little is known about its physicochemical, structural, thermal, and rheological properties. It also represents a promising and relatively unexplored area of study on polymers and encapsulating materials in the current research context [7]. It is widely available in Peru and produces pods of high economic value due to the tannins in the leaflets and the gum in the endosperm of the seeds [13]. Conversely, various starches, proteins, maltodextrins, and gums are employed as wall materials for encapsulating diverse bioactive compounds using various techniques [1,14–17]. Understanding this gum's structure and physicochemical properties can contextualise it within current encapsulant options, identifying its advantages and possible areas for improvement. This approach contributes significantly to advancing the field by expanding the repertoire of available materials and promoting sustainable solutions [18,19].

While tara gum holds encapsulating potential, its behaviour under physicochemical, structural, thermal, and rheological analyses remains to be thoroughly explored. This study comprehensively characterises fundamental properties: yield, moisture, water activity, hygroscopicity, bulk density, colour, ζ-potential, particle size, polydispersity, structure, elemental composition, functional groups, thermal stability, and flow behaviour. The results provide a rigorous understanding of the essential properties of tara gum and confirm its potential as an encapsulating material for multiple industrial applications. The main objective is to lay a solid scientific foundation to use this local resource in encapsulation, promoting regional circular economy and sustainability.

2. Materials and Methods

2.1. Materials

The tara pods used were kindly provided by farmers in the district of Talavera, province of Andahuaylas, Apurimac Region in Perú (coordinates 13°40′17.11″ S, 73°25′0.39″ W). Likewise, a commercial organic sample of tara gum, produced by the company Associated Mills SAC, was used for comparative purposes. We also used absolute ethanol (Scharlau, Sentmenat, Spain), sodium chloride (Spectrum Chemical Mfg. Corp, Bathurst, NB, Canada), and potassium bromide IR-grade (Thermo Fisher Scientific, Garfield, NJ, USA).

2.2. Obtaining the Tara Gum

A total of 120 g of germ-free tara seeds were placed in 800 mL of distilled water and stirred using an M6 thermomagnetic stirrer (CAT, Ballrechten-Dottingen, Germany) at 70 °C for 12 h to obtain a viscous substance. The extract was then filtered, purified, and mixed with 96% ethanol in a 1:1 ratio. This process was carried out to precipitate the viscous substance [20,21].

Subsequently, the purified extract was separated into three parts and subjected to a drying process using three methods. In the first method, a mini spray dryer B-290 (BÜCHI Labortechnik AG, Flawil, Switzerland) was used, with an inlet temperature of 100 °C and a pumping rate of 20%. The second method used forced convection oven drying FED 115 (BINDER, Tuttlingen, Germany) at 60 °C for 24 h. In the third method, a vacuum oven VD56 (Binder, Tuttlingen, Germany) was used at 30 °C and a pressure of 10 mBar.

Finally, the material obtained was collected in low-density polyethylene bags and stored in a desiccator at 20 °C for later use. Likewise, a commercial sample of tara gum produced by the company Associated Mills SAC was acquired and used for comparative purposes, which was obtained through the mechanical processes of the separation and milling of the endosperm from the tara seed, being considered a natural and gluten-free product, used as a thickener in the food industry. An experimental flow diagram is shown in Figure 1.

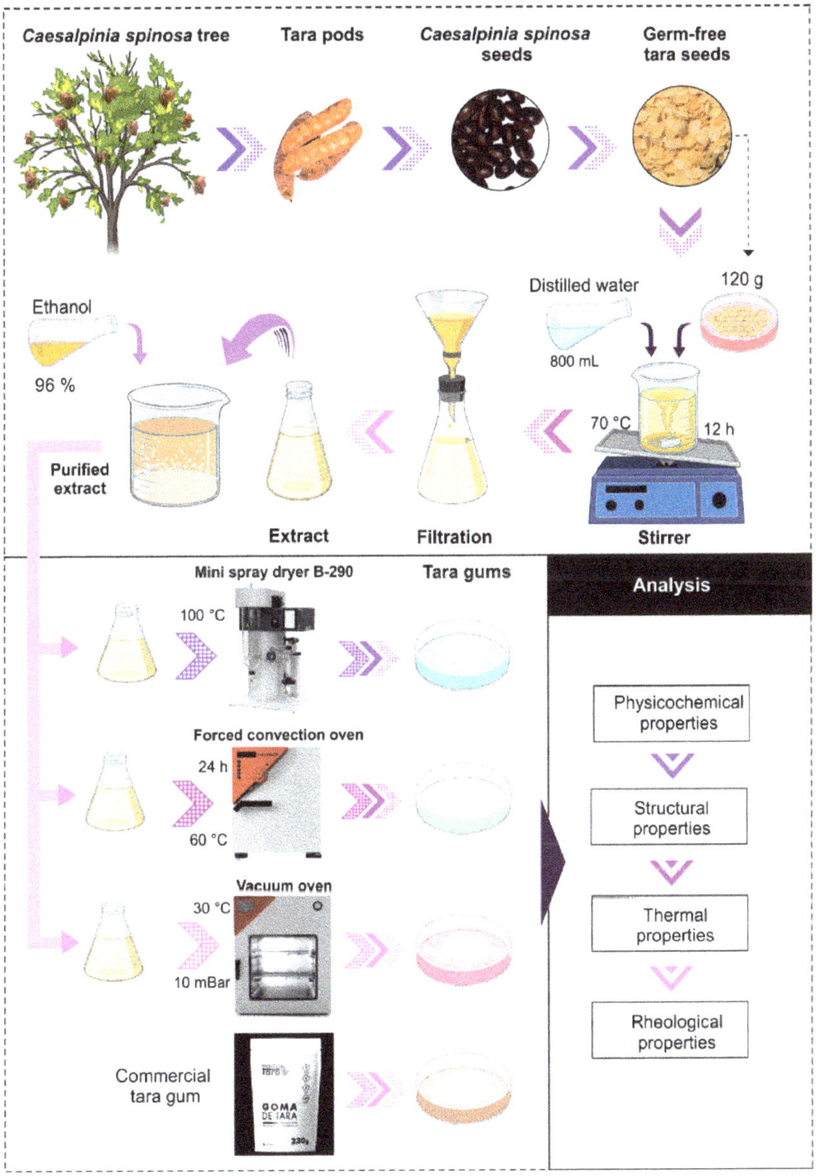

Figure 1. Experimental flow diagram.

2.3. Yield

The yield was calculated based on the initial mass of the tara seeds used, and the final mass obtained after the extraction and purification was used to determine the efficiency of obtaining tara gum. The yield was expressed as a percentage and determined using the ratio between the initial and final mass, according to the following equation [22]:

$$Y\ (\%) = \frac{mi}{mf} \times 100 \quad (1)$$

where $Y\ (\%)$ is the yield in percent, mi is the initial mass (g), and mf is the final mass (g).

2.4. Moisture

Moisture was determined according to the methodology of AOAC 950.10. A total of 2 g of sample was weighed into watch capsules and placed inside a forced convection oven FED 115 (BINDER, Tuttlingen, Germany) at a controlled temperature of 105 °C until a constant weight was reached. Moisture content was expressed as the percentage ratio of the sample weight after drying to the sample weight before drying [23].

2.5. Water Activity (Aw)

To determine the Aw of the tara gum samples, specialised equipment known as a water activity meter was used. In particular, the HygroPalm23-AW model from Rotronic (Bassersdorf, Switzerland) was used. Before the measurements, the equipment was calibrated according to the manufacturer's instructions. This ensured the reliability of the Aw readings as the meter was adjusted with standard solutions. Tara gum samples were placed in the reading chamber of the equipment, which measures the equilibrium relative humidity on the sample. This relative humidity value is converted to Aw according to the product characteristic curves stored in the meter's memory [24].

2.6. Hygroscopicity

A total of 1 g of the sample was dispersed in 100 mL of a saturated NaCl solution (with a relative humidity of 75%) in an airtight container. The temperature was kept constant at 25 °C. After seven days, the samples were weighed again, and the results were expressed as a percentage increase in mass [25]:

$$\%H = \left(\frac{m3 - m2}{m2 - m1}\right) \times 100 \quad (2)$$

where $\%H$ is the mass increase in percent, $m1$ is the weight of the empty Petri dish, $m2$ is the weight of the Petri dish + sample, and $m3$ is the weight of the Petri dish + sample after seven days.

2.7. Bulk Density

The displaced volume method was used to determine the bulk density of the tara gum samples. This consists of weighing a known amount of the powdered sample and placing it in a 10 mL graduated cylinder. The graduated cylinder was gently tapped on a smooth, firm surface to settle the powder. The tapping removed the voids between the particles. Finally, the volume displaced by the sample after tapping the graduated cylinder was recorded. With the known mass and the measured volume, the bulk density (g/mL) was calculated using the mass/volume ratio [26].

2.8. Colour Analysis

A CR-5 colorimeter from Konica Minolta (Tokyo, Japan) was used, which allows the rapid and objective measurement of different colorimetric parameters based on the interaction of the sample with a standard light source. Before the measurements, the colorimeter was calibrated with standard black and white reference plates to ensure reliable readings. Then, powdered tara gum samples were successfully placed inside the reading cell, and the lightness values (L^*), chroma a^*, and chroma b^* were recorded [27,28].

2.9. ζ-Potential

The dynamic laser light scattering technique was used to evaluate colloidal stability using ζ-potential measurements. Powdered samples (20 mg) were dispersed in ultrapure water (50 mL) and subjected to ultrasonication for 60 s before readings to ensure adequate particle distribution. Measurements were performed with a Zetasizer ZSU3100 (Malvern Instruments, Worcestershire, UK) at 25 °C using a specific DTS1080 cell to read the ζ-potential of the dispersions. This electrokinetic analysis allowed the prediction of the stability of the polymer suspensions due to the magnitude of the repulsive forces between the particles [29].

2.10. Particle Size and Polydispersity

To determine the particle size and polydispersity using a laser light scattering method, the sample particles were dispersed in isopropanol and subjected to ultrasound for 30 s. Then, a 600 nm helium–neon laser was incident on the sample, and the resulting diffraction pattern was measured using a Mastersizer 3000 instrument (Malvern Instruments, Worcestershire, UK) [30]. The polydispersity was determined using the amplitude index [31,32] according to the following relationship:

$$Span\ index = \frac{D\ (90) - D\ (10)}{D\ (50)} \tag{3}$$

where $D\ (10)$, $D\ (50)$, and $D\ (90)$ correspond to the diameters relative to 10%, 50%, and 90% of the cumulative size distribution.

2.11. SEM-EDS Analysis

Morphological and elemental analyses were carried out using a scanning electron microscope (SEM) Prism E model (Thermo Fisher, Waltham, MA, USA). To prepare the tara gum samples, they were arranged on carbon adhesive tape, ensuring uniform distribution. Photomicrographs were captured at 30 kV, 2500×, and 500× magnification under low-vacuum conditions, allowing a detailed evaluation of the surface morphology. In addition, the elemental analysis was carried out using X-ray energy dispersion (EDS) [33].

2.12. FTIR Analysis

The functional groups of the tara gums were analysed using a Fourier transform spectrophotometer (FTIR) model Nicolet IS50 (ThermoFisher, Waltham, MA, USA). Tablets were prepared with 1 mg of sample and 99 mg of KBr. The mixture was crushed and passed through a press at 10 tons [34]. For the readings, the transmission module was used in the range of 400 cm^{-1} to 4000 cm^{-1}, making use of a resolution of 8 cm^{-1}, 32 scans were carried out to ensure an accurate capture of the spectral information [26].

2.13. Thermal Analysis

A total of 10 mg of the sample was utilised for thermal stability analysis via thermogravimetry (TGA). The measurement was performed using a thermal analyser (TA Instrument, New Castle, DE, USA) with a heating rate of 10 °C/min. A differential scanning calorimeter (DSC2500, TA Instrument, New Castle, DE, USA) was also used with 2 mg of the sample. The temperature range was set from 0 to 250 °C, with a heating rate of 10 °C/min in a nitrogen atmosphere [22].

2.14. Rheological Analysis

The experimental samples were continuously tested through an MCR 702e MultiDrive rotational rheometer (Anton Paar GmbH, Austria). Flow curve measurements were determined using the CC27 concentric cylinder with a diameter of 26.661 mm and a length of 39.996, with a cup diameter of 29.914 mm at a shear rate of 1 to 300 s^{-1}. They were then analysed using shear stress models for fluids through the power law, Bingham Plastic, and Herschel–Bulkley models [35]. Table 1 shows the details of the rheological models.

Table 1. Rheological models for non-Newtonian fluids.

Model	Equation	Parameters
Power Law	$\tau = k\gamma^n$	k, n
Bingham Plastic	$\tau = \tau_y + n_B\gamma$	τ_y, n_B
Herschel–Bulkley	$\tau = \tau_y + k_H\gamma^n$	τ_y, k_H, n

Note: τ is the yield strength (Pa), γ is the shear rate (s^{-1}), k is the consistency index (Pa*s^n), n is the behavioural index, τ_y is the yield strength (Pa), n_B is the plastic viscosity (Pa*s), and k_H is the consistency index (Pa*s^n).

2.15. Statistical Analysis

Tukey's multiple comparison test was used to evaluate possible significant differences between groups, with a confidence level of 95%, after analysing variance (ANOVA). This analysis allowed the variations between groups to be examined and statistical disparities to be accurately determined. All statistical analyses and graphical representations were realised using OriginPro 2024 software from Origin Lab Corporation, based in Northampton, MA, USA. This methodological approach provided a rigorous and comprehensive evaluation of the differences between groups, ensuring a robust interpretation of the results obtained.

3. Results and Discussions

3.1. Physical and Chemical Properties of Tara Gum

Table 2 presents the physicochemical properties of spray-dried tara gum (GA), forced convection oven-dried tara gum (GE), vacuum oven-dried tara gum (GV), and commercial tara gum (GC) for comparison. In terms of yield, the vacuum oven drying method (GV) yielded the highest value at 45.21%, indicating optimistic process efficiency. Moisture content ranged from 8.63% to 12.55% across different samples, with these values being considered adequate as low moisture promotes material stability [36]. Water activity (Aw) values ranged from 0.39 to 0.41, a desirable characteristic for encapsulating agents due to increased resistance to microbial growth and biochemical deterioration reactions [37]. Other authors reported water activity values of 0.55 for almond gum and 0.53 for gum arabic [38]. Hygroscopicity values below 13.41% were found, a fundamental value, considering that the wall materials must present inert properties toward the active ingredients of the core, which must also be non-hygroscopic [39]. Hygroscopicity is a critical factor affecting the stability of food products [32]. Powders with values above 15.1% are considered highly hygroscopic [40]. On the other hand, powders with hygroscopicity values below 10% are considered non-hygroscopic and between 10.1% and 15% slightly hygroscopic, determined at a relative humidity of 75% [41]. Also, hygroscopicity values were reported for gum arabic between 26% and 40%, where fine particles (37 μm) with the highest value of 40% presented a high capacity to retain and adsorb water [42]. Concerning bulk density, values between 0.43 and 0.76 g/mL were found, which coincide with the densities of xanthan gum (0.46 g/mL) [43] and gum arabic (0.50 g/mL) [37].

The lightness values (L^*) indicated that the spray-dried tara gum sample (GA) was the lightest, indicating a better preservation of the original characteristics through this technique, while the oven-dried samples (GE and GV) showed some darkening; similar values were obtained for gum arabic [37]. Regarding the tendency toward reddish (positive a^* values) or greenish (negative a^* values) tones, the GA sample presented a slightly reddish tone, unlike the rest; the commercial sample (GC) was the one with the highest greenish intensity. In the case of gum arabic, the value recorded for this parameter was 2.48 [37]. Yellow tones (positive b^* values) were observed in all samples; other authors have reported values of 2.48 to 3.39 and 11.62 to 14.29 for gum arabic and almond gum, respectively [37].

The zeta potential values were negative in all tara gum samples due to ionisable acidic groups in their structures [22,44]. The zeta potential represents the potential difference between the compact ion layer and the dispersant medium, determining the repulsive force between the particles and their colloidal stability [45]. All the gums presented moderate stability between −16.57 and −23.77 millivolts [46].

Table 2. Physical and chemical properties of tara gum obtained using different methods.

Properties	GA		GE		GV		GC	
	$\bar{x}$ ±SD	*	$\bar{x}$ ±SD	*	$\bar{x}$ ±SD	*	$\bar{x}$ ±SD	*
Yield (%)	12.18 ± 0. 45	a	42.38 ± 0.44	b	45.21 ± 0.60	c	51.37 ± 0.19	d
Moisture (%)	8.63 ± 0. 35	a	11.42 ± 0.40	bc	12.55 ± 0.48	c	10.86 ± 0.53	b
Aw	0.37 ± 0.01	a	0.39 ± 0.01	b	0.41 ± 0.01	b	0.40 ± 0.01	b
Hygroscopicity (%)	10.51 ± 0. 41	a	12.29 ± 0.43	b	13.41 ± 0.36	c	11.42 ± 0.16	d
Bulk Density (g/mL)	0.43 ± 0.01	a	0.71 ± 0.01	b	0.68 ± 0.01	c	0.76 ± 0.01	d
L^*	91.02 ± 0.01	a	77.82 ± 0.01	b	79.62 ± 0.01	c	89.75 ± 0.01	d
a^*	0.06 ± 0.01	a	−0.33 ± 0.05	b	−0.17 ± 0.04	c	−0.41 ± 0.01	b
b^*	3.52 ± 0.05	a	4.23 ± 0.05	b	6.01 ± 0.05	c	9.83 ± 0.02	d
ζ-Potential (mV)	−16.57 ± 0.31	a	−22.29 ± 0.38	b	−22.97 ± 0.32	bc	−23.77 ± 0.29	c
Particle Size (μm)	3.46 ± 0.01	a	30.94 ± 0.79	b	115.80 ± 1.64	c	139.60 ± 0.89	d
Polydispersity	1.77 ± 0.01	a	2.88 ± 0.09	b	3.09 ± 0.05	c	1.54 ± 0.02	d

Note: $\bar{x}$ is the arithmetic mean, SD is the standard deviation, and different letters indicate significant differences per row evaluated for triplicates at 5% significance (*), the background color is the reference of the three chromatic coordinates. Spray-dried tara gum (GA), forced convection oven-dried tara gum (GE), vacuum oven-dried tara gum (GV), and commercial tara gum (GC).

The data indicate that tara gum is favourable for developing spray-drying encapsulation processes, avoiding the agglomeration of the encapsulating particles [22,30,47]. In food applications, adequate colloidal stability improves the interaction with the charged components of the matrix. The results are similar to those of commercial samples of tara gum used in spray-drying encapsulation processes [30,47]. The negative values suggest its usefulness in food and pharmaceutical products, promoting favourable interactions with its negative charge [5]. In addition to zeta potential, pH, ionic strength, and additives influence colloidal stability, requiring a more comprehensive approach for optimising the functional properties of these gums in various food and pharmaceutical applications [48].

The GA sample obtained by spray drying had the most petite particle sizes with a D10 of 1.71 μm, D50 of 3.46 μm, and D90 of 7.84 μm. This indicates that 90% of the particles were below 7.84 μm, with a mean of 3.46 μm and a minimum of 1.71 μm. This narrow distribution of sizes less than 10 microns makes the GA sample the most suitable for forming the walls of microcapsules, whose sizes vary between 1 and 100 μm. In contrast, the GV sample obtained by vacuum drying exhibited the largest sizes with a D90 of 375.40 μm and an average of 115.80 μm. Considering that the size of the nanostructures is located between the range of 1–1000 nm, this sample would not be helpful as a wall material for the nanoencapsulation process, for which the GA and GE samples with a D50 below 50 μm seem to be more viable. Regarding polydispersity, the GA sample showed the lowest index of 1.77, while GE and GV were the most polydisperse with indexes of 2.88 and 3.09, respectively. This reflects that spray drying and convection oven drying methods give rise to more homogeneous distributions of particle sizes [1,49].

Polydispersity constitutes an important indicator that describes the homogeneity of particle size distributions [45] for tara gum samples. This shows that most of the samples studied presented heterogeneous size distributions with significant variations between the most extensive and most minor fractions, according to the parameters D90 and D10, respectively. However, it should be noted that the tara gum sample obtained by spray drying (GA) exhibited the lowest polydispersity index of 1.77. This indicates that using this technique makes it possible to obtain the narrowest and most homogeneous distribution of particle sizes compared to the other methods evaluated. Higher polydispersity values were obtained for xanthan gum with 9.70 [50]. The results show that the tara gum obtained by spray drying has the most suitable characteristics to be used as a wall material in micro- and nanoencapsulation processes.

The results shown above show that tara gum's physicochemical properties vary according to the drying method used. Spray-dried tara gum presented the best characteristics for use as an encapsulant. Spray drying allows for the quick drying of droplets and the formation of tiny, uniform spherical particles [51,52]. This method prevents browning and thermal degradation with convection and vacuum drying [53,54]. Therefore, choosing the drying method is a crucial factor that must be carefully considered during the production of gums for encapsulation applications [55]. However, further research is necessary to optimise the drying parameters to obtain gums with ideal physicochemical properties. In conclusion, this study provides valuable information to guide the selection and production of tara gums as wall materials in encapsulation processes.

3.2. SEM-EDS Analysis

Figure 2 shows microphotographs of the tara gum samples, which presented variability in shape and size. Sample GA was observed at a magnification of 2500× (scale bar of 50 μm), and GE, GV, and GC were observed at 500× (scale bar of 400 μm). The GA sample showed a homogeneous surface without large visible pores and with mostly spherical shapes. Concerning GE, a primarily smooth topography with defined particles was observed. At the same time, GV exhibited an irregular surface, with numerous pores distributed over the particles, and some angular edges were distinguished, denoting fragility. GC showed a heterogeneous appearance with some aggregated particles, in which a wide distribution of particle sizes can be distinguished.

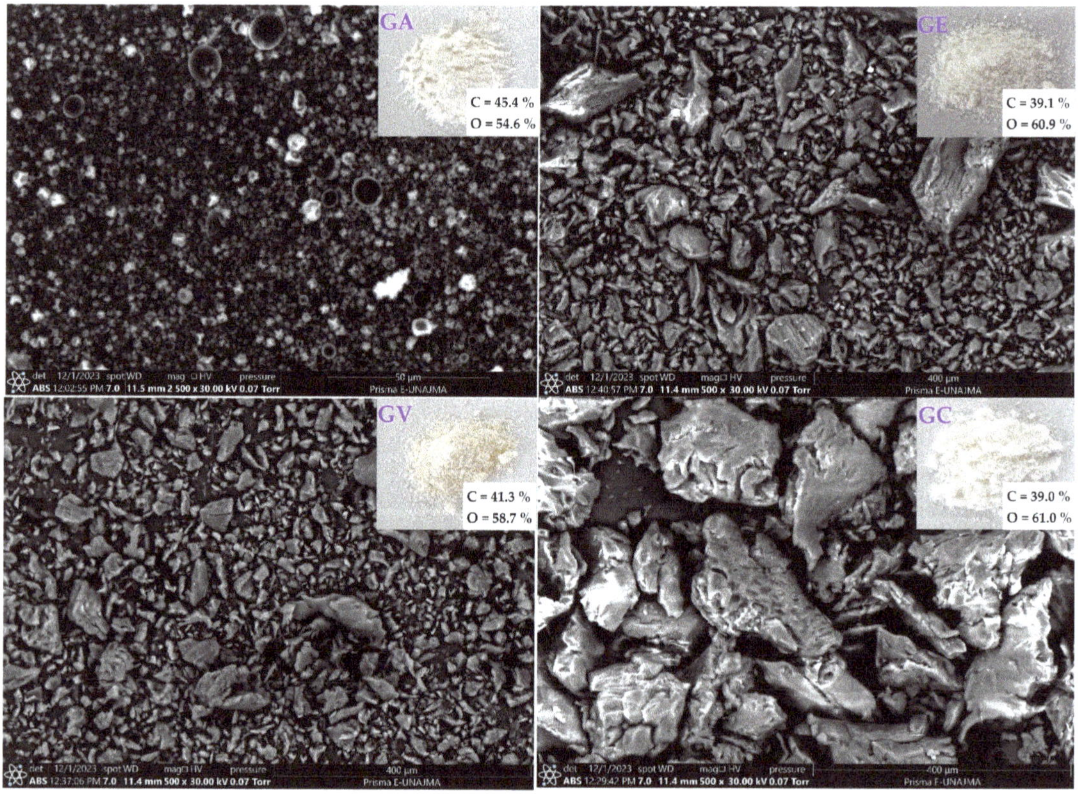

Figure 2. Scanning electron microscopy analysis on spray-dried tara gum (GA), forced convection-dried tara gum (GE), vacuum-dried tara gum (GV), and commercial tara gum (GC).

GA presented particles with a smaller size and a more homogeneous shape; the sizes of the particles can influence the dissolution rate of powdered gums [56], which in turn directly influences the intrinsic viscosity and molecular mass [57]. Generally, the dissolution rate of polysaccharide powders increases with a reduction in particle size [37,56]. Likewise, the exudates' gums have various shapes and sizes [37,57].

Likewise, the C and O surface contents were analysed via energy-dispersive X-ray spectroscopy (EDS). It was observed that the carbon content varied between 39% and 45.4%, and the oxygen content varied between 54.6% and 61%. No significant differences in the C and O compositions were observed between the gums. EDS analysis confirmed that the different tara gum samples possess a similar composition of carbon and oxygen, which are chemical elements in biopolymers used as wall materials [26,58,59]. This verifies that the obtained gums are suitable with wall materials in micro- and nanoencapsulation processes. The different drying methods used do not alter this desired composition.

SEM and EDS analyses show that the drying method affects the morphology of tara gum. Spray drying achieves homogeneous spherical particles ideal for encapsulation, as water's rapid evaporation fixes the droplets' spherical shape [60]. Convective and vacuum methods cause irregularities due to higher thermal exposure [61]. The C and O composition remains stable, indicating that the chemical nature does not change. The drying conditions must be optimised to obtain the desired microstructure while preserving the composition.

3.3. FTIR Analysis

The infrared spectra of the gums obtained through spray drying, a forced convection oven, a vacuum oven, and commercially are shown in Figure 3. The four samples presented similar absorption patterns, indicating a similar chemical composition based on polysaccharides. Characteristic bands of polysaccharides such as hydroxyl (–OH), carbonyl (C=O), and C–O and O–H bonds were observed. Validated methodology was used to analyse and interpret the infrared (IR) spectra [62].

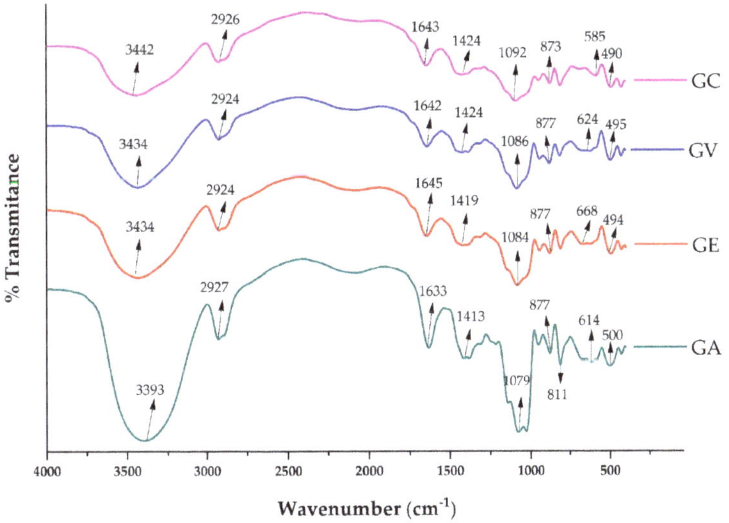

Figure 3. FTIR spectra of commercial tara gum (GC), vacuum-dried tara gum (GV), forced convection-dried tara gum (GE), and spray-dried tara gum (GA).

The IR spectra of the tara gum samples showed a broad absorption band at 3393–3442 cm^{-1} due to the stretching frequency of the OH group [63,64]. The presence of –OH stretching could be attributable to sugars such as galactose and arabinose [5]. The absorption band between 2924 and 2926 cm^{-1} is attributable to the stretching of the CH of the alkyl group [37,65], and those located between 1413 and 1424 cm^{-1} are due to the bending of the CH of the methyl group [64,65]. Likewise, the absorption bands located between 1633 and 1643 cm^{-1} are attributable to the asymmetric stretching of the carboxylate ion [65].

The bands located between 1079 and 1092 cm^{-1} corresponded to the presence of galactans [37]. Finally, the bands lying between 494 and 668 cm^{-1} were attributed to skeletal mode vibrations of the pyranose rings [37]. The FTIR analysis showed that tara gum's chemical composition is preserved regardless of the drying method. The characteristic bands of functional groups are maintained, indicating that the polysaccharide structure is not significantly altered.

3.4. Thermal Analysis

Graphical representations of the TG curves of all the samples of the tare gums are shown in Figure 4a–d. It can be noted that the samples exhibited similar thermal behaviours, evidencing the presence of mainly two events.

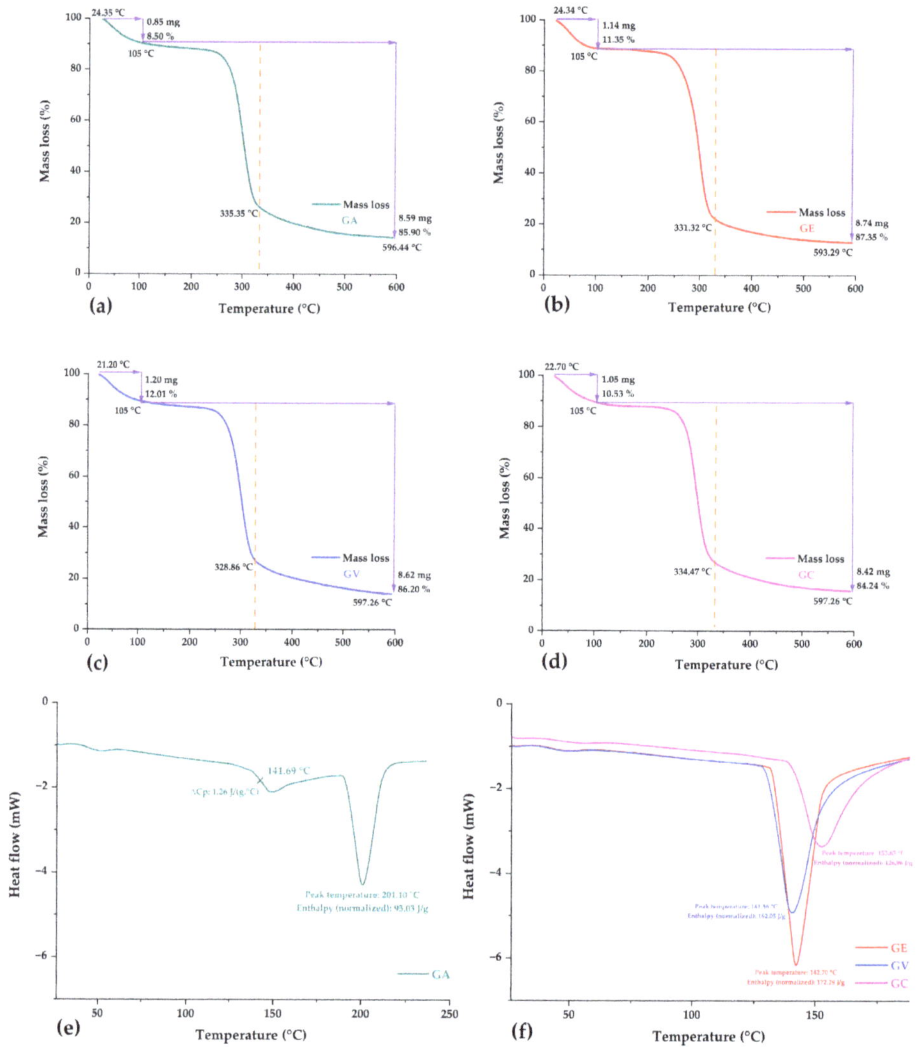

Figure 4. Thermogravimetric analysis: (**a**) spray-dried tara gum GA, (**b**) oven-dried tara gum GE, (**c**) vacuum-dried tara gum GV, (**d**) commercial tara gum GC, (**e**) DSC curves in GA, (**f**) DSC curves in GE, GV, and GC.

The first event occurred between 21.20 °C and 105 °C, with a mass loss between 8.50% and 12.01%, similar to the moisture content determined in the samples, which would correspond to the evaporation of free water due to the hydrophilic nature of the functional groups of the polysaccharides present in the polymeric matrix, this decrease is attributed to the onset of hydrogen bond breaking and the evaporation of water molecules, in addition to the elimination of other thermolabile compounds of low molecular weight [66,67].

A second event occurred between 105 °C and around 600 °C (with peaks around 328.86 °C and 335.35 °C), which was associated with a mass loss of between 84.24% and 87.35%, which is related to the thermal decomposition of the carbohydrates present in the tara gum [67,68]. At high temperatures, other organic compounds were eliminated in the gums, culminating in obtaining final residues [59,69,70].

Figure 4e illustrates the DSC analysis performed on all samples; in the specific case of the spray-dried tara gum (GA), an elevated glass transition temperature of 141.69 °C was obtained as a slight change in the slope of the curve [71]. This finding offers valuable insights into phase transitions and structural modifications within this material [33,72]. Glass transition temperature (Tg) is a critical parameter that influences the functional properties and encapsulation behaviours of polymeric materials, thus impacting their potential applications. A higher Tg can facilitate elevated temperatures during encapsulation processes, which is crucial for developing specific applications [22,30,47]. Similar glass transition temperatures have been reported for various polymeric wall materials used in spray-drying processes, as was the case for maltodextrin (155.34 °C) [72]; also, Kurozawa and Deschamps reported a Tg of 160 °C for maltodextrin [73,74], arabic gum (139.81 °C), native potato starch (138.26 °C) [22], quinoa starch (139.21 °C) [75], and spray-dried tara gum (157.7 °C) [22]. Glass transition temperatures below these values in the microcapsules and nanocapsules would indicate that encapsulation was successfully achieved [33,72].

In the cases of the GE, GV, and GC gums (Figure 4f), endothermic melting peaks of 142.70 °C, 141.36 °C, and 153.63 °C were observed. In addition, an absence of change in heat capacity between the initial and final state was observed, which was a limitation for determining the glass transition temperature in the present study. Therefore, further analysis is required for a complete understanding of the thermal processes present in these gums, as it is known that the glass transition temperatures of natural gums are usually high [22,30,47,75].

3.5. Rheological Properties

Figure 5 shows the non-Newtonian rheological behaviours, with a nonlinear relationship between shear stress and strain rate. The GA sample recorded the highest maximum shear stress but one of the lowest maximum shear rates compared to the rest. The GE gum showed moderate values of these rheological parameters compared to the other samples. Sample GV showed similar strain rates to GA but significantly lowered maximum shear stress. Finally, GC exhibited the lowest levels of shear stress and maximum shear rate relative to GA, GE, and GV.

Within the common non-Newtonian nature, GC presented the most favourable rheological profile considering the magnitude of the experimentally determined parameters. Similar behaviours were reported for aqueous solutions of mucilage isolated from Opuntia ficus indica [76], which could be correlated with the shear rate using the power law model; likewise, locust bean gum showed non-Newtonian flow behaviour at speeds of high cut [77].

In Figure 6, the apparent viscosity is not constant; as the temperature increases from 40 to 80 °C, the viscosities of all the samples decrease, with downwardly shifted curves being observed at higher shear rates. The observed behaviour is because increasing temperature reduces the intermolecular forces. This trend occurs because it contributes to increased intermolecular forces of attraction [5]. Similar behaviours were reported for Pithecellobium dulce, in which the viscosity of the gum decreased with increasing temperature, and the effect was more pronounced at temperatures below 50 °C [5]. The temperature study revealed that the tara gum samples under study show shear thinning behaviours at temperatures of 80 °C.

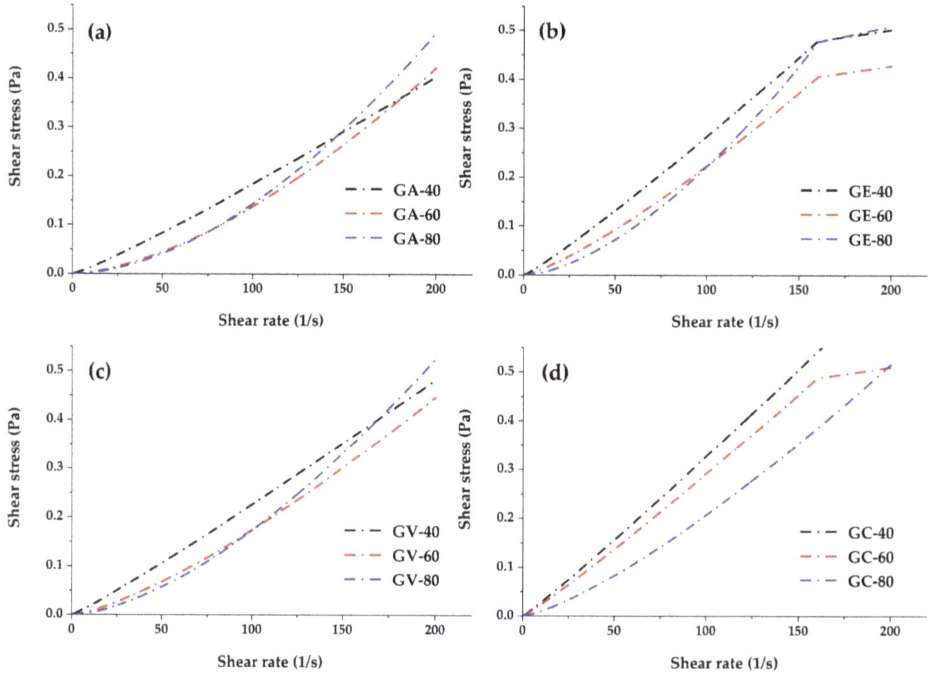

Figure 5. Shear rate and shear stress curves of spray-dried tara gum GA (**a**), oven-dried tara gum GE (**b**), oven vacuum-dried tara gum GV (**c**), and commercial tara gum GC (**d**) at 40, 60, and 80 °C.

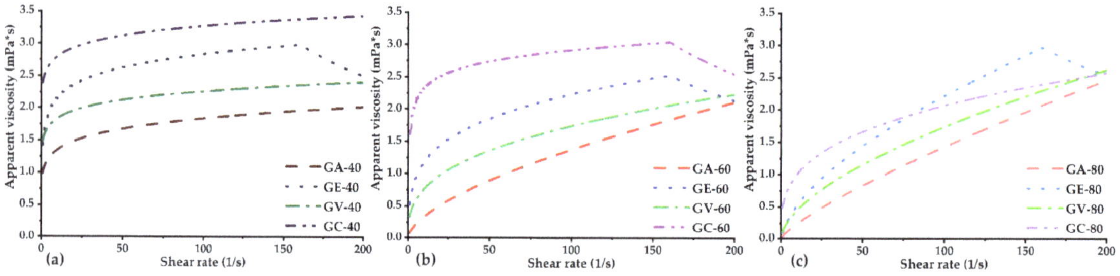

Figure 6. Apparent viscosity and shear stress curves at (**a**) 40 °C, (**b**) 60 °C, and (**c**) 80 °C.

Apparent viscosity was observed to increase with the shear rate, this behaviour being typical in polymer solutions such as gums [78]. This phenomenon, observed in colloidal suspensions such as starch–water mixtures, is attributed to dissipative hydrodynamic interactions between particles, which induce the formation of hydroclusters [79].

This phenomenon, called dilatancy, causes changes in the structure and alignment of the polymer chains at high shear rates, increasing the interactions between chains and, therefore, the apparent viscosity of the fluid. This phenomenon explains the non-Newtonian behaviour and is a determining factor in the rheology of solutions of rubbers and other polymers [80]. The GE gum at all its temperatures and GC gum at 60 °C exhibit a particular rheological behaviour. An increase in apparent viscosity is observed at low shear rates, known as shear thickening or dilatation. However, as the shear rate exceeds

150 1/s, the apparent viscosity begins to decrease, a phenomenon called shear thinning. This behaviour is called rheopectic or rheopexy, where the viscosity versus shear rate curve forms a hysteresis loop [81]. A similar rheopectic behaviour has been reported for a maltodextrin-based thickener [82].

3.6. *Rheological Analysis*

The rheological models of the tara gum samples, obtained through atomisation, by forced convection drying, under vacuum, and as a commercial gum, are shown in Table 3. Non-Newtonian pseudoplastic rheological behaviour was observed ($n < 1$) for all gums at different temperatures. Most of the experimental data for tara gum fit the Herschel–Bulkley model very well, with R^2 values greater than 0.9743. However, the initial shear stresses for GA, GE, GV, and GC were very close to zero; therefore, the power law model was the one that fit best with R^2 values greater than 0.9739. This model describes the flow curves well, especially at high strain rates [83]. Xanthan gum, which is widely used as an encapsulating agent, fits the power law model [43,84], as did gum of diutan [84] and carboxymethylated tara gum [6], and has been used on numerous occasions to adjust the relationship between the apparent viscosity and the shear rate of biopolymer solutions [84].

Table 3. Rheological analysis.

Model	Parameters	GA			GE			GV			GC		
		40 °C	60 °C	80 °C	40 °C	60 °C	80 °C	40 °C	60 °C	80 °C	40 °C	60 °C	80 °C
Power Law	k ($\times 10^{-4}$ Pa·s^n)	9.9045	0.7980	0.3872	14.20	5.2710	0.9830	14.90	3.3455	1.1252	0.0024	16.10	4.8530
	n	1.1344	1.6180	1.7884	1.1073	1.2642	1.6141	1.0908	1.3583	1.5944	1.0684	1.0866	1.3152
	R^2	0.9739	0.9854	0.9913	0.9912	0.9816	0.9884	0.9902	0.9796	0.9938	0.9965	0.9967	0.9926
Bingham Plastic	τ_y(Pa)	0	0	0	0	0	0	0	0	0	0	0	0
	n_B(Pa·s^n)	0.0020	0.002	0.0023	0.0025	0.0021	0.0024	0.0024	0.0022	0.0025	0.0034	0.0026	0.0025
	R^2	0.9626	0.8578	0.8081	0.9826	0.9472	0.8628	0.9828	0.9244	0.8726	0.9944	0.9880	0.9422
Herschel–Bulkley	τ_y(Pa)	0.0093	0.0166	0.0119	0.0034	0.0185	0.0215	0	0.0221	0.0219	0.0027	0.0035	0.0141
	k_H ($\times 10^{-4}$ Pa·s^n)	6.8532	0.2850	0.1993	0.0013	2.2715	0.3197	0.0017	1.1315	0.3804	0.0023	14.60	2.8672
	n	1.2010	1.8081	1.9070	1.1250	1.4179	1.8217	1.0672	1.5573	1.7947	1.0780	1.1040	1.4111
	R^2	0.9743	1	1	0.9912	1	1	0.9892	1	0.9967	0.9966	0.9938	1

Note: τ is the yield strength (Pa), γ is the shear rate (s^{-1}), k is the consistency index (Pa·s^n), n is the behavioural index, τ_y is the yield strength (Pa), n_B is the plastic viscosity (Pa*s), k_H is the consistency index (Pa·s^n), and R^2 is the correlation coefficient. GA—spray-dried tara gum, GE—forced convection dried tara gum, GV—vacuum-dried tara gum, and GC—commercial tara gum (GC).

3.7. *Result Overview*

A principal component analysis (PCA) of the physicochemical properties studied allowed for establishing relationships between these complex variables [85,86], providing a comprehensive view of their interactions. This methodology facilitated the identification of significant trends [47], which could help design and optimise obtaining tara gum with specific properties for use in the food, pharmaceutical, and cosmetic industries [30].

Figure 7 shows the PCA study, and it can be seen that spray-dried tara gum is associated with properties such as ζ-potential, luminosity L*, and chroma a*. These correlations suggest a possible interdependence between these properties, which could influence the stability and whiteness of tara gum. On the other hand, tara gums dried using a forced convection oven and vacuum oven (GE and GV) are more related to high values of moisture, Aw, hygroscopicity, and polydispersity, which would indicate that these gums are more prone to deterioration, and they present more significant heterogeneity in particle size. Finally, commercial tara gum (GC) is more related to high values of yield, particle size, apparent density, and chroma b*, which would indicate that in economic terms of the process, it would be more efficient, but on the contrary, due to the size of the particle, it would not be the most appropriate to use in the micro- and nanoencapsulation of compounds. Based on the results, spray-dried tara gum has the best physicochemical, structural, thermal, and rheological properties.

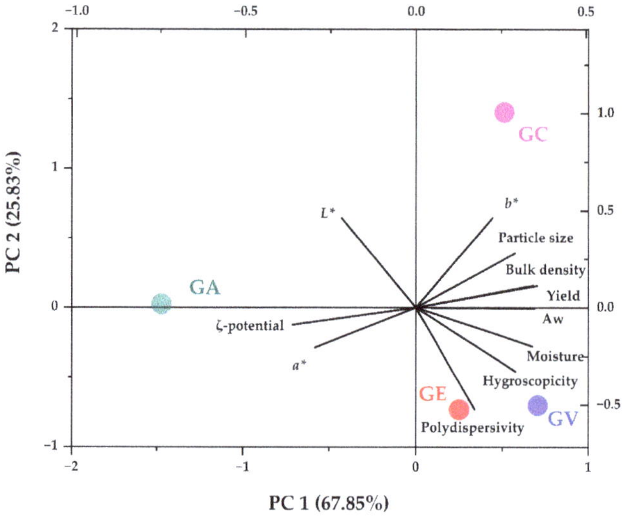

Figure 7. PCA study.

The Pearson correlogram (Figure 8) shows a significant positive correlation between hygroscopicity and moisture (r = 0.96), attributable to the more significant number of hydrophilic sites in the samples with higher residual water content [30]; a positive correlation was also observed between yield and bulk density (r = 0.99) since higher yields are obtained in denser and more compact particles [47]. On the other hand, significant negative correlations were found between yield and ζ-potential (r = −1.00), chroma a* and bulk density (r = −0.96), ζ-potential and bulk density (r = −0.98), and polydispersity and lightness (r = −0.96), which could be attributed to competitive interactions between the evaluated parameters. However, further studies are required to clarify the mechanisms involved. In conclusion, the correlation analysis revealed significant associations between the physicochemical properties of the studied gums.

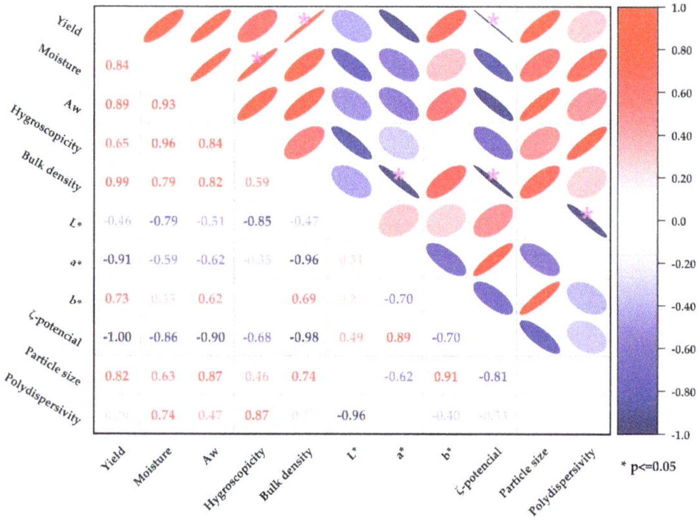

Figure 8. Pearson correlogram.

4. Conclusions

Tara gum emerges as a promising natural polymer for applications as a wall material in encapsulation processes. The present work provides a comprehensive characterisation of tara gum in terms of physicochemical, structural, thermal, and rheological properties relevant to its application as a wall material. The tara gum obtained through different drying methods showed moisture contents between 8.63% and 12.55%, water activity between 0.37 and 0.41, hygroscopicity less than 13.41%, and apparent density between 0.43 and 0.76 g/mL. The FTIR analysis confirmed the presence of polysaccharides.

The spray-dried tara gum presented a better preservation of the original colour, smaller particle size, and better thermal properties. All samples showed non-Newtonian pseudoplastic rheological behaviour, adequately described by the power law. The apparent viscosity decreased with increasing temperature, showing shear thinning at 80 °C. The results position tara gum as a viable material for encapsulation, with favourable characteristics compared to other encapsulating agents. It is recommended that research be continued aimed at practical applications of encapsulation with tara gum as a wall material in the food and pharmaceutical industries.

Author Contributions: Conceptualisation, E.M.-M., C.A.L.-S., D.F.C.H. and D.C.-Q.; methodology, J.C.A.-Q., G.D.l.C., M.L.H.-C., U.R.Q.-Q. and R.L.-A.; software, M.L.H.-C., W.C.C.P. and U.R.Q.-Q.; validation, R.L.-A. and W.C.C.P.; formal analysis, R.S.-L., U.R.Q.-Q. and D.F.C.H.; investigation, M.L.H.-C., E.M.-M., C.A.L.-S., J.C.A.-Q., G.D.l.C., R.L.-A. and D.C.-Q.; data curation, J.C.A.-Q., M.L.H.-C. and U.R.Q.-Q.; writing—original draft preparation, C.A.L.-S., E.M.-M., U.R.Q.-Q., G.D.l.C. and W.C.C.P.; writing—review and editing, R.S.-L., U.R.Q.-Q. and D.F.C.H.; supervision, R.L.-A. and D.F.C.H.; project administration, R.S.-L. All authors have read and agreed to the published version of the manuscript.

Funding: This work was funded in part by the Research Group on Nutraceuticals and Biomaterials of the Universidad Nacional José María Arguedas.

Institutional Review Board Statement: Not applicable.

Data Availability Statement: They are available in the same article.

Acknowledgments: The authors acknowledge the Food Nanotechnology Research Laboratory of the Universidad Nacional José María Arguedas.

Conflicts of Interest: The authors declare no conflicts of interest.

References

1. Taheri, A.; Jafari, S.M. Gum-based nanocarriers for the protection and delivery of food bioactive compounds. *Adv. Colloid Interface Sci.* **2019**, *269*, 277–295. [CrossRef]
2. Halahlah, A.; Piironen, V.; Mikkonen, K.S.; Ho, T.M. Polysaccharides as wall materials in spray-dried microencapsulation of bioactive compounds: Physicochemical properties and characterization. *Crit. Rev. Food Sci. Nutr.* **2022**, *63*, 6983–7015. [CrossRef]
3. Coimbra, P.P.S.; Cardoso, F.d.S.N.; Goncalves, E.C.B.d.A. Spray-drying wall materials: Relationship with bioactive compounds. *Crit. Rev. Food Sci. Nutr.* **2021**, *61*, 2809–2826. [CrossRef]
4. Mukherjee, K.; Dutta, P.; Badwaik, H.R.; Saha, A.; Das, A.; Giri, T.K. Food Industry applications of Tara Gum and its modified forms. *Food Hydrocoll. Health* **2022**, *3*, 100107. [CrossRef]
5. Chaudhari, B.B.; Annapure, U.S. Physiochemical and rheological characterization of Pithecellobium dulce (Roxb.) benth gum exudate as a potential wall material for the encapsulation of rosemary oil. *Carbohydr. Polym. Technol. Appl.* **2020**, *1*, 100005. [CrossRef]
6. Santos, M.B.; dos Santos, C.H.C.; de Carvalho, M.G.; de Carvalho, C.W.P.; Garcia-Rojas, E.E. Physicochemical, thermal and rheological properties of synthesized carboxymethyl tara gum (Caesalpinia spinosa). *Int. J. Biol. Macromol.* **2019**, *134*, 595–603. [CrossRef]
7. Murga-Orrillo, H.; Lobo, F.D.A.; Santos Silva Amorim, R.; Fernandes Silva Dionisio, L.; Nuñez Bustamante, E.; Chu-Koo, F.W.; López, L.A.A.; Arévalo-Hernández, C.O.; Abanto-Rodriguez, C. Increased Production of Tara (Caesalpinia spinosa) by Edaphoclimatic Variation in the Altitudinal Gradient of the Peruvian Andes. *Agronomy* **2023**, *13*, 646. [CrossRef]
8. Santander, S.; Aoki, M.; Hernandez, J.; Pombo, M.; Moins-Teisserenc, H.; Mooney, N.; Fiorentino, S. Galactomannan from Caesalpinia spinosa induces phenotypic and functional maturation of human dendritic cells. *Int. Immunopharmacol.* **2011**, *11*, 652–660. [CrossRef]

9. Ramakrishnan, Y.; Adzahan, N.M.; Yusof, Y.A.; Muhammad, K. Effect of wall materials on the spray drying efficiency, powder properties and stability of bioactive compounds in tamarillo juice microencapsulation. *Powder Technol.* **2018**, *328*, 406–414. [CrossRef]
10. Samborska, K.; Poozesh, S.; Barańska, A.; Sobulska, M.; Jedlińska, A.; Arpagaus, C.; Malekjani, N.; Jafari, S.M. Innovations in spray drying process for food and pharma industries. *J. Food Eng.* **2022**, 110960. [CrossRef]
11. Maisuthisakul, P.; Gordon, M.H. Influence of polysaccharides and storage during processing on the properties of mango seed kernel extract (microencapsulation). *Food Chem.* **2012**, *134*, 1453–1460. [CrossRef]
12. Outuki, P.M.; de Francisco, L.M.B.; Hoscheid, J.; Bonifácio, K.L.; Barbosa, D.S.; Cardoso, M.L.C. Development of arabic and xanthan gum microparticles loaded with an extract of Eschweilera nana Miers leaves with antioxidant capacity. *Colloids Surf. A Physicochem. Eng. Asp.* **2016**, *499*, 103–112. [CrossRef]
13. Aguilar-Galvez, A.; Noratto, G.; Chambi, F.; Debaste, F.; Campos, D. Potential of tara (Caesalpinia spinosa) gallotannins and hydrolysates as natural antibacterial compounds. *Food Chem.* **2014**, *156*, 301–304. [CrossRef]
14. Rahaiee, S.; Assadpour, E.; Faridi Esfanjani, A.; Silva, A.S.; Jafari, S.M. Application of nano/microencapsulated phenolic compounds against cancer. *Adv. Colloid Interface Sci.* **2020**, *279*, 102153. [CrossRef]
15. Zabot, G.L.; Schaefer Rodrigues, F.; Polano Ody, L.; Vinícius Tres, M.; Herrera, E.; Palacin, H.; Córdova-Ramos, J.S.; Best, I.; Olivera-Montenegro, L. Encapsulation of Bioactive Compounds for Food and Agricultural Applications. *Polymers* **2022**, *14*, 4194. [CrossRef]
16. Galves, C.; Galli, G.; Kurozawa, L. Potato protein: Current review of structure, technological properties, and potential application on spray drying microencapsulation. *Crit. Rev. Food Sci. Nutr.* **2023**, *63*, 6564–6579. [CrossRef]
17. Liang, X.; Chen, L.; McClements, D.J.; Peng, X.; Xu, Z.; Meng, M.; Jin, Z. Bioactive delivery systems based on starch and its derivatives: Assembly and application at different structural levels. *Food Chem.* **2024**, *432*, 137184. [CrossRef]
18. Hamdani, A.M.; Wani, I.A.; Bhat, N.A. Sources, structure, properties and health benefits of plant gums: A review. *Int. J. Biol. Macromol.* **2019**, *135*, 46–61. [CrossRef]
19. Amiri, M.S.; Mohammadzadeh, V.; Yazdi, M.E.; Barani, M.; Rahdar, A.; Kyzas, G.Z. Plant-Based Gums and Mucilages Applications in Pharmacology and Nanomedicine: A Review. *Molecules* **2021**, *26*, 1770. [CrossRef]
20. Vilaró Luna, P. Obtención y Caracterización de Gomas Provenientes de Semillas de Especies Nativas del Género Prosopis. 2019. Available online: https://www.colibri.udelar.edu.uy/jspui/handle/20.500.12008/21333 (accessed on 13 March 2024).
21. Al-Shammari, B.; Al-Ali, R.; Al-Sahi, A. Electrical, characterization and functional properties of extract gum (Tniqonella *Foenum graecum* L.) from Fenugreek seeds. In Proceedings of the IOP Conference Series: Earth and Environmental Science, Kerbala City, Iraq, 17–18 November 2019; p. 012060.
22. Ligarda-Samanez, C.A.; Moscoso-Moscoso, E.; Choque-Quispe, D.; Ramos-Pacheco, B.S.; Arévalo-Quijano, J.C.; Cruz, G.D.; Huamán-Carrión, M.L.; Quispe-Quezada, U.R.; Gutiérrez-Gómez, E.; Cabel-Moscoso, D.J.; et al. Native Potato Starch and Tara Gum as Polymeric Matrices to Obtain Iron-Loaded Microcapsules from Ovine and Bovine Erythrocytes. *Polymers* **2023**, *15*, 3985. [CrossRef]
23. Horwitz, W. Official methods of analysis of AOAC International. In *Agricultural Chemicals, Contaminants, Drugs*; Horwitz, W., Ed.; AOAC International: Gaithersburg, MD, USA, 2010; Volume I.
24. Choque-Quispe, D.; Ligarda-Samanez, C.A.; Huamán-Rosales, E.R.; Aguirre Landa, J.P.; Agreda Cerna, H.W.; Zamalloa-Puma, M.M.; Álvarez-López, G.J.; Barboza-Palomino, G.I.; Alzamora-Flores, H.; Gamarra-Villanueva, W. Bioactive Compounds and Sensory Analysis of Freeze-Dried Prickly Pear Fruits from an Inter-Andean Valley in Peru. *Molecules* **2023**, *28*, 3862. [CrossRef] [PubMed]
25. Moreira, G.É.G.; Costa, M.G.M.; de Souza, A.C.R.; de Brito, E.S.; de Medeiros, M.d.F.D.; de Azeredo, H.M. Physical properties of spray dried acerola pomace extract as affected by temperature and drying aids. *LWT-Food Sci. Technol.* **2009**, *42*, 641–645. [CrossRef]
26. Ligarda-Samanez, C.A.; Choque-Quispe, D.; Moscoso-Moscoso, E.; Huamán-Carrión, M.L.; Ramos-Pacheco, B.S.; Peralta-Guevara, D.E.; De la Cruz, G.; Martínez-Huamán, E.L.; Arévalo-Quijano, J.C.; Muñoz-Saenz, J.C. Obtaining and characterizing andean multi-floral propolis nanoencapsulates in polymeric matrices. *Foods* **2022**, *11*, 3153. [CrossRef]
27. Ligarda-Samanez, C.A.; Palomino-Rincón, H.; Choque-Quispe, D.; Moscoso-Moscoso, E.; Arévalo-Quijano, J.C.; Huamán-Carrión, M.L.; Quispe-Quezada, U.R.; Muñoz-Saenz, J.C.; Gutiérrez-Gómez, E.; Cabel-Moscoso, D.J.; et al. Bioactive Compounds and Sensory Quality in Chips of Native Potato Clones (Solanum tuberosum spp. andigena) Grown in the High Andean Region of PERU. *Foods* **2023**, *12*, 2511. [CrossRef]
28. Ramos-Pacheco, B.S.; Choque-Quispe, D.; Ligarda-Samanez, C.A.; Solano-Reynoso, A.M.; Palomino-Rincón, H.; Choque-Quispe, Y.; Peralta-Guevara, D.E.; Moscoso-Moscoso, E.; Aiquipa-Pillaca, Á.S. Effect of Germination on the Physicochemical Properties, Functional Groups, Content of Bioactive Compounds, and Antioxidant Capacity of Different Varieties of Quinoa (Chenopodium quinoa Willd.) Grown in the High Andean Zone of Peru. *Foods* **2024**, *13*, 417. [CrossRef]
29. Zhang, T.; Li, X.; Xu, J.; Shao, J.; Ding, M.; Shi, S. Preparation, characterization, and evaluation of breviscapine nanosuspension and its freeze-dried powder. *Pharmaceutics* **2022**, *14*, 923. [CrossRef]
30. Ligarda-Samanez, C.A.; Choque-Quispe, D.; Moscoso-Moscoso, E.; Huamán-Carrión, M.L.; Ramos-Pacheco, B.S.; De la Cruz, G.; Arévalo-Quijano, J.C.; Muñoz-Saenz, J.C.; Muñoz-Melgarejo, M.; Quispe-Quezada, U.R. Microencapsulation of Propolis and Honey Using Mixtures of Maltodextrin/Tara Gum and Modified Native Potato Starch/Tara Gum. *Foods* **2023**, *12*, 1873. [CrossRef]

31. Jedlińska, A.; Samborska, K.; Wieczorek, A.; Wiktor, A.; Ostrowska-Ligęza, E.; Jamróz, W.; Skwarczyńska-Maj, K.; Kiełczewski, D.; Błażowski, Ł.; Tułodziecki, M. The application of dehumidified air in rapeseed and honeydew honey spray drying-Process performance and powders properties considerations. *J. Food Eng.* **2019**, *245*, 80–87. [CrossRef]
32. Napiórkowska, A.; Szpicer, A.; Wojtasik-Kalinowska, I.; Perez, M.D.T.; González, H.D.; Kurek, M.A. Microencapsulation of juniper and black pepper essential oil using the coacervation method and its properties after freeze-drying. *Foods* **2023**, *12*, 4345. [CrossRef]
33. Ligarda-Samanez, C.A.; Choque-Quispe, D.; Moscoso-Moscoso, E.; Palomino-Rincón, H.; Taipe-Pardo, F.; Landa, J.P.A.; Arévalo-Quijano, J.C.; Muñoz-Saenz, J.C.; Quispe-Quezada, U.R.; Huamán-Carrión, M.L. Nanoencapsulation of Phenolic Extracts from Native Potato Clones (Solanum tuberosum spp. andigena) by Spray Drying. *Molecules* **2023**, *28*, 4961. [CrossRef]
34. Ren, Z.; Huang, X.; Shi, L.; Liu, S.; Yang, S.; Hao, G.; Qiu, X.; Liu, Z.; Zhang, Y.; Zhao, Y.; et al. Characteristics and potential application of myofibrillar protein from golden threadfin bream (Nemipterus virgatus) complexed with chitosan. *Int. J. Biol. Macromol.* **2023**, *240*, 124380. [CrossRef]
35. Quispe-Chambilla, L.; Pumacahua-Ramos, A.; Choque-Quispe, D.; Curro-Pérez, F.; Carrión-Sánchez, H.M.; Peralta-Guevara, D.E.; Masco-Arriola, M.L.; Palomino-Rincón, H.; Ligarda-Samanez, C.A. Rheological and functional properties of dark chocolate with partial substitution of peanuts and sacha inchi. *Foods* **2022**, *11*, 1142. [CrossRef]
36. Piñón-Balderrama, C.I.; Leyva-Porras, C.; Terán-Figueroa, Y.; Espinosa-Solís, V.; Álvarez-Salas, C.; Saavedra-Leos, M.Z. Encapsulation of active ingredients in food industry by spray-drying and nano spray-drying technologies. *Processes* **2020**, *8*, 889. [CrossRef]
37. Bashir, M.; Haripriya, S. Assessment of physical and structural characteristics of almond gum. *Int. J. Biol. Macromol.* **2016**, *93*, 476–482. [CrossRef]
38. Bouaziz, F.; Koubaa, M.; Neifar, M.; Zouari-Ellouzi, S.; Besbes, S.; Chaari, F.; Kamoun, A.; Chaabouni, M.; Chaabouni, S.E.; Ghorbel, R.E. Feasibility of using almond gum as coating agent to improve the quality of fried potato chips: Evaluation of sensorial properties. *LWT-Food Sci. Technol.* **2016**, *65*, 800–807. [CrossRef]
39. Vijeth, S.; Heggannavar, G.B.; Kariduraganavar, M.Y. Encapsulating wall materials for micro-/nanocapsules. *Microencapsul.-Process. Technol. Ind. Appl.* **2019**, 1–19. [CrossRef]
40. Zotarelli, M.F.; da Silva, V.M.; Durigon, A.; Hubinger, M.D.; Laurindo, J.B. Production of mango powder by spray drying and cast-tape drying. *Powder Technol.* **2017**, *305*, 447–454. [CrossRef]
41. Schuck, P.; Jeantet, R.; Dolivet, A. *Analytical Methods for Food and Dairy Powders*; John Wiley & Sons: Hoboken, NJ, USA, 2012.
42. Rosland Abel, S.E.; Yusof, Y.A.; Chin, N.L.; Chang, L.S.; Mohd Ghazali, H.; Manaf, Y.N. Characterisation of physicochemical properties of gum arabic powder at various particle sizes. *Food Res.* **2020**, *4*, 107–115. [CrossRef]
43. Lee, H.; Yoo, B. Agglomerated xanthan gum powder used as a food thickener: Effect of sugar binders on physical, microstructural, and rheological properties. *Powder Technol.* **2020**, *362*, 301–306. [CrossRef]
44. Goff, H.D.; Guo, Q. The role of hydrocolloids in the development of food structure. *Handb. Food Struct. Dev.* **2019**, *18*, 1.
45. Lunardi, C.N.; Gomes, A.J.; Rocha, F.S.; De Tommaso, J.; Patience, G.S. Experimental methods in chemical engineering: Zeta potential. *Can. J. Chem. Eng.* **2021**, *99*, 627–639. [CrossRef]
46. Schramm, L.L. *Emulsions, Foams, Suspensions, and Aerosols: Microscience and Applications*; John Wiley & Sons: Hoboken, NJ, USA, 2014.
47. Ligarda-Samanez, C.A.; Moscoso-Moscoso, E.; Choque-Quispe, D.; Palomino-Rincón, H.; Martínez-Huamán, E.L.; Huamán-Carrión, M.L.; Peralta-Guevara, D.E.; Aroni-Huamán, J.; Arévalo-Quijano, J.C.; Palomino-Rincón, W.; et al. Microencapsulation of Erythrocytes Extracted from Cavia porcellus Blood in Matrices of Tara Gum and Native Potato Starch. *Foods* **2022**, *11*, 2107. [CrossRef]
48. Cano-Sarmiento, C.; Téllez-Medina, D.I.; Viveros-Contreras, R.; Cornejo-Mazón, M.; Figueroa-Hernández, C.Y.; García-Armenta, E.; Alamilla-Beltrán, L.; García, H.S.; Gutiérrez-López, G.F. Zeta potential of food matrices. *Food Eng. Rev.* **2018**, *10*, 113–138. [CrossRef]
49. Huamaní-Meléndez, V.J.; Mauro, M.A.; Darros-Barbosa, R. Physicochemical and rheological properties of aqueous Tara gum solutions. *Food Hydrocoll.* **2021**, *111*, 106195. [CrossRef]
50. Sikora, M.; Kowalski, S.; Tomasik, P. Binary hydrocolloids from starches and xanthan gum. *Food Hydrocoll.* **2008**, *22*, 943–952. [CrossRef]
51. Díaz-Montes, E. Wall Materials for Encapsulating Bioactive Compounds via Spray-Drying: A Review. *Polymers* **2023**, *15*, 2659. [CrossRef]
52. Samborska, K.; Boostani, S.; Geranpour, M.; Hosseini, H.; Dima, C.; Khoshnoudi-Nia, S.; Rostamabadi, H.; Falsafi, S.R.; Shaddel, R.; Akbari-Alavijeh, S.; et al. Green biopolymers from by-products as wall materials for spray drying microencapsulation of phytochemicals. *Trends Food Sci. Technol.* **2021**, *108*, 297–325. [CrossRef]
53. Calín-Sánchez, Á.; Lipan, L.; Cano-Lamadrid, M.; Kharaghani, A.; Masztalerz, K.; Carbonell-Barrachina, Á.A.; Figiel, A. Comparison of Traditional and Novel Drying Techniques and Its Effect on Quality of Fruits, Vegetables and Aromatic Herbs. *Foods* **2020**, *9*, 1261. [CrossRef] [PubMed]
54. Guiné, R. The drying of foods and its effect on the physical-chemical, sensorial and nutritional properties. *Int. J. Food Eng.* **2018**, *2*, 93–100. [CrossRef]
55. Labuschagne, P. Impact of wall material physicochemical characteristics on the stability of encapsulated phytochemicals: A review. *Food Res. Int.* **2018**, *107*, 227–247. [CrossRef]

56. Cui, S.W. *Food Carbohydrates: Chemistry, Physical Properties, and Applications*; CRC Press: Boca Raton, FL, USA, 2005.
57. Pachuau, L.; Lalhlenmawia, H.; Mazumder, B. Characteristics and composition of Albizia procera (Roxb.) Benth gum. *Ind. Crops Prod.* **2012**, *40*, 90–95. [CrossRef]
58. Adsare, S.R.; Annapure, U.S. Microencapsulation of curcumin using coconut milk whey and Gum Arabic. *J. Food Eng.* **2021**, *298*, 110502. [CrossRef]
59. Medina-Torres, L.; Calderas, F.; Nunez Ramírez, D.; Herrera-Valencia, E.; Bernad Bernad, M.; Manero, O. Spray drying egg using either maltodextrin or nopal mucilage as stabilizer agents. *J. Food Sci. Technol.* **2017**, *54*, 4427–4435. [CrossRef] [PubMed]
60. Hosseini, H.; Yaghoubi Hamgini, E.; Jafari, S.M. Chapter 9—Spray drying of starches and gums. In *Spray Drying for the Food Industry*; Jafari, S.M., Samborska, K., Eds.; Woodhead Publishing: Sawston, UK, 2024; pp. 243–274.
61. Fathi, F.; Ebrahimi, S.N.; Matos, L.C.; Oliveira, M.B.P.P.; Alves, R.C. Emerging drying techniques for food safety and quality: A review. *Compr. Rev. Food Sci. Food Saf.* **2022**, *21*, 1125–1160. [CrossRef]
62. Nandiyanto, A.B.D.; Oktiani, R.; Ragadhita, R. How to read and interpret FTIR spectroscope of organic material. *Indones. J. Sci. Technol.* **2019**, *4*, 97–118. [CrossRef]
63. Akcicek, A.; Bozkurt, F.; Akgül, C.; Karasu, S. Encapsulation of olive pomace extract in rocket seed gum and chia seed gum nanoparticles: Characterization, antioxidant activity and oxidative stability. *Foods* **2021**, *10*, 1735. [CrossRef] [PubMed]
64. Yahoum, M.M.; Toumi, S.; Tahraoui, H.; Lefnaoui, S.; Kebir, M.; Amrane, A.; Assadi, A.A.; Zhang, J.; Mouni, L. Formulation and Evaluation of Xanthan Gum Microspheres for the Sustained Release of Metformin Hydrochloride. *Micromachines* **2023**, *14*, 609. [CrossRef] [PubMed]
65. Ahuja, M.; Kumar, A.; Singh, K. Synthesis, characterization and in vitro release behavior of carboxymethyl xanthan. *Int. J. Biol. Macromol.* **2012**, *51*, 1086–1090. [CrossRef] [PubMed]
66. Ballesteros, L.F.; Ramirez, M.J.; Orrego, C.E.; Teixeira, J.A.; Mussatto, S.I. Encapsulation of antioxidant phenolic compounds extracted from spent coffee grounds by freeze-drying and spray-drying using different coating materials. *Food Chem.* **2017**, *237*, 623–631. [CrossRef]
67. Pant, K.; Thakur, M.; Chopra, H.K.; Nanda, V. Encapsulated bee propolis powder: Drying process optimization and physicochemical characterization. *LWT* **2022**, *155*, 112956. [CrossRef]
68. Jafari, Y.; Sabahi, H.; Rahaie, M. Stability and loading properties of curcumin encapsulated in Chlorella vulgaris. *Food Chem.* **2016**, *211*, 700–706. [CrossRef]
69. Castro-López, C.; Espinoza-González, C.; Ramos-González, R.; Boone-Villa, V.D.; Aguilar-González, M.A.; Martínez-Ávila, G.C.; Aguilar, C.N.; Ventura-Sobrevilla, J.M. Spray-drying encapsulation of microwave-assisted extracted polyphenols from Moringa oleifera: Influence of tragacanth, locust bean, and carboxymethyl-cellulose formulations. *Food Res. Int.* **2021**, *144*, 110291. [CrossRef]
70. Zhang, H.; Gong, T.; Li, J.; Pan, B.; Hu, Q.; Duan, M.; Zhang, X. Study on the Effect of Spray Drying Process on the Quality of Microalgal Biomass: A Comprehensive Biocomposition Analysis of Spray-Dried S. acuminatus Biomass. *BioEnergy Res.* **2022**, *15*, 320–333. [CrossRef]
71. Leyva-Porras, C.; Cruz-Alcantar, P.; Espinosa-Solís, V.; Martínez-Guerra, E.; Piñón-Balderrama, C.I.; Compean Martínez, I.; Saavedra-Leos, M.Z. Application of Differential Scanning Calorimetry (DSC) and Modulated Differential Scanning Calorimetry (MDSC) in Food and Drug Industries. *Polymers* **2020**, *12*, 5. [CrossRef] [PubMed]
72. Pashazadeh, H.; Zannou, O.; Ghellam, M.; Koca, I.; Galanakis, C.M.; Aldawoud, T.M.S. Optimization and Encapsulation of Phenolic Compounds Extracted from Maize Waste by Freeze-Drying, Spray-Drying, and Microwave-Drying Using Maltodextrin. *Foods* **2021**, *10*, 1396. [CrossRef] [PubMed]
73. Kurozawa, L.E.; Park, K.J.; Hubinger, M.D. Effect of maltodextrin and gum arabic on water sorption and glass transition temperature of spray dried chicken meat hydrolysate protein. *J. Food Eng.* **2009**, *91*, 287–296. [CrossRef]
74. Descamps, N.; Palzer, S.; Roos, Y.H.; Fitzpatrick, J.J. Glass transition and flowability/caking behaviour of maltodextrin DE 21. *J. Food Eng.* **2013**, *119*, 809–813. [CrossRef]
75. Ligarda-Samanez, C.A.; Choque-Quispe, D.; Moscoso-Moscoso, E.; Pozo, L.M.F.; Ramos-Pacheco, B.S.; Palomino-Rincón, H.; Gutiérrez, R.J.G.; Peralta-Guevara, D.E. Effect of Inlet Air Temperature and Quinoa Starch/Gum Arabic Ratio on Nanoencapsulation of Bioactive Compounds from Andean Potato Cultivars by Spray-Drying. *Molecules* **2023**, *28*, 7875. [CrossRef] [PubMed]
76. Medina-Torres, L.; Brito-De La Fuente, E.; Torrestiana-Sanchez, B.; Katthain, R. Rheological properties of the mucilage gum (Opuntia ficus indica). *Food Hydrocoll.* **2000**, *14*, 417–424. [CrossRef]
77. Hussain, R.; Singh, A.; Vatankhah, H.; Ramaswamy, H.S. Effects of locust bean gum on the structural and rheological properties of resistant corn starch. *J. Food Sci. Technol.* **2017**, *54*, 650–658. [CrossRef]
78. Dolez, P.I.; Mlynarek, J. 22—Smart materials for personal protective equipment: Tendencies and recent developments. In *Smart Textiles and Their Applications*; Koncar, V., Ed.; Woodhead Publishing: Oxford, UK, 2016; pp. 497–517.
79. Wagner, N.J.; Brady, J.F. Shear thickening in colloidal dispersions. *Phys. Today* **2009**, *62*, 27–32. [CrossRef]
80. Rapp, B.E. *Microfluidics: Modeling, Mechanics and Mathematics*; Elsevier: Amsterdam, The Netherlands, 2022.
81. Xu, Y. *Stabbing Resistance of Soft Ballistic Body Armour Impregnated with Shear Thickening Fluid*; The University of Manchester: Manchester, UK, 2017.

82. Dewar, R.J.; Joyce, M.J. The thixotropic and rheopectic behaviour of maize starch and maltodextrin thickeners used in dysphagia therapy. *Carbohydr. Polym.* **2006**, *65*, 296–305. [CrossRef]
83. Díaz, R. Reología aplicada a sistemas alimentarios. *Editor. Grupo Compás.* **2018**, *18*, 8080.
84. Gao, X.; Huang, L.; Xiu, J.; Yi, L.; Zhao, Y. Evaluation of Viscosity Changes and Rheological Properties of Diutan Gum, Xanthan Gum, and Scleroglucan in Extreme Reservoirs. *Polymers* **2023**, *15*, 4338. [CrossRef] [PubMed]
85. Ligarda-Samanez, C.A.; Choque-Quispe, D.; Palomino-Rincón, H.; Ramos-Pacheco, B.S.; Moscoso-Moscoso, E.; Huamán-Carrión, M.L.; Peralta-Guevara, D.E.; Obregón-Yupanqui, M.E.; Aroni-Huamán, J.; Bravo-Franco, E.Y.; et al. Modified Polymeric Biosorbents from Rumex acetosella for the Removal of Heavy Metals in Wastewater. *Polymers* **2022**, *14*, 2191. [CrossRef] [PubMed]
86. Vítězová, M.; Jančiková, S.; Dordević, D.; Vítěz, T.; Elbl, J.; Hanišáková, N.; Jampílek, J.; Kushkevych, I. The possibility of using spent coffee grounds to improve wastewater treatment due to respiration activity of microorganisms. *Appl. Sci.* **2019**, *9*, 3155. [CrossRef]

polymers

MDPI

Article

Stability and Biaxial Behavior of Fresh Cheese Coated with Nanoliposomes Encapsulating Grape Seed Tannins and Polysaccharides Using Immersion and Spray Methods

Angela Monasterio [1], Emerson Núñez [2], Valeria Verdugo [1] and Fernando A. Osorio [1,*]

[1] Department of Food Science and Technology, Technological Faculty, University of Santiago-Chile (USACH), Av. El Belloto 3735, Estación Central, Santiago 9170022, Chile; angela.monasterio@usach.cl (A.M.); valeria.verdugo@usach.cl (V.V.)
[2] Department of Fruit Production and Enology, School of Agricultural and Natural Systems, Pontificia Universidad Católica de Chile, Av. Vicuña Mackenna 4860, Macul, Santiago 7820436, Chile; ennunez@uc.cl
* Correspondence: fernando.osorio@usach.cl

Abstract: In the food industry context, where fresh cheese stands out as a highly perishable product with a short shelf life, this study aimed to extend its preservation through multi-layer edible coatings. The overall objective was to analyze the biaxial behavior and texture of fresh cheese coated with nanoliposomes encapsulating grape seed tannins (NTs) and polysaccharides (hydroxypropyl methylcellulose; HPMC and kappa carrageenan; KC) using immersion and spray methods, establishing comparisons with uncoated cheeses and commercial samples, including an accelerated shelf-life study. NT, HPMC, and KC were employed as primary components in the multi-layer edible coatings, which were applied through immersion and spray. The results revealed significant improvements, such as a 20% reduction in weight loss and increased stability against oxidation, evidenced by a 30% lower peroxide index than the uncoated samples. These findings underscore the effectiveness of edible coatings in enhancing the quality and extending the shelf life of fresh cheese, highlighting the innovative application of nanoliposomes and polysaccharide blends and the relevance of applying this strategy in the food industry. In conclusion, this study provides a promising perspective for developing dairy products with improved properties, opening opportunities to meet market demands and enhance consumer acceptance.

Keywords: biaxial viscosity; texture; cheese; nanoliposomes; tannins

check for updates

Citation: Monasterio, A.; Núñez, E.; Verdugo, V.; Osorio, F.A. Stability and Biaxial Behavior of Fresh Cheese Coated with Nanoliposomes Encapsulating Grape Seed Tannins and Polysaccharides Using Immersion and Spray Methods. *Polymers* **2024**, *16*, 1559. https://doi.org/10.3390/polym16111559

Academic Editors: César Leyva-Porras and Zenaida Saavedra-Leos

Received: 25 April 2024
Revised: 22 May 2024
Accepted: 29 May 2024
Published: 31 May 2024

1. Introduction

Fresh cheese, whether made from cow or goat milk, stands out as one of the highly perishable non-liquid foods, characterized by a short shelf life of 7 to 9 days [1,2] due to its high water and nutrient content leading to increased syneresis, decreased pH, and lipolysis [3]. The loss of quality, flavor, and texture indicates deterioration in cheeses, and this alteration can occur due to various factors, such as temperature and humidity, as well as internal factors inherent in the food composition [4]. One way to prevent food deterioration is by applying edible coatings, defined as a thin layer of edible materials applied directly to the surface of food, acting as a natural barrier against external conditions [5]. To address intrinsic factors affecting cheese deterioration, molecules with antioxidant and antimicrobial properties can be incorporated into edible coatings to enhance their functionality.

Condensed tannins or proanthocyanidins are phenolic compounds with high antioxidant capacity that provide various health benefits [6]. Techniques have been developed to incorporate these properties into different foods for therapeutic, structural, and preservative purposes. However, astringency and bitterness are the main organoleptic properties of these compounds [7]. The perception of these properties in the mouth is directly related to the ability of tannins to interact with salivary proteins, forming aggregates [8]; this makes their direct incorporation into foods challenging. Nanotechnology represents an innovative and versatile alternative to reduce unpleasant tastes in the mouth, produced either by the interaction between food components and saliva components or by the food's nature [9,10].

Nanoliposomes are carrier systems for bioactive compounds widely used in various industries, such as cosmetics, pharmaceuticals, and food, as they exhibit excellent biocompatibility with biological membranes due to their amphiphilic nature [11]. Recently, lipid- and polysaccharide-based edible coatings have proven to be an effective strategy to improve product quality [12]. Biopolymers are considered non-toxic compounds, easy to modify and optimize, with some, such as hydroxypropyl methylcellulose and kappa carrageenan, providing structure to food matrices, thus improving their texture [13].

Coatings can be applied to or inside foods using various methods [14–16]. Immersion is the most common coating method for fresh products, ensuring good uniformity on complex and rough surfaces. It is also simple and cost-effective. On the other hand, spraying increases the liquid's surface by forming small droplets and dispersing them through a set of nozzles onto the food surface [17,18].

The percentage increase in the weight of the samples coated by immersion was expressed in Equation (1).

$$Weight\ gain\ (\%) = \frac{Initial\ weight - Final\ weight}{Initial\ weight} \times 100 \quad (1)$$

For a steady nozzle centrifugal device, the atomization process, characterized by the mean drop diameter, D, is governed by the following physical parameters [19].

- L: The characteristic dimension of the atomizer, nozzle hole diameter.
- U: The initial relative velocity of the injected suspension and ambient air.
- σ: The surface tension of the coating.
- ρ_s, ρ_a: The densities of the coating suspension and air, respectively.
- μ_s, μ_a: The dynamic viscosities of the coating suspension and air.

After applying the Buckingham Pi theorem, the following relationship (Equation (2)) is obtained:

$$\frac{D}{L} = f\left(R_e, W_e, \frac{\rho_a}{\rho_s}, \frac{\mu_a}{\mu_s}\right) \quad (2)$$

With four dimensionless numbers, the first term corresponds to the Reynolds number, which represents the ratio of inertial force to viscous force:

$$R_e = \frac{\rho U L}{\mu} \quad (3)$$

The second term represents the Weber number, which corresponds to the ratio of inertial force to surface tension force:

$$W_e = \frac{U^2 \rho L}{\sigma} \quad (4)$$

The third and fourth terms represent the density and viscosity ratios of the air and coating suspension used, respectively.

Combining the first two terms in Equation (2) to eliminate velocity allows obtaining another dimensionless number, the Ohnesorge number (O_h), which highlights the relative importance of interfacial viscous and surface tension.

$$O_h = {W_e}^{0.5} {R_e}^{-1} = \frac{\mu}{(\rho \sigma L)^{0.5}} \quad (5)$$

Studies on determining texture in cheeses include lubricated uniaxial and biaxial compression tests, which determine extensional and compressional flows [20]. Young's modulus and viscosity are essential parameters obtained through lubricated compression tests to define the texture of fresh cheeses (FCs) coagulated with commercial acids [21]. Lazárková et al. [22] evaluated the texture of white cheese in brine using stress–strain graphs corrected against Hencky deformation. Giliel et al. [23] used a double compression test to determine the hardness, elasticity, cohesion, and chewiness parameters in cheeses coated with a film-forming solution based on Ziziphus joazeiro fruit pulp.

Biaxial compression tests help determine the texture of hard, semi-hard, and soft cheeses. They are based on mechanical properties and conducted using specialized equipment. In these tests, the shape of the sample is maintained. At the same time, the contact area with the plate increases as the height of the sample decreases, thereby determining the cheese's behavior in terms of stress and deformation [24,25].

The biaxial extension velocity distribution may be expressed in terms of the Hencky stress, $\dot{\varepsilon}_h$, as follows:

$$u_z = -2\dot{\varepsilon}_B z = \dot{\varepsilon}_h z \quad (6)$$

$$u_r = \dot{\varepsilon}_B r = \frac{\dot{\varepsilon}_h r}{2} \quad (6a)$$

$$u_\theta = 0 \quad (6b)$$

where

u_z = velocity in the z-direction [ms^{-1}];
$\dot{\varepsilon}_B = \dot{\varepsilon}_h/2$ = biaxial extensional strain rate [s^{-1}];
u_r = velocity in the r-direction [ms^{-1}];
z = axial coordinate [m];
$\dot{\varepsilon}_h$ = Hencky strain rate [s^{-1}];
u_θ = velocity in the θ-direction [ms^{-1}].

The compression stress in the vertical direction is as follows:

$$d\dot{\varepsilon}_h = -\frac{dh}{h} \quad (7)$$

where h = the height that separates the plates [m].

The biaxial extensional strain rate (also called radial extension rate) is equal to half of the Hencky vertical strain rate [24]:

$$\dot{\varepsilon}_B = \left(\frac{1}{2}\right)\dot{\varepsilon}_h = \left(-\frac{1}{2h}\right)\frac{dh}{dt} = \frac{u_z}{2(ho - u_z t)} \quad (8)$$

where h_o = initial height of the sample [m]; h = height [m]; t = time [s].

The extensional viscosity is calculated from the stretching stress and the strain rate.

$$\eta_B = f(t) = \frac{\sigma_B}{\dot{\varepsilon}_B} = \frac{\sigma_B\, 2(ho - u_z t)}{u_z} \quad (9)$$

where

η_B = biaxial extensional viscosity [Pa s];
σ_B = biaxial compression stress [Pa];
$\dot{\varepsilon}_B$ = biaxial extensional strain rate.

The biaxial compression stress (σ_B) is obtained from the following expression:

$$\sigma_B = \frac{Fh}{\pi {R_o}^2 ho} \quad (10)$$

where

R_o = initial radius of the sample [m];
F = force [N].

Moreover, and as a complement to the mentioned studies, the composition of the food is a relevant factor for result analysis, as the content of proteins, fat, water, and carbohydrates affects the rheological behavior and, therefore, the textural characteristics [26]. In this context, the objective of this research was to analyze the biaxial behavior and texture of fresh cheese coated with nanoliposomes encapsulating grape seed tannins (NTs) and polysaccharides (HPMC and KC) using immersion and spray methods to establish comparisons with uncoated cheeses and commercial samples, including an accelerated shelf-life study.

Studying texture in coated cheeses is fundamental as it is considered a quality attribute that influences sensory perception, consumer acceptability, and the cheeses' characterization, classification, and processing. Moreover, it provides valuable information for manufacturers and consumers and contributes to the continuous improvement of dairy products. The potential benefits associated with the research encompass a general improvement in product quality, strengthening its structure, and adding antioxidant capacity to the matrix, leading to new products, improvements in food processing, and product shelf life.

2. Materials and Methods

2.1. Materials

Condensed tannin powder with a medium degree of polymerization of 2.5 $\pm$ 0.2, a galloylation degree of 15.5 $\pm$ 1.1, and an average molecular weight of 784 $\pm$ 61 was used, sourced from the Enology Laboratory at the Pontifical Catholic University of Chile. Dimerco Comercial Ltd. provided soy lecithin (Santiago, Chile). Glycerol PA (purity >99%) and sodium thiosulfate were purchased from Sigma Aldrich (St. Louis, MO, USA). Ethanol PA from Merck, HPLC-grade methanol ($\geq$99.9%), Trifluoroacetic acid (TFA) from Merck (Darmstadt, Germany), and Milli-Q water were used as solvents. Kappa carrageenan (Carragel PGU 5289, Gelymar®, Santiago, Chile) and HPMC (Methocel E19, Dow Wolff Cellulosics, Bomlitz, Germany) were employed. The liquid rennet CHY-MAX® M200 for FC preparation was obtained from DILACO (Santiago, Chile).

2.2. Experimental Design

To determine the optimal concentrations of Tween-80 and GLY to be added in the preparation of nanoliposomes encapsulating tannins (NTs), a multi-level factorial design with 18 experimental runs was conducted in two blocks [27]. The factors, levels, and evaluated response variables are presented in Table 1.

Table 1. Multi-level factorial design *.

Factors	Low	High	Levels	Units	Answer	Units
Tween-80	1	1	3	[%]	ρ	[kg/m^3]
GLY	5	5	3	[%]	Y	[mN/m]
					OD	[Dimensionless]

* Software STATGRAPHICS Centurion XVI (v.16.1.03), 2022. GLY: glycerol; ρ: density; Y: surface tension; OD: optical density.

In this experimental design, three levels were considered for the concentrations of each factor. The evaluated response variables were density, surface tension, and optical density. The design matrix was generated using the STATGRAPHICS Centurion XVI software (v. 16.1.03), and the experiments were conducted in random order to minimize bias. Subsequently, multiple optimizations of the response variables were performed using the desirability function, where all the variables were maximized. This approach aimed to enhance the encapsulating material per unit volume, potentially improving the retention capacity of active compounds. Simultaneously, higher surface tension leads to the formation of smaller droplets, favoring the uniform dispersion of coating components [28].

Additionally, Tween-80 and GLY concentrations were selected based on their critical roles in emulsion stability and nanoliposome formation. The decision to focus on these two factors was supported by previous research indicating their significant impact on nanoliposome properties [29]. While other formulation components were not included in the factorial design, their levels were kept constant based on preliminary experiments and literature review. This ensured a focused investigation of the main factors affecting nanoliposome characteristics [30]. This approach allowed for a more efficient exploration of the key parameters influencing the encapsulation process.

2.3. Nanoliposomes Encapsulating Tannins (NTs)

The formation of NT is based on the methodology described by Jafari et al. [31] with some specific modifications. An oily phase was prepared, composed of Tween-80, grape seed oil (OG), GLY, and Milli-Q water. Simultaneously, an aqueous phase containing tannins in suspension (TS) in citrate buffer (0.1 M at pH 3) was prepared. Subsequently, the aqueous phase was gradually incorporated into the oily phase using a programmable syringe pump to control the flow and prevent the formation of aggregates. After incorporation, the mixture was homogenized using an Ultraturrax (IKA T18, Staufen, Germany) at 10,000 rpm for 3 min. Then, it underwent 5 min of sonication to obtain liposomes using an ultrasonic cell disruptor (HIELSCHER UP100H, Teltow, Germany, max. 100 W) with an MS7 Micro tip 7 sonotrode (7 mm in diameter, 120 mm in length, 130 W/cm^2 acoustic power density) working at 50% amplitude. Empty nanoliposomes (ENs) were used as a control. The resulting liposomal suspension was stored at 4 °C to preserve its stability, and finally, the efficiency of encapsulation and stability analyses were conducted before its use as the main ingredient in multilayer edible coatings (MECs).

2.4. Encapsulation Efficiency (EE)

The methodology described by Babazadeh et al. [32] was employed to determine EE. Two milliliters (2 mL) of NT were centrifuged at 4 °C and 14,000 rpm for 1 h (Hanil Scientific Inc. Supra R22, Gimpo, Republic of Korea) to separate the suspended tannins. The supernatant was filtered through 0.22 μm syringe filters and deposited in borosilicate vials for analysis via UHPLC (Thermo Scientific Dionex UltiMate 3000, Waltham, MA, USA). A C18 column (5 μm, 250 × 4.6 mm, Perkin Elmer, Shelton, CT, USA) and a UV detector were used for tannin analysis. The mobile phases and gradients employed were as described by Bianchi et al. [33]. Epicatechin monomer was used as the standard for tannin quantification, with a calibration curve (5–500 μg/mL, R^2 = 0.999). Epicatechin detection was performed at 280 nm. The EE in percentage was determined in triplicate and calculated using Equation (11).

$$EE = \frac{[TE] - [TL]}{[TE]} \cdot 100 \tag{11}$$

where

$[TE]$: The initial concentration of encapsulated tannins, [mg of epicatechin/g of sample].
$[TL]$: The concentration of free tannins in suspension, [mg of epicatechin/g of sample].

2.5. Stability Study of NT

The stability study of NT was conducted using a LITESIZER 500 particle analyzer (Anton Paar, Graz, Austria). The analysis included determining the particle size, polydispersity index, diffusion coefficient, zeta potential, conductivity, and transmittance for each sample over a 15-day storage period at 4 °C. A backscatter measurement angle of 175° was employed [34,35].

2.6. Film-Forming Suspensions Based on Polysaccharides (FSs)

HPMC and KC were used as polysaccharides to prepare film-forming suspensions. Milli-Q water was heated to the dissolution temperature of each compound and stirred magnetically at 1400 rpm for 15 min. After this time, the polysaccharide was slowly and gradually added in a thin stream until complete dissolution. Once the polymer was dissolved, it was tempered to 20 °C and dispersed using an Ultraturrax at 10,000 rpm for

5 min. The resulting film-forming suspension was degassed in an ultrasound bath and stored under refrigeration [36].

2.7. Formation of Multilayer Edible Coatings (MECs)

In this section, we employed the experimental designs previously conducted by our research group to systematically vary the concentrations of components and optimize relevant variables for creating multilayer edible coatings (MECs).

The selected NT from the experimental design and the FS-HPMC and FS-KC were used for the formation of the MEC. The first coating comprised 65% NT and 35% HPMC, designated as coating A. The second coating was prepared with 75% NT and 25% KC, designated as coating B. Both A and B were prepared using magnetic stirring at 1400 rpm and 20 °C for 1 h. After this time, the mixture was homogenized using an Ultraturrax at 3000 rpm for 15 s [37].

2.8. Preparation of Fresh Cheese

The methodology described by Nemati et al. [38] was employed to produce fresh cheese with some modifications. Semi-skimmed cow's milk was pasteurized at 75 °C for 15 s. After this thermal process, the milk was cooled to 35 ± 2 °C, and CHY-MAX® M200 liquid rennet, derived from Aspergillus niger, and calcium chloride were added. This inoculation was allowed to rest for 30 min. Afterwards, the coagulum was cut into approximately 2 cm cubes, and sodium chloride was added to salt the mixture. The draining and molding process took place in a cheese vat measuring 10×10 cm^2, with weight applied, at refrigeration temperature for 24 h.

The commercial samples of goat cheese (GC) and vegan cheese substitute (VC), used for texture and biaxial behavior comparisons, were purchased from a local supermarket in Estación Central, Santiago, Chile.

2.9. Comparison with Commercial Samples

This study compares our experimental cheese samples with the commercial samples of goat cheese and a vegan cheese substitute. This comparison highlights nanoliposome coatings' potential advantages and unique properties on the experimental samples. While it is acknowledged that the commercial samples differ significantly in composition and were not coated with liposomes, this comparison is still valid and provides meaningful insights for several reasons.

Firstly, commercial goat cheese and vegan cheese substitutes represent standard benchmarks in the market. By comparing our experimental samples to these established products, we can evaluate how the liposome coatings influence key properties such as texture, shelf life, and nutritional content relative to products consumers are already familiar with. This helps contextualize the experimental cheese's performance within the broader market.

Secondly, although the commercial samples were not coated with liposomes, the comparison allows us to identify specific areas where the liposome coatings offer clear benefits. For example, improvements in the stability and controlled release of bioactive compounds in the experimental samples can be directly contrasted with the properties of the commercial products; this helps demonstrate the added value liposome technology can bring to cheese production.

Finally, the study aims to pave the way for future research and development. Including the commercial samples without liposome coatings provides a baseline against which future studies could compare results when similar coatings are applied to commercial products. Future research could involve applying nanoliposome coatings to commercial cheeses and vegan substitutes, allowing for a more direct comparison and validation of the observed benefits.

2.10. Proximate Composition

The proximate analysis of the cheeses included the determination of moisture content through the gravimetric method described in A.O.A.C. 925.45 standard, protein content using the Kjeldahl method as proposed by A.O.A.C. 990.03 standard, lipid or fat content through the alkaline hydrolysis methodology described in A.O.A.C. 996.06 standard, and the total ash content of the samples determined in a muffle furnace at 500 °C according to A.O.A.C. 923.03 standard [39–42]. Carbohydrates were calculated by difference from the previously obtained data and expressed as a percentage [43].

2.11. Dipping Coating Process

Pieces of fresh cheese were immersed in the coating suspension for 5 s to apply the MEC to the samples. Subsequently, the coated cheeses were placed on mesh screens to allow drying at room temperature (20 °C) for 1 h. Afterwards, they were transferred to individual aluminum trays with perforations and stored under refrigeration conditions at a constant temperature of 4 °C for six days, following the procedure outlined by Vasiliauskaite et al. [44].

2.12. Spray Coating Process

In forming an edible layer on the surface of fresh cheese, an experimental approach based on a pilot spray system was followed, as previously described by Silva-Vera et al. [45]. The liquid flow was controlled using a rotameter connected to the supply line to carry out this task. The MEC needed for layer formation was prepared in a sealed 2.5-L tank. A spraying device from Spraying System S.S. Co (model VA67255–60° S.S., Glendale Heights, IL, USA) and air atomizing caps from the same brand (model VF2850–SS) were used to optimize the process. The experimental tests were conducted at 5 [L/h], 50 [kPa] pressure, and 0.3 [m] height. This combination of variables was applied to the cheese samples for 20 s. After application, the samples were initially stored at room temperature (20 °C) for 1 h for the formation and setting of the edible layer, followed by storage under refrigeration conditions (4° C).

2.13. Lubricated Compression Test

A lubricated compression test was conducted to determine the biaxial behavior of the samples to minimize friction and ensure extensional deformation. Rectangular prisms of cheese measuring 5 cm in length × 2 cm in width × 1.5 cm in height were cut and positioned between two parallel plates lubricated with vegetable oil, attached to a universal testing machine (Zwick/Roell BDO-FBO.5T5 Texture Analyzer, Ulm, Germany). The samples underwent biaxial deformation at a constant deformation rate (1 mm/s) and were compressed to 85% of their initial height. The obtained data files were transferred to Microsoft Excel® (v.2404/2019) for subsequent curve fitting [46].

2.14. Weight Loss (WL)

The percentage weight loss of the cheeses was determined by subtracting the initial weight of the cheese from the weights measured at various intervals during storage under accelerated conditions (30 °C), as per Equation (12) [47].

$$WL = \frac{[W_i] - [W_t]}{[W_i]} \cdot 100 \quad (12)$$

where

W_i: The initial weight of the cheese.
W_t: The weight of the cheese at time t.

2.15. Peroxide Index (PI)

The cheese samples with coatings were subjected to accelerated storage conditions at 30 °C for 6 days, following the methodology described by Sánchez-González and Pérez [48]. The peroxide index was determined according to the UNE-ISO 3960-2017 [49] standard for oils and fats of animal and vegetable origin. A 0.3 [g] sample was weighed and dissolved in 30 [mL] of acetic acid–chloroform solution (3:2). Subsequently, 3 [g] of potassium iodide and 500 [µL] of distilled water were added, and the mixture was stirred for 1 min. Then, 30 [mL] of distilled water and 1.5 [mL] of a 1% starch solution were incorporated as an indicator. Titration was carried out with 0.001 N sodium thiosulfate until a color change was observed. Blank titration was performed, and the peroxide index was calculated according to Equation (13).

$$PI = \frac{(V_M - V_B) \cdot N \cdot 1000}{m} \quad (13)$$

where

PI: Peroxide index expressed in Meq. O_2/kg of cheese.
V_M: The volume of sodium thiosulfate spent on the sample.
V_B: The volume of sodium thiosulfate spent on the blank.
N: The normality of sodium thiosulfate.
m: The mass of cheese used.

2.16. Statistical Analysis

All the assays outlined in this manuscript were conducted in triplicate, and the experimental data obtained were expressed as mean ± standard deviation. Differences among three or more groups were assessed using ANOVA tests, followed by Tukey's comparison tests with a confidence level of 95% ($p < 0.05$) to establish statistical significance [50]. All the statistical analyses were performed using the STATGRAPHICS Centurion XVI software, v.16.1.03 (StatPoint Technologies, Inc., Warrenton, VA, USA).

3. Results

3.1. Multi-Level Factorial Design

During our research, we implemented a multi-level factorial design to optimize nanoliposome composition (NT) for grape seed oil formulation. Figure 1 displays the estimated response surface detailing the optimal combination of Tween-80 and GLY. The results revealed that the optimal concentration of Tween-80 was 2.6%, while that of GLY was 1% relative to the amount of grape seed oil. These proportions were critical for maximizing the nanoliposome's efficacy and stability. Under these optimal conditions, a desirability value of 0.55 was attained, indicating an ideal combination that maximizes the nanoliposome's desired properties. These findings underscore the significance of meticulous component optimization in nanoliposome formulation, emphasizing efficiency in encapsulating bioactive compounds and ensuring the quality and stability of the final product.

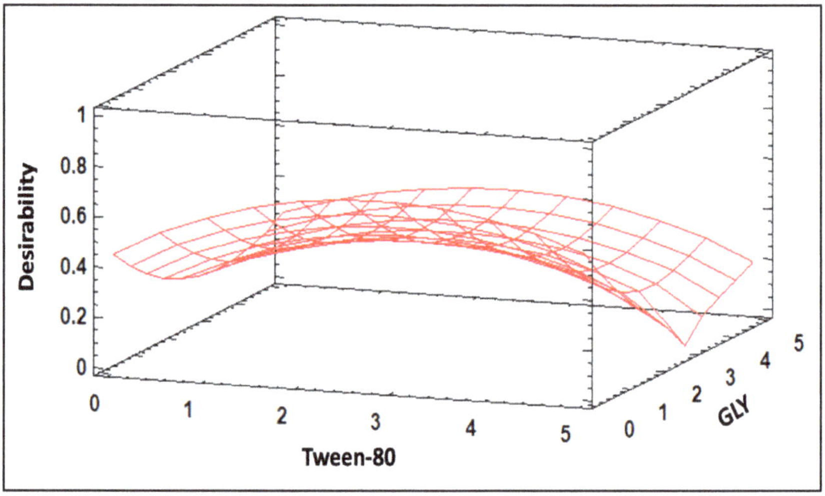

Figure 1. Estimated response surface for NT ingredients.

3.2. Encapsulation Efficiency (EE)

Within the scope of our research, we assessed the EE for the epicatechin monomer in NT derived from grape seed tannin powder, Tween-80, and glycerol. The results unveil a significant EE, registering at 76 ± 0.03 [%]. This metric attests to the effectiveness of the encapsulation process, showcasing the nanoliposomes' capability to retain and safeguard the epicatechin monomer during formulation.

The notable 76% EE underscores the efficiency of the specific combination of the components employed in the nanoliposome synthesis. As determined in the earlier multi-level factorial design, the presence of Tween-80 and glycerol optimizes the desirability function and contributes significantly to the encapsulation efficiency. The nanoliposomes' ability to encapsulate the epicatechin monomer with such efficacy is pivotal for ensuring the stability and bioavailability of the compound across diverse applications, spanning from the food industry to pharmaceuticals.

These findings reinforce the relevance of precise formulation in nanoliposome synthesis for the efficient encapsulation of bioactive compounds, thus promoting their successful application across various scientific and industrial domains.

3.3. Stability Study of NT

Table 2 shows the comprehensive results of a stability study conducted on nanoliposomes (NTs) throughout a 15-day refrigerated storage period. The mean particle size (MPS), polydispersity index (PDI), diffusion coefficient, Z-potential, conductivity, and transmittance parameters play crucial roles in determining the stability and behavior of the nanoparticles in storage conditions. The results provide valuable insights into the dynamic evolution of NT characteristics over time, shedding light on their potential applications in diverse fields.

Table 2. Stability study of NT during the refrigerated storage period *.

Days	MPS [nm]	PDI	Diffusion Coefficient [μm²/s]	Z-Potential [mV]	Conductivity [mS/cm]	Transmittance [%]
0	450 ± 3 d	0.299 ± 0.10 d	1.088 ± 0.04 a	30.47 ± 1.89 b	2.98 ± 0.04 c	0.069 ± 0.01 a
1	434 ± 2 c	0.248 ± 0.04 b	1.129 ± 0.05 a	25.58 ± 2.65 a	2.94 ± 0.09 c	0.071 ± 0.02 a
2	356 ± 5 c	0.260 ± 0.05 c	1.375 ± 0.04 b	36.31 ± 2.32 b	2.83 ± 0.08 bc	0.066 ± 0.03 a
3	337 ± 3 b	0.265 ± 0.04 c	1.453 ± 0.07 c	29.38 ± 3.12 ab	2.62 ± 0.02 b	0.071 ± 0.01 a
4	349 ± 4 c	0.238 ± 0.09 ab	1.405 ± 0.05 c	42.66 ± 6.57 c	2.61 ± 0.07 b	0.071 ± 0.02 a
5	345 ± 3 c	0.226 ± 0.09 a	1.420 ± 0.02 c	40.05 ± 2.62 c	2.62 ± 0.09 b	0.069 ± 0.02 a
6	331 ± 5 a	0.259 ± 0.03 c	1.479 ± 0.06 c	28.59 ± 3.03 ab	1.84 ± 0.06 a	0.067 ± 0.04 a
7	319 ± 3 a	0.218 ± 0.05 a	1.537 ± 0.06 d	47.89 ± 5.23 c	2.86 ± 0.03 bc	0.062 ± 0.04 a
8	320 ± 4 a	0.216 ± 0.08 a	1.528 ± 0.01 d	27.93 ± 1.02 ab	2.61 ± 0.07 b	0.065 ± 0.01 a
9	324 ± 3 a	0.254 ± 0.02 c	1.511 ± 0.02 d	27.86 ± 1.14 ab	3.01 ± 0.03 c	0.066 ± 0.03 a
10	329 ± 4 a	0.263 ± 0.06 c	1.227 ± 0.09 a	28.57 ± 0.65 ab	2.69 ± 0.07 b	0.063 ± 0.03 a
11	321 ± 3 a	0.220 ± 0.04 a	1.528 ± 0.09 d	38.03 ± 1.72 a	2.74 ± 0.09 b	0.065 ± 0.02 a
12	324 ± 3 a	0.240 ± 0.06 b	1.432 ± 0.02 c	30.19 ± 0.29 ab	3.02 ± 0.02 c	0.063 ± 0.02 a
13	327 ± 2 a	0.243 ± 0.04 b	1.451 ± 0.04 c	32.65 ± 2.46 b	3.02 ± 0.01 c	0.064 ± 0.01 a
14	329 ± 3 a	0.241 ± 0.03 b	1.585 ± 0.09 d	34.74 ± 1.62 b	2.99 ± 0.03 c	0.064 ± 0.01 a
15	332 ± 2 a	0.242 ± 0.05 b	1.476 ± 0.02 c	37.96 ± 0.08 c	3.00 ± 0.03 c	0.063 ± 0.02 a

* Results are presented as means ± standard deviation. Different letters in the same column indicate significant differences ($p < 0.05$). MPS: mean particle size; PDI: polydispersity index.

The data indicate that nanoliposomes exhibit significant stability over the 15 days, with only minor fluctuations in critical parameters such as MPS, PDI, and Z-potential. This stability is crucial for their potential use in various industries. For instance, in the pharmaceutical industry, stable nanoliposomes can be used as drug delivery systems, where consistent particle size and charge are essential for the effective and controlled release of therapeutic agents [51]. The cosmetic industry can also benefit from these findings, as stable nanoliposomes can enhance the delivery of active ingredients in skincare products, improving their efficacy and shelf life [52]. Additionally, in the food industry, nanoliposome stability can be leveraged to encapsulate and protect sensitive bioactive compounds, such as vitamins and antioxidants, ensuring their sustained release and enhancing the nutritional profile of food products [53].

These applications underscore the importance of understanding and optimizing nanoliposome stability under various storage conditions. Future research could explore the long-term stability of these nanocarriers and their behavior under different environmental stresses, further broadening their potential industrial applications.

3.4. Proximate Analysis

Table 3 presents the proximate composition of the FC coated with MEC based on NT and polysaccharides such as HPMC and KC, alongside the commercial GC and VC samples. Significant variations in moisture, proteins, lipids, ashes, and carbohydrates are observed among the samples, indicating the coatings' differential impact on the cheeses' nutritional composition. These results provide a crucial starting point for the detailed discussion on how the properties of the coatings, especially those incorporating nanoliposomes, influence the composition and characteristics of fresh cheeses and their relevance in comparison to commercial and vegan products.

Table 3. Proximate composition of coated cheeses with MEC and studied commercial cheeses *.

Sample	Moisture [%]	Protein [%]	Lipids [%]	Ashes [%]	Carbohydrates [%]
FC	57.8 ± 0.4 ^{d}	17.9 ± 1.3 ^{c}	3.6 ± 0.2 ^{ab}	3.1 ± 0.1 ^{c}	17.5 ± 1.5 ^{b}
FC-NT/HPMC	54.9 ± 0.3 ^{c}	19.8 ± 0.3 ^{d}	3.0 ± 0.3 ^{a}	2.8 ± 0.1 ^{b}	19.4 ± 0.4 ^{b}
FC-NT/KC	54.2 ± 0.3 ^{c}	16.1 ± 0.2 ^{b}	4.6 ± 0.1 ^{b}	2.6 ± 0.1 ^{a}	22.4 ± 0.2 ^{c}
GC	47.9 ± 0.2 ^{a}	22.2 ± 0.2 ^{e}	25.2 ± 0.8 ^{d}	3.1 ± 0.1 ^{c}	1.7 ± 0.8 ^{a}
VC	49.3 ± 0.4 ^{b}	2.1 ± 0.1 ^{a}	7.4 ± 0.3 ^{c}	7.2 ± 0.1 ^{d}	33.9 ± 0.4 ^{d}

* Results are presented as means ± standard deviation. Different letters in the same column indicate significant differences ($p < 0.05$). FC: control fresh cheese; FC-NT/HPMC: fresh cheese coated with nanoliposomes encapsulating grape seed tannins and hydroxypropyl methylcellulose; FC-NT/KC: fresh cheese coated with nanoliposomes encapsulating grape seed tannins and kappa carrageenan; GC: goat cheese; VC: vegan cheese substitute.

3.5. Biaxial Behavior of Cheeses Coated with MEC

Figures 2 and 3 depict the results of the biaxial behavior of the cheese samples, derived from the Hencky stress–strain equations when a compressive force is applied in the vertical direction.

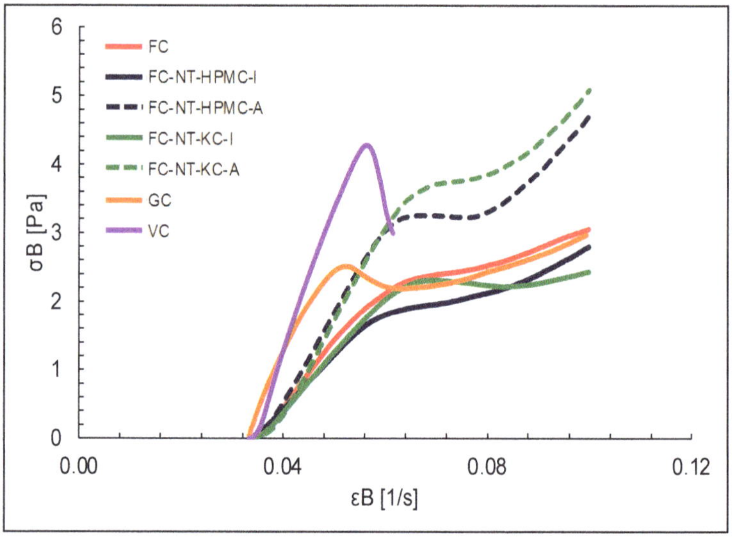

Figure 2. Biaxial stress rate versus biaxial extensional strain rate for different types of cheese.

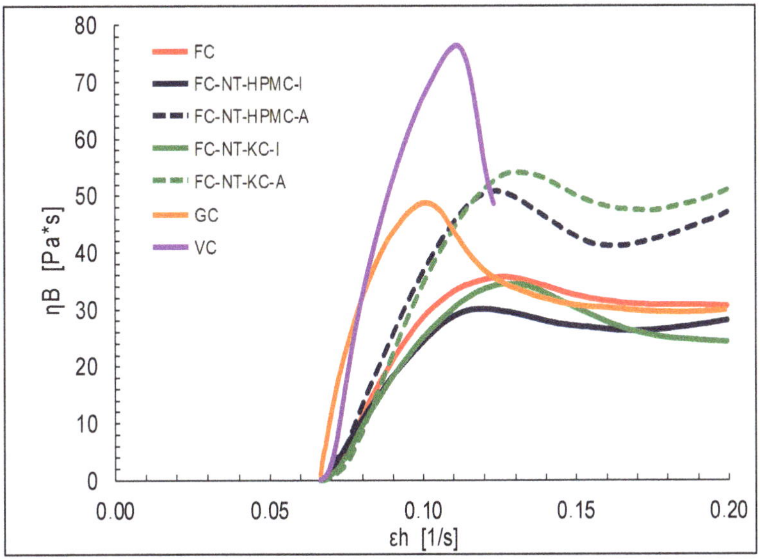

Figure 3. Biaxial extensional viscosity versus Hencky strain rate for different types of cheese.

The obtained data provide a detailed insight into how various factors influence the texture of the cheese samples subjected to biaxial compression. Firstly, the influence of the coating type becomes apparent, with the NT-HPMC-coated samples requiring a higher force for deformation compared to those coated with NT-KC.

To model biaxial behavior, a linear fit was applied to the compressive stress curves [Pa] against time θ [s], represented in Equations (14) and (15), where θ represents the difference between the compression time and the stabilization of the test. This time adjustment corresponds to the difference between the actual compression time (t) and, as visualized in Figure 4, an initial time threshold (t_0) passing through the origin.

$$\sigma = a\theta \tag{14}$$

$$\sigma = a(t - t_0) \tag{15}$$

where

σ: Compressive stress [Pa].
a: The slope of the line [Pa/s].
$\theta = (t - t_0)$: Difference between the compression time and the initial time threshold [s].

The results of fitting the biaxial behavior curves for different cheese types at a compression rate of 1 mm/s are summarized in Table 4.

Table 4. Fitting results of biaxial behavior curves of cheese samples *.

Sample	Velocity [mm/s]	A [Pa/s]	R^2	t_0 [s]	θ $(t - t_0) < t_R$ [s]
FC	1	862.43	0.962	0.02	$\theta < (8.32 - t_0)$
FC-NT-HPMC-I	1	685.21	0.958	0.02	$\theta < (8.40 - t_0)$
FC-NT-HPMC-A	1	1036.8	0.919	0.02	$\theta < (8.10 - t_0)$
FC-NT-KC-I	1	769.06	0.941	0.02	$\theta < (8.45 - t_0)$
FC-NT-KC-A	1	1083.2	0.895	0.02	$\theta < (8.50 - t_0)$
GC	1	1385.8	0.995	0.02	$\theta < (5.96 - t_0)$
VC	1	1664.8	0.946	0.02	$\theta < (7.28 - t_0)$

* Results are presented as means ± standard deviation. FC: control fresh cheese; FC-NT-HPMC: fresh cheese coated with nanoliposomes that encapsulate grape seed tannins and hydroxypropylmethylcellulose; FC-NT-KC: fresh cheese coated with nanoliposomes that encapsulate grape seed tannins and kappa carrageenan; GC: goat cheese; VC: vegan cheese substitute.

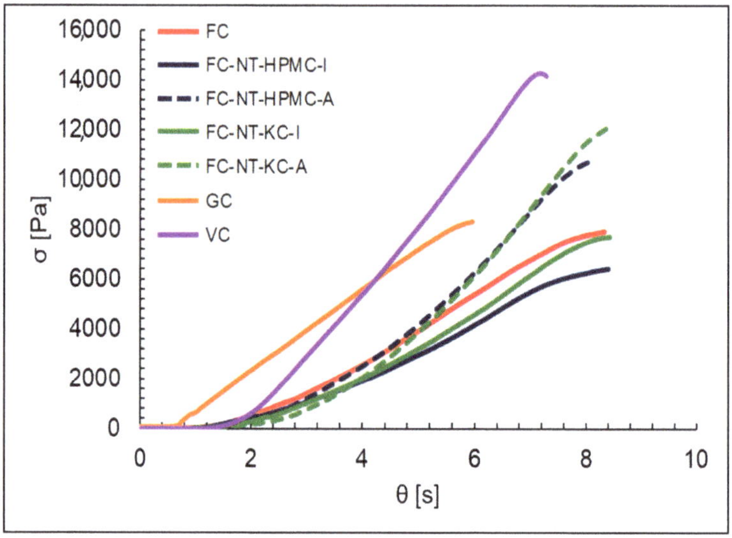

Figure 4. Modeling of the biaxial behavior for different types of cheese at a compression speed of 1 mm/s.

3.6. Quality Parameters in Fresh Cheese

Figure 5 reveals information on weight loss in the different cheese samples over six-day storage under accelerated conditions. The FC sample showed a gradual weight loss, which is expected due to natural dehydration during ripening. The low weight loss value suggests that the uncoated FC maintains predictable weight loss over time.

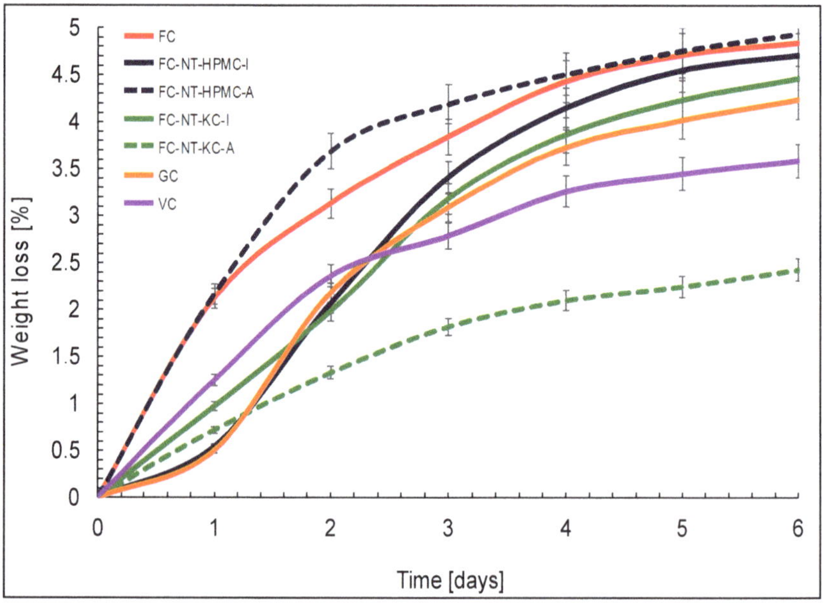

Figure 5. Weight loss for samples of cheeses coated with MEC and commercial cheeses versus storage time under accelerated conditions.

Furthermore, the peroxide index indicates the amount of peroxides (oxidizing compounds) present in a sample. It is commonly used as an indicator of the rancidity or oxidation of a product. The results presented in Table 5 provide a detailed insight into how the peroxide index evolves in the different cheese samples over a six-day storage period under accelerated conditions.

Table 5. Peroxide index expressed in Meq. O_2/kg of cheese in a storage period under accelerated conditions *.

	Days						
Sample	0	1	2	3	4	5	6
FC	2.2 ± 0.2 [b]	3.2 ± 0.2 [b]	3.7 ± 0.3 [b]	5.1 ± 0.4 [b]	5.7 ± 0.4 [c]	6.5 ± 0.2 [c]	9.1 ± 0.4 [c]
FC-NT-HPMC-I	0.7 ± 0.3 [a]	0.8 ± 0.2 [a]	1.4 ± 0.2 [a]	1.7 ± 0.3 [a]	1.9 ± 0.2 [a]	1.8 ± 0.5 [a]	2.8 ± 0.5 [ab]
FC-NT-HPMC-A	0.8 ± 0.5 [a]	1.4 ± 0.2 [a]	1.6 ± 0.2 [a]	1.7 ± 0.3 [a]	1.8 ± 0.2 [a]	1.9 ± 0.4 [a]	2.2 ± 0.4 [a]
FC-NT-KC-I	1.0 ± 0.3 [a]	1.4 ± 0.5 [a]	1.8 ± 0.2 [a]	2.0 ± 0.3 [a]	2.1 ± 0.2 [a]	3.0 ± 0.3 [b]	3.1 ± 0.5 [ab]
FC-NT-KC-A	1.2 ± 0.2 [ab]	1.3 ± 0.3 [a]	1.9 ± 0.4 [a]	2.1 ± 0.2 [a]	2.3 ± 0.3 [a]	3.2 ± 0.4 [b]	3.9 ± 0.6 [b]
GC	2.2 ± 0.4 [b]	3.0 ± 0.3 [b]	3.4 ± 0.2 [b]	4.2 ± 0.5 [b]	4.7 ± 0.3 [b]	6.2 ± 0.2 [c]	9.4 ± 0.4 [c]
VC	2.1 ± 0.5 [b]	3.0 ± 0.3 [b]	3.7 ± 0.3 [b]	4.8 ± 0.4 [b]	6.9 ± 0.2 [d]	7.9 ± 0.5 [d]	9.7 ± 0.3 [c]

* Results are presented as means ± standard deviation. Different letters in the same column indicate significant differences ($p < 0.05$). FC: control fresh cheese; FC-NT-HPMC: fresh cheese coated with nanoliposomes that encapsulate grape seed tannins and hydroxypropylmethylcellulose; FC-NT-KC: fresh cheese coated with nanoliposomes that encapsulate grape seed tannins and kappa carrageenan; GC: goat cheese; VC: vegan cheese substitute.

4. Discussion

4.1. Multi-Level Factorial Design

The observed outcome underscores the pivotal roles played by both Tween-80 and GLY in optimizing the desirability function value. Glycerol, known for its humectant and plasticizing properties, emerges as a critical component [54]. Its inclusion enhances the film-forming capacity and texture of the MEC. Glycerol's humectant properties contribute to moisture retention, a feature particularly beneficial for preserving freshness in cheese products. Moreover, its plasticizing characteristics augment the overall flexibility and pliability of the MEC, contributing to improved film integrity. These attributes are of paramount importance for the successful application of MEC in fresh cheese production, where factors such as texture, film-forming capacity, and moisture retention play crucial roles in determining the quality and shelf life of the product. Thus, the synergy between Tween-80 and glycerol emerges as a strategic formulation approach, optimizing both functional and organoleptic attributes for enhanced applicability in fresh cheese manufacturing.

4.2. Encapsulation Efficiency (EE)

The observed enhancement in EE with the application of ultrasound suggests a notable facilitation of the encapsulation process for smaller units of condensed tannins, as highlighted by the findings of Rosales et al. [55]. Ultrasound, known for its ability to induce controlled cavitation and enhance mass transfer, likely plays a pivotal role in promoting the encapsulation of these smaller tannin units, resulting in the heightened efficiency observed in our study.

Furthermore, the influence of citric acid, a tricarboxylic acid, on encapsulation is noteworthy. As a crystalline solid under normal conditions, citric acid's high solubility in water becomes a critical factor in forming encapsulation complexes with epicatechin compounds during the NT fabrication process, as discussed by Munin et al. [56]. The hydroxyl group's positioning at position 3 in epicatechin introduces subtle molecular interactions within the encapsulation matrix. These interactions, along with the three-dimensional structure of the molecule, potentially contribute to variations in EE, as proposed by Li et al. [57].

Regarding the interaction between lipidic particles (NTs) and carbohydrate polymers, it is essential to acknowledge that these interactions can be complex and somewhat unpredictable due to the heterogeneous nature of the nanoliposome matrix. However, these intricate interactions may play a crucial role in imparting stability to the coating.

The reported EE values align with prior studies demonstrating high EE for nanoliposomes loaded with diverse bioactive compounds, such as alpha-linolenic acid (ALA), an omega-3-rich fatty acid, and polyphenols like naringin and naringenin [58–60]. The consistency in EE values across different studies underscores the robustness of the nanoliposome platform for encapsulating a broad range of bioactive compounds, showcasing its potential for diverse applications in the delivery of therapeutic and nutritional agents.

4.3. Stability Study of NT

The stability study of NT revealed several dynamic interactions between its components, shedding light on their role in maintaining the nanoliposome's stability. The observed fluctuations in key parameters provide valuable insights into the complex interplay of factors influencing nanoliposome behavior during refrigerated storage.

The mean particle size (MPS) fluctuations observed in the initial days, followed by stabilization around 331 ± 5 [nm] by day 6 (as shown in Table 2), may be indicative of an equilibrium point reached in the aggregation–disaggregation processes of the nanoparticles [61]. The dynamic nature of nanoliposome interactions can result in variations in MPS, and the observed stabilization could suggest a balance between aggregation and disaggregation phenomena [62].

Furthermore, the interactions between lipidic particles (NTs) and carbohydrate polymers, such as HPMC and KC, which constitute the multilayer edible coatings (MECs), play a crucial role in coated nanoliposome stability. The hydrophobic–hydrophilic balance between the lipidic particles and the polysaccharide matrix influences the nanoliposome's overall stability by affecting their aggregation–disaggregation dynamics and surface charge properties [63].

The polydispersity index (PDI) values, consistently within a moderate range, indicate the relatively uniform dispersion of particle sizes. The stability in PDI implies that the nanoliposomes maintained a relatively homogenous size distribution over the refrigerated storage period [64].

The peak in the diffusion coefficient on day 2, indicating enhanced particle movement or diffusion, provides valuable insights into the mobility of nanoliposomes. This understanding can significantly impact their behavior in various applications such as drug delivery or catalysis [65].

The Z-potential values, representing the surface charge of NT, declined until day eight before stabilizing. Changes in Z-potential can influence the stability of colloidal systems, and the observed variations suggest alterations in the electrostatic interactions governing nanoliposome stability [66]. Further investigations into the factors influencing surface charge dynamics could enhance our understanding of nanoliposome behavior.

The conductivity fluctuations, which reached a minimum on day 7, could potentially indicate alterations in the ionic strength or conductivity of the surrounding media. These variations might be attributed to changes in the nanoliposome's environment, such as interactions with the storage medium or potential surface modifications [67].

The transmittance values remained stable, indicating the maintenance of optical clarity [68]; this is crucial for applications involving transparent or translucent materials, and the consistent transmittance values suggest the preservation of the NT optical properties over the storage period.

These results suggest a dynamic interplay of factors influencing nanoliposome stability during refrigerated storage. The observed fluctuations in key parameters highlight the need for a nuanced understanding of the specific interactions between the components of the MEC and their effects on the coated nanoliposome stability. Further mechanistic insights into these interactions will enhance our understanding of nanoliposome behavior and facilitate the development of more stable and efficient nanocarrier systems for various applications.

4.4. Proximate Analysis

The proximate composition of the analyzed cheeses reveals statistically significant differences ($p < 0.05$) among the studied samples. The uncoated control cheese (FC) exhibits a high moisture content (57.81% ± 0.40), characteristic of its fresh and unripened nature, retaining a higher water content during its manufacturing process [69]. In contrast, GC displays the lowest moisture content (47.92% ± 0.24). This lower moisture content is likely due to the different properties of rennet curds from goat's milk compared to cow's milk used in FC, which typically results in a firmer texture [70]. The specific ripening times for the commercial cheese samples should also be considered, as they also influence moisture content and texture. For instance, the commercial GC typically undergoes a ripening period of approximately 4 to 6 months, compared to the FC, which does not undergo a significant ripening process [71].

Regarding protein content, GC shows the highest concentration (22.17% ± 0.22), attributable to the higher proportion of proteins typically found in goat milk compared to the cow's milk used in FC [72]. The coated cheese FC-NT/HPMC also exhibits a significant ($p < 0.05$) protein content (19.82% ± 0.32), potentially linked to the coating and the components used in its production.

Goat cheese stands out for its high lipid content (25.17% ± 0.77), characteristic of goat cheeses rich in fat, contributing to its creamy texture and flavor [73,74]. The coated cheese FC-NT/KC also shows a significant ($p < 0.05$) lipid content (4.61% ± 0.07), possibly influenced by grape seed tannins and the properties of kappa carrageenan in retaining fat in the cheese.

The ash content in cheese is related to the presence of minerals and inorganic compounds. Both FC and GC display a notable ash content, possibly due to the mineral composition of the milk used. In contrast, the coated cheeses FC-NT/HPMC and FC-NT/KC show lower ash levels. The reduction in ash content could be due to the interaction between the coating materials and the mineral components in the cheese. The polysaccharides in the coatings might bind to certain minerals, thereby reducing their availability or solubility, which can lead to lower ash content in the final product [75].

Furthermore, the cheese substitute (VC) stands out for its high carbohydrate content (33.98% ± 0.37), attributable to the use of plant-based ingredients instead of dairy [76]. These variations in the proximate composition of the cheeses indicate how the nature of ingredients and manufacturing processes influence the composition and, consequently, the physical properties of cheeses.

The introduction of MEC made with NT and polysaccharides such as HPMC and KC seems to have variable effects on the composition, suggesting the need for further research to understand better its influence on the quality and nutritional value of cheeses.

4.5. Biaxial Behavior of Cheeses Coated with MEC

The results provide a detailed insight into how various factors influence the texture of the biaxially compressed cheese samples. Firstly, the apparent influence of the type of coating is highlighted, where the samples coated with NT-HPMC required a significantly higher deformation force compared to those coated with NT-KC [77,78]. This discrepancy is attributed to the intrinsic properties of the polysaccharides used in the coatings, showing that HPMC tends to generate more rigid coatings in contrast to KC, which provides more elastic coatings. Additionally, carrageenan in the coatings appears to accelerate the deformation rate, possibly due to its gelling properties and water-retention capacity [79].

Another finding relates to the coating application method, where the samples coated by immersion require a higher deformation force than those coated by spraying [80]. This difference is mainly attributed to the coating thickness, as immersion results in a thicker layer that demands additional force for deformation. In this regard, the application method emerges as a critical factor in the rheological behavior of the samples.

Compared to laboratory-made cheeses, A significant difference is observed in the biaxial extensional stress between the commercial cheeses GC and VC [81,82]. This disparity could result from variations in the dairy and non-dairy matrices used, the production process, and the days of maturation before acquiring commercial samples.

Analyzing the biaxial extensional viscosity of coated and control cheeses reveals a pseudoplastic behavior in several samples, suggesting a decrease in viscosity as Hencky's deformation rate increases [20]. This behavior is attributed to the coatings' structure and composition, which are more pronounced in the samples coated by immersion. The observed differences between the immersion and spray-coated samples are associated with the uniformity of the coating and the formation of a stable biopolymer network [83].

Furthermore, the commercial cheeses GC and VC exhibit higher biaxial extensional viscosity, which is attributed to the proximal composition of the samples. Higher lipid and carbohydrate contents provide structure to the food, requiring more effort for deformation.

The analysis of compression stress highlights the unique response of each cheese type. The control FC shows a slope of 862.43 [Pa/s], indicating a relatively quick response to compression [84]. Moreover, the high coefficient of determination (R^2) of 0.962 suggests an excellent fit quality of the linear model. This could be related to the typical soft and moist texture of this type of cheese, which deforms and yields to compression more quickly.

On the other hand, GC exhibits the highest slope, with a value of 1385.8 [Pa/s] and an R^2 of 0.995; this indicates a swift response to compression and excellent model fit quality. Due to their prolonged maturation process, the dense and firm texture of goat cheeses could be related to this rapid response, as the cheese effectively resists compression [85]. VC shows a similar behavior, suggesting a quick response and a good model fit. The texture of vegan cheeses generally resembles that of harder cheeses, which could explain this fast response to compression [86].

The cheeses coated by immersion with NT and polysaccharides FC-NT/HPMC-I and FC-NT/KC-I also show considerable slopes of 685.21 and 769.06 [Pa/s], respectively. These results suggest that nanoliposome and tannin coatings can influence the mechanical response of cheeses, possibly related to the interaction of these ingredients in the cheese matrix [22]. In the case of the cheeses coated by spray with NT and polysaccharides FC-NT-HPMC-A and FC-NT-KC-A, the slopes indicate even faster responses. These results could be related to the specific properties of nanoliposomes and KC, affecting coated cheeses' texture and compression response. The results suggest that immersion application may moderate the mechanical response of cheese; on the contrary, spraying presents a faster mechanical response, thus providing the cheese with a coating texture that resists compression more firmly [87].

Overall, these findings underscore the importance of carefully considering the composition of coatings, the type of carrageenan, the coating application method, and the type of cheese in the formulation of food products. A deeper understanding of texture is essential for developing food products with desired properties and ensuring the product's quality. These findings provide valuable guidelines for optimizing formulations and processes in the food industry.

4.6. Quality Parameters in Fresh Cheese

The samples coated with HPMC, either by immersion (FC-NT-HPMC-I) or spraying (FC-NT-HPMC-A), exhibit significantly less weight loss compared to the uncoated control (FC). However, it is observed that immersion results in slightly higher weight loss than spraying, which may be due to increased moisture retention in the spray-coated samples [88]. The low weight loss values in both the NT-HPMC-coated samples suggest that applying NT-HPMC provides consistency in moisture retention over time. Similarly, samples coated with KC, either by immersion (FC-NT-KC-I) or spraying (FC-NT-KC-A), also experience less weight loss compared to the control cheese (FC). The choice of application method again influences the magnitude of weight loss, with immersion resulting in slightly higher loss than spraying.

Lastly, the samples from commercial cheeses (GC and VC) exhibited weight loss over the six days. However, this weight loss is significantly lower than the control cheese (FC).

This difference could be related to variations in commercial cheeses' proximal composition and manufacturing process compared to FC.

In summary, these findings underscore the effectiveness of coatings with NT-HPMC and NT-KC in reducing weight loss compared to the uncoated cheese [89].

In the case of FC, a progressive increase in the peroxide index is observed over time. This increase suggests that lipids present in FC are undergoing an oxidation process. The variability in peroxide values, indicated by the rising standard deviation, suggests that oxidation may vary between individual samples, possibly due to differences in lipid quantity, oxygen levels, or cheese storage conditions [90].

NT-HPMC and NT-KC coatings, applied by immersion and spraying, clearly demonstrate antioxidant effects in reducing peroxide formation in cheese during storage. This antioxidant effect is attributed to the ability of these coatings to act as physical barriers, limiting the exposure of cheese lipids to environmental oxygen and the presence of tannins, which are gradually released from the coating [91]. The cheese coated with NT-HPMC stands out for maintaining consistent antioxidant protection over time, suggesting constant effectiveness. On the other hand, NT-KC is also effective but shows more significant variability in peroxide values and standard deviations, which could be due to specific sample factors. The choice between NT-HPMC and NT-KC may depend on the desired consistency in antioxidant protection and other specific factors related to the product or manufacturing process.

Furthermore, the samples from commercial cheeses GC and VC also exhibit lower peroxide indices than FC, suggesting that these commercial cheeses have been formulated or processed to reduce peroxide formation. This reduction could be attributed to selecting ingredients (additives) and manufacturing methods that minimize exposure of cheese lipids to oxygen. The low variability in peroxide values and lower standard deviations indicate that these commercial cheeses maintain consistent quality and are less prone to oxidation.

5. Conclusions

During the analysis of the proximal composition, significant differences ($p < 0.05$) were observed in the contents of moisture, proteins, lipids, ash, and carbohydrates between cheeses coated with nanoliposomes (FC-NT/HPMC and FC-NT/KC) and commercial cheeses (GC and VC). Particularly noteworthy is the significantly high protein content of 19.82% $\pm$ 0.32 in the FC-NT/HPMC coated cheese, indicating a potential influence of the coating and its components on its production.

The analysis of biaxial behavior revealed that coating, application method, and cheese type significantly influenced texture and compression response, with biaxial extensional viscosity values ranging from 1036 to 1083 [Pa·s]. Furthermore, MEC exhibited antioxidant properties by reducing peroxide formation by 75% compared to the uncoated cheese, limiting weight loss to below 5% during refrigerated storage.

In terms of quality parameters, the nanoliposome coatings (NT-HPMC and NT-KC) proved effective in significantly reducing ($p < 0.05$) weight loss and acted as antioxidant barriers, limiting peroxide formation during storage. This antioxidant effect was attributed to the ability of these coatings to act as physical barriers and the progressive release of tannins. The commercial cheeses also showed less susceptibility to oxidation, emphasizing the importance of formulation and processing in the final product's quality.

These results underscore the complexity of the interaction between coatings, cheese composition, and processing methods, emphasizing the need to carefully consider these factors to develop food products with desired properties and ensure quality.

Author Contributions: Conceptualization, F.A.O.; data curation, A.M., E.N. and V.V.; formal analysis, F.A.O.; funding acquisition, F.A.O.; investigation, A.M. and E.N.; methodology, A.M., V.V. and F.A.O.; project administration, F.A.O.; software, E.N.; supervision, F.A.O.; writing—original draft, Angela Monasterio; writing—review and editing, A.M. and F.A.O. All authors have read and agreed to the published version of the manuscript.

Funding: The Research and Development Chilean National Agency funded this research (ANID) FONDECYT Regular project 1241250.

Institutional Review Board Statement: Not applicable.

Data Availability Statement: The data presented in this study are available on request from the corresponding author. The data are not publicly available because it belongs to an ongoing project.

Acknowledgments: A.M. thanks the Research and Development Chilean National Agency (ANID, Chile) for the support through the National Doctoral Scholarship 21200051-2020. F.A.O. thanks the Anamin Group for their support with the equipment to carry out the particle stability study.

Conflicts of Interest: The authors declare no conflicts of interest. The funders had no role in the study's design; in the collection, analyses, or interpretation of data; in the writing of the manuscript; or in the decision to publish the results.

References

1. Barukčić, I.; Ščetar, M.; Marasović, I.; Lisak Jakopović, K.; Galić, K.; Božanić, R. Evaluation of quality parameters and shelf life of fresh cheese packed under modified atmosphere. *J. Food Sci. Technol.* **2020**, *57*, 2722–2731. [CrossRef] [PubMed]
2. Vasiliauskaite, A.; Mileriene, J.; Songisepp, E.; Rud, I.; Muizniece-Brasava, S.; Ciprovica, I.; Axelsson, L.; Lutter, L.; Aleksandrovas, E.; Tammsaar, E.; et al. Application of edible coating based on liquid acid whey protein concentrate with indigenous *Lactobacillus helveticus* for acid-curd cheese quality improvement. *Foods* **2022**, *11*, 3353. [CrossRef] [PubMed]
3. Duarte, R.V.; Casal, S.; Gomes, A.M.; Delgadillo, I.; Saraiva, J.A. Nutritional and quality evaluation of hyperbaric stored fresh cheeses. *Food Chem. Adv.* **2023**, *2*, 100212. [CrossRef]
4. Sun, Q.; Yin, S.; He, Y.; Cao, Y.; Jiang, C. Biomaterials and Encapsulation Techniques for Probiotics: Current Status and Future Prospects in Biomedical Applications. *Nanomaterials* **2023**, *13*, 2185. [CrossRef] [PubMed]
5. Baghi, F.; Gharsallaoui, A.; Dumas, E.; Ghnimi, S. Advancements in biodegradable active films for food packaging: Effects of nano/microcapsule incorporation. *Foods* **2022**, *11*, 760. [CrossRef] [PubMed]
6. Resat Atilgan, M.; Bayraktar, O. Procyanidins. In *Handbook of Food Bioactive Ingredients: Properties and Applications*; Springer International Publishing: Cham, Switzerland, 2022; pp. 1–43.
7. Soares, S.; Brandão, E.; Guerreiro, C.; Soares, S.; Mateus, N.; De Freitas, V. Tannins in food: Insights into the molecular perception of astringency and bitter taste. *Molecules* **2020**, *25*, 2590. [CrossRef] [PubMed]
8. Pires, M.A.; Pastrana, L.M.; Fuciños, P.; Abreu, C.S.; Oliveira, S.M. Sensorial perception of astringency: Oral mechanisms and current analysis methods. *Foods* **2020**, *9*, 1124. [CrossRef] [PubMed]
9. Bleotu, C.; Mambet, C.; Matei, L.; Dragu, L.D. Improving wine quality and safety through nanotechnology applications. In *Nanoengineering in the Beverage Industry*; Academic Press: New York, NY, USA, 2020; pp. 437–458.
10. Marques, M.S.; Lima, L.A.; Poletto, F.; Contri, R.V.; Guerreiro IC, K. Nanotechnology for the treatment of paediatric diseases: A review. *J. Drug Deliv. Sci. Technol.* **2022**, *75*, 103628. [CrossRef]
11. Lv, J.M.; Ismail, B.B.; Ye, X.Q.; Zhang, X.Y.; Gu, Y.; Chen, J.C. Ultrasonic-assisted nanoencapsulation of kiwi leaves proanthocyanidins in liposome delivery system for enhanced biostability and bioavailability. *Food Chem.* **2023**, *416*, 135794. [CrossRef]
12. Yousuf, B.; Sun, Y.; Wu, S. Lipid and lipid-containing composite edible coatings and films. *Food Rev. Int.* **2022**, *38*, 574–597. [CrossRef]
13. Zhang, S.; Qamar, S.A.; Junaid, M.; Munir, B.; Badar, Q.; Bilal, M. Algal Polysaccharides-Based Nanoparticles for Targeted Drug Delivery Applications. *Starch-Stärke* **2022**, *74*, 2200014. [CrossRef]
14. Zhong, Y.; Cavender, G.; Zhao, Y. Investigation of different coating application methods on the performance of edible coatings on Mozzarella cheese. *LWT-Food Sci. Technol.* **2014**, *56*, 1–8. [CrossRef]
15. Jafarzadeh, S.; Salehabadi, A.; Nafchi, A.M.; Oladzadabbasabadi, N.; Jafari, S.M. Cheese packaging by edible coatings and biodegradable nanocomposites; improvement in shelf life, physicochemical and sensory properties. *Trends Food Sci. Technol.* **2021**, *116*, 218–231. [CrossRef]
16. Miazaki, J.B.; dos Santos, A.R.; de Freitas, C.F.; Stafussa, A.P.; Mikcha JM, G.; de Cássia Bergamasco, R.; da Silva Scapim, M.R. Edible coatings and application of photodynamics in ricotta cheese preservation. *LWT* **2022**, *165*, 113697. [CrossRef]
17. Senturk Parreidt, T.; Schmid, M.; Müller, K. Effect of dipping and vacuum impregnation coating techniques with alginate based coating on physical quality parameters of cantaloupe melon. *J. Food Sci.* **2018**, *83*, 929–936. [CrossRef]
18. Suhag, R.; Kumar, N.; Petkoska, A.T.; Upadhyay, A. Film formation and deposition methods of edible coating on food products: A review. *Food Res. Int.* **2020**, *136*, 109582. [CrossRef]
19. Andrade, R.D.; Skurtys, O.; Osorio, F.A. Atomizing spray systems for application of edible coatings. *Compr. Rev. Food Sci. Food Saf.* **2012**, *11*, 323–337. [CrossRef]
20. Kontou, V.; Dimitreli, G.; Raphaelides, S.N. Elongational flow studies of processed cheese spreads made from traditional greek cheese varieties. *LWT* **2019**, *107*, 318–324. [CrossRef]
21. Skarlatos, L.; Marinopoulou, A.; Petridis, A.; Raphaelides, S.N. Texture attributes of acid coagulated fresh cheeses as assessed by instrumental and sensory methods. *Int. Dairy J.* **2021**, *114*, 104939. [CrossRef]

22. Lazárková, Z.; Šopík, T.; Talár, J.; Purevdorj, K.; Salek, R.N.; Buňková, L.; Černíková, M.; Novotný, M.; Pachlová, V.; Němečková, I.; et al. Quality evaluation of white brined cheese stored in cans as affected by the storage temperature and time. *Int. Dairy J.* **2021**, *121*, 105105. [CrossRef]
23. Giliel, L.R.; de Souza, O.F.; de Medeiros, T.K.F.; de Oliveira, J.P.F.; de Medeiros, R.S.; de Albuquerque, P.B.S.; de Souza, M.P. Quality and safety of the Coalho cheese using a new edible coating based on the *Ziziphus joazeiro* fruit pulp. *Future Foods* **2021**, *4*, 100089. [CrossRef]
24. Steffe, J. *Rheological Methods in Food Process Engineering*, 2nd ed.; Freeman Press: Dallas, TX, USA, 1996.
25. Osorio, F.; Gahona, E.; Alvarez, F. Water absorption effects on biaxial extensional viscosity of wheat flour dough. *J. Texture Stud.* **2003**, *34*, 147–157. [CrossRef]
26. Guzman, L.E.; Tejada, C.; de la Ossa, Y.J.; Rivera, C.A. Análisis comparativo de perfiles de textura de quesos frescos de leche de cabra y vaca. *Biotecnol. Sect. Agropecu. Agroind.* **2015**, *13*, 139. [CrossRef]
27. Neter, J.; Wasserman, W.; Kunter, M. *Applied Linear Statistical Models*, 2nd ed.; Richard D. Irwin, Inc.: Homewood, IL, USA, 1985; pp. 491–505.
28. Ho, J.Y.; Rabbi, K.F.; Sett, S.; Wong, T.N.; Miljkovic, N. Dropwise condensation of low surface tension fluids on lubricant-infused surfaces: Droplet size distribution and heat transfer. *Int. J. Heat Mass Transf.* **2021**, *172*, 121149. [CrossRef]
29. Lopez-Polo, J.; Monasterio, A.; Cantero-López, P.; Osorio, F.A. Combining edible coatings technology and nanoencapsulation for food application: A brief review with an emphasis on nanoliposomes. *Food Res. Int.* **2021**, *145*, 110402. [CrossRef] [PubMed]
30. Monasterio, A.; Osorio, F.A. Physicochemical Properties of Nanoliposomes Encapsulating Grape Seed Tannins Formed with Ultrasound Cycles. *Foods* **2024**, *13*, 414. [CrossRef] [PubMed]
31. Jafari, S.M.; Vakili, S.; Dehnad, D. Production of a functional yogurt powder fortified with nanoliposomal vitamin D through spray drying. *Food Bioprocess Technol.* **2019**, *12*, 1220–1231. [CrossRef]
32. Babazadeh, A.; Ghanbarzadeh, B.; Hamishehkar, H. Phosphatidylcholine-rutin complex as a potential nanocarrier for food applications. *J. Funct. Foods* **2017**, *33*, 134–141. [CrossRef]
33. Bianchi, S.; Kroslakova, I.; Mayer, I. Determination of molecular structures of condensed tannins from plant tissues using HPLC-UV combined with thiolysis and MALDI-TOF mass spectrometry. *Bio-Protocol* **2016**, *6*, e1975. [CrossRef]
34. Dai, Y.; Ma, Y.; Liu, X.; Gao, R.; Min, H.; Zhang, S.; Hu, S. Formation Optimization, Characterization and Antioxidant Activity of Auricularia auricula-judae Polysaccharide Nanoparticles Obtained via Antisolvent Precipitation. *Molecules* **2022**, *27*, 5807037. [CrossRef]
35. Ghadiri, N.; Mirghazanfari, S.M.; Hadi, V.; Hadi, S.; Mohammadimehr, M.; Ardestani, M.M.; Talatappeh, H.D.; Mohajeri, M. Physicochemical properties and antioxidant activity of polyvinyl alcohol orally disintegrating films containing sweet almond oil nanoemulsion. *J. Food Meas. Charact.* **2023**, *17*, 4045–4059. [CrossRef]
36. Andrade-Pizarro, R.D.; Skurtys, O.; Osorio-Lira, F. Effect of cellulose nanofibers concentration on mechanical, optical, and barrier properties of gelatin-based edible films. *Dyna* **2015**, *82*, 219–226. [CrossRef]
37. Monasterio, A.; Núñez, E.; Brossard, N.; Vega, R.; Osorio, F.A. Mechanical and Surface Properties of Edible Coatings Elaborated with Nanoliposomes Encapsulating Grape Seed Tannins and Polysaccharides. *Polymers* **2023**, *15*, 3774. [CrossRef]
38. Nemati, V.; Hashempour-Baltork, F.; Sadat Gharavi-Nakhjavani, M.; Feizollahi, E.; Marangoni Júnior, L.; Mirza Alizadeh, A. Application of a Whey Protein Edible Film Incorporated with Cumin Essential Oil in Cheese Preservation. *Coatings* **2023**, *13*, 1470. [CrossRef]
39. *A.O.A.C. 925.45*; Loss on drying (moisture) in sugars. AOAC International Publisher: Rockville, MD, USA, 2016.
40. *A.O.A.C. 990.03*; Protein (Crude) in Animal Feed, Combustion Method. AOAC International Publisher: Rockville, MD, USA, 2016.
41. *A.O.A.C. 996.06*; Fat (Total, Saturated and Unsaturated) in Foods, Hydrolytic Extraction Gas Chromatographic Method. AOAC International Publisher: Rockville, MD, USA, 2016.
42. *A.O.A.C. 923.03*; Ash of flour. Direct method. AOAC International Publisher: Rockville, MD, USA, 2016.
43. A.O.A.C. *The Official Methods of Analysis of AOAC International*, 20th ed.; AOAC International Publisher: Rockville, MD, USA, 2016.
44. Vasiliauskaite, A.; Mileriene, J.; Kasparaviciene, B.; Aleksandrovas, E.; Songisepp, E.; Rud, I.; Serniene, L. Screening for Antifungal Indigenous Lactobacilli Strains Isolated from Local Fermented Milk for Developing Bioprotective Fermentates and Coatings Based on Acid Whey Protein Concentrate for Fresh Cheese Quality Maintenance. *Microorganisms* **2023**, *11*, 557. [CrossRef] [PubMed]
45. Silva-Vera, W.; Zamorano-Riquelme, M.; Rocco-Orellana, C.; Vega-Viveros, R.; Giménez-Castillo, B.; Silva-Weiss, A.; Osorio-Lira, F. Study of spray system applications of edible coating suspensions based on hydrocolloids containing cellulose nanofibers on grape surface (*Vitis vinifera* L.). *Food Bioprocess Technol.* **2018**, *11*, 1575–1585. [CrossRef]
46. Bousi, C.; Sismanidou, O.X.; Marinopoulou, A.; Raphaelides, S. Effect of sugar addition on the elongational and shearing deformation behavior of sesame paste systems. *LWT* **2022**, *153*, 112479. [CrossRef]
47. Elfadaly, S.S.; Mattar, A.A.; Sorour, M.A.; Karam-Allah AA, K.; Soliman, T.N. Preparation and Characterization of Whey Protein Edible Coating with Potato and Mango Peels extract: Application in Processed Cheese. *Egypt. J. Chem.* **2023**, *66*, 401–416.
48. Sánchez-González, J.A.; Pérez Cueva, J.A. Vida útil sensorial del queso mantecoso por pruebas aceleradas. *Sci. Agropecu.* **2016**, *7*, 215–222. [CrossRef]
49. *UNE-ISO 3960-2017*; Animal and vegetable fats and oils—Determination of peroxide value—Iodometric (visual) endpoint determination. Asociación Española de Normalización: Madrid, Spain, 2017.

50. Nanda, A.; Mohapatra, B.B.; Mahapatra, A.P.K.; Mahapatra, A.P.K.; Mahapatra, A.P.K. Multiple comparison test by Tukey's honestly significant difference (HSD): Do the confident level control type I error. *Int. J. Stat. Appl. Math.* **2021**, *6*, 59–65. [CrossRef]
51. Baek, Y.; Jeong, E.W.; Lee, H.G. Encapsulation of resveratrol within size-controlled nanoliposomes: Impact on solubility, stability, cellular permeability, and oral bioavailability. *Colloids Surf. B Biointerfaces* **2023**, *224*, 113205. [CrossRef]
52. Sethi, M.; Rana, R.; Sambhakar, S.; Chourasia, M.K. Nanocosmeceuticals: Trends and Recent Advancements in Self Care. *AAPS PharmSciTech* **2024**, *25*, 51. [CrossRef]
53. Arratia-Quijada, J.; Nuño, K.; Ruíz-Santoyo, V.; Andrade-Espinoza, B.A. Nano-encapsulation of probiotics: Need and critical considerations to design new non-dairy probiotic products. *J. Funct. Foods* **2024**, *116*, 106192. [CrossRef]
54. Giosafatto, C.V.L.; Porta, R. Advanced Biomaterials for Food Edible Coatings. *Int. J. Mol. Sci.* **2023**, *24*, 9929. [CrossRef]
55. Rosales TK, O.; da Silva FF, A.; Bernardes, E.S.; Paulo Fabi, J. Plant-derived polyphenolic compounds: Nanodelivery through polysaccharide-based systems to improve the biological properties. In *Critical Reviews in Food Science and Nutrition*; Taylor & Francis: Oxford, UK, 2023; pp. 1–25.
56. Munin, A.; Edwards-Lévy, F. Encapsulation of natural polyphenolic compounds: A review. *Pharmaceutics* **2011**, *3*, 793–829. [CrossRef] [PubMed]
57. Li, N.; Yang, J.; Wang, C.; Wu, L.; Liu, Y. Screening bifunctional flavonoids of anti-cholinesterase and anti-glucosidase by in vitro and in silico studies: Quercetin, kaempferol and myricetin. *Food Biosci.* **2023**, *51*, 102312. [CrossRef]
58. Joseph, T.M.; Kar Mahapatra, D.; Esmaeili, A.; Piszczyk, Ł.; Hasanin, M.S.; Kattali, M.; Thomas, S. Nanoparticles: Taking a unique position in medicine. *Nanomaterials* **2023**, *13*, 574. [CrossRef] [PubMed]
59. Choudhary, P.; Dutta, S.; Moses, J.A.; Anandharamakrishnan, C. Nanoliposomal encapsulation of chia oil for sustained delivery of α-linolenic acid. *Int. J. Food Sci. Technol.* **2021**, *56*, 4206–4214. [CrossRef]
60. Chen, M.; Li, R.; Gao, Y.; Zheng, Y.; Liao, L.; Cao, Y.; Li, J.; Zhou, W. Encapsulation of hydrophobic and low-soluble polyphenols into nanoliposomes by ph-driven method: Naringenin and naringin as model compounds. *Foods* **2021**, *10*, 963. [CrossRef]
61. Bian, T.; Gardin, A.; Gemen, J.; Houben, L.; Perego, C.; Lee, B.; Klajn, R. Electrostatic co-assembly of nanoparticles with oppositely charged small molecules into static and dynamic superstructures. *Nat. Chem.* **2021**, *13*, 940–949. [CrossRef]
62. D'Onofrio, M.; Munari, F.; Assfalg, M. Alpha-synuclein—nanoparticle interactions: Understanding, controlling and exploiting conformational plasticity. *Molecules* **2020**, *25*, 5625. [CrossRef] [PubMed]
63. Lúcio, M.; Lopes, C.M.; Fernandes, E.; Gonęalves, H.; Oliveira, M.E.C.R. Organic nanocarriers for brain drug delivery. In *Nanoparticles for Brain Drug Delivery*; Jenny Stanford Publishing: Dubai, United Arab Emirates, 2021; pp. 75–160.
64. Fan, C.; Feng, T.; Wang, X.; Xia, S.; Swing, C.J. Liposomes for encapsulation of liposoluble vitamins (A, D, E and K): Comparation of loading ability, storage stability and bilayer dynamics. *Food Res. Int.* **2023**, *163*, 112264. [CrossRef] [PubMed]
65. Saravanakumar, S.M.; Cicek, P.V. Microfluidic mixing: A physics-oriented review. *Micromachines* **2023**, *14*, 1827. [CrossRef] [PubMed]
66. Souri, P.; Emamifar, A.; Davati, N. Physical and antimicrobial properties of nano-ZnO-loaded nanoliposomes prepared by thin layer hydration-sonication and heating methods. *Food Bioprocess Technol.* **2023**, *16*, 1822–1836. [CrossRef]
67. Siva, S.; Bodkhe, G.A.; Cong, C.; Kim, S.H.; Kim, M. Electrohydrodynamic-printed ultrathin Ti3C2Tx-MXene field-effect transistor for probing aflatoxin B1. *Chem. Eng. J.* **2024**, *479*, 147492. [CrossRef]
68. Medina, M.; Nuñez, M. Cheeses from ewe and goat milk. In *Cheese*; Academic Press: Cambridge, MA, USA, 2017; pp. 1069–1091.
69. Li, Y.; Wang, F.; Xu, J.; Wang, T.; Zhan, J.; Ma, R.; Tian, Y. Improvement in the optical properties of starch coatings via chemical-physical combination strategy for fruits preservation. *Food Hydrocoll.* **2023**, *137*, 108405. [CrossRef]
70. Nájera, A.I.; Nieto, S.; Barron LJ, R.; Albisu, M. A review of the preservation of hard and semi-hard cheeses: Quality and safety. *Int. J. Environ. Res. Public Health* **2021**, *18*, 9789. [CrossRef] [PubMed]
71. Jia, R.; Zhang, F.; Song, Y.; Lou, Y.; Zhao, A.; Liu, Y.; Wang, B. Physicochemical and textural characteristics and volatile compounds of semihard goat cheese as affected by starter cultures. *J. Dairy Sci.* **2021**, *104*, 270–280. [CrossRef] [PubMed]
72. Stergiadis, S.; Nørskov, N.P.; Purup, S.; Givens, I.; Lee, M.R. Comparative nutrient profiling of retail goat and cow milk. *Nutrients* **2019**, *11*, 2282. [CrossRef]
73. Deshwal, G.K.; Ameta, R.; Sharma, H.; Singh, A.K.; Panjagari, N.R.; Baria, B. Effect of ultrafiltration and fat content on chemical, functional, textural and sensory characteristics of goat milk-based Halloumi type cheese. *LWT* **2020**, *126*, 109341. [CrossRef]
74. Pawlos, M.; Znamirowska-Piotrowska, A.; Kowalczyk, M.; Zaguła, G.; Szajnar, K. Possibility of Using Different Calcium Compounds for the Manufacture of Fresh Acid Rennet Cheese from Goat's Milk. *Foods* **2023**, *12*, 3703. [CrossRef]
75. Gruskiene, R.; Bockuviene, A.; Sereikaite, J. Microencapsulation of bioactive ingredients for their delivery into fermented milk products: A review. *Molecules* **2021**, *26*, 4601. [CrossRef]
76. Benevides, S.D.; Wurlitzer, N.J.; Dionisio, A.P.; dos Santos Garruti, D.; Nunes GM, V.C.; Chagas, B.A.; de Sousa PH, M. Alternative Protein and Fiber-based Cheese and Hamburger Analogues: Meeting Consumer Demand for Differentiated Plant-based Products. *Chem. Eng. Trans.* **2023**, *102*, 25–30.
77. Xiao, Q. Coating and film-forming properties. In *Food Hydrocolloids: Functionalities and Applications*; Springer: Singapore, 2021; pp. 267–306.
78. Manzoor, M.; Singh, J.; Bandral, J.D.; Gani, A.; Shams, R. Food hydrocolloids: Functional, nutraceutical and novel applications for delivery of bioactive compounds. *Int. J. Biol. Macromol.* **2020**, *165*, 554–567. [CrossRef]

79. Zhu, Y.; Bhandari, B.; Prakash, S. Tribo-rheometry behaviour and gel strength of κ-carrageenan and gelatin solutions at concentrations, pH and ionic conditions used in dairy products. *Food Hydrocoll.* **2018**, *84*, 292–302. [CrossRef]
80. Carvalho, A.F.; Kulyk, B.; Fernandes, A.J.; Fortunato, E.; Costa, F.M. A review on the applications of graphene in mechanical transduction. *Adv. Mater.* **2022**, *34*, 2101326. [CrossRef]
81. Chatziantoniou, S.E.; Thomareis, A.S.; Kontominas, M.G. Effect of different stabilizers on rheological properties, fat globule size and sensory attributes of novel spreadable processed whey cheese. *Eur. Food Res. Technol.* **2019**, *245*, 2401–2412. [CrossRef]
82. Atik, D.S.; Huppertz, T. Melting of natural cheese: A review. *Int. Dairy J.* **2023**, *142*, 105648. [CrossRef]
83. Siyar, Z.; Motamedzadegan, A.; Mohammadzadeh Milani, J.; Rashidinejad, A. The effect of the liposomal encapsulated saffron extract on the physicochemical properties of a functional ricotta cheese. *Molecules* **2021**, *27*, 120. [CrossRef]
84. Lv, Z.; Chen, J.; Holmes, M. Human capability in the perception of extensional and shear viscosity. *J. Texture Stud.* **2017**, *48*, 463–469. [CrossRef] [PubMed]
85. Burgos, L.S.; Pece Azar NB, D.C.; Maldonado, S. Textural, rheological and sensory properties of spreadable processed goat cheese. *Int. J. Food Stud.* **2020**, *9*, 62–74. [CrossRef]
86. Briceño-Ahumada, Z.; Mikhailovskaya, A.; Staton, J.A. The role of continuous phase rheology on the stabilization of edible foams: A review. *Phys. Fluids* **2022**, *34*, 031302. [CrossRef]
87. Hesarinejad, M.A.; Lorenzo, J.M.; Rafe, A. Influence of gelatin/guar gum mixture on the rheological and textural properties of restructured ricotta cheese. *Carbohydr. Polym. Technol. Appl.* **2021**, *2*, 100162. [CrossRef]
88. Mouzakitis, C.K.; Sereti, V.; Matsakidou, A.; Kotsiou, K.; Biliaderis, C.G.; Lazaridou, A. Physicochemical properties of zein-based edible films and coatings for extending wheat bread shelf life. *Food Hydrocoll.* **2022**, *132*, 107856. [CrossRef]
89. Salehi, F. Effect of coatings made by new hydrocolloids on the oil uptake during deep-fat frying: A review. *J. Food Process. Preserv.* **2020**, *44*, e14879. [CrossRef]
90. Rinaldi, S.; Palocci, G.; Di Giovanni, S.; Iacurto, M.; Tripaldi, C. Chemical characteristics and oxidative stability of buffalo mozzarella cheese produced with fresh and frozen curd. *Molecules* **2021**, *26*, 1405. [CrossRef]
91. Kaur, J.; Singh, J.; Rasane, P.; Gupta, P.; Kaur, S.; Sharma, N.; Sowdhanya, D. Natural additives as active components in edible films and coatings. *Food Biosci.* **2023**, *53*, 102689. [CrossRef]

polymers

MDPI

Article

Exploring the Equilibrium State Diagram of Maltodextrins across Diverse Dextrose Equivalents

Zenaida Saavedra-Leos [1], Anthony Carrizales-Loera [1], Daniel Lardizábal-Gutiérrez [2], Laura Araceli López-Martínez [3] and César Leyva-Porras [2,*]

[1] Multidisciplinary Academic Unit, Altiplano Region (COARA), Autonomous University of San Luis Potosi, Carretera a Cedral km 5+600, Matehuala 78700, Mexico; zenaida.saavedra@uaslp.mx (Z.S.-L.); a275849@alumnos.uaslp.mx (A.C.-L.)

[2] Advanced Materials Research Center (CIMAV), Miguel de Cervantes 120, Complejo Industrial Chihuahua, Chihuahua 31136, Mexico; daniel.lardizabal@cimav.edu.mx

[3] Academic Coordination of the Western High Plateau Region, Autonomous University of San Luis Potosi, Salinas de Hidalgo 78600, Mexico; araceli.lopez@uaslp.mx

* Correspondence: cesar.leyva@cimav.edu.mx; Tel.: +52-(614)-4391100 (ext. 2011)

Abstract: This study investigates the equilibrium state diagram of maltodextrins with varying dextrose equivalents (DE 10 and 30) for quercetin microencapsulation. Using XRD, SEM, and optical microscopy, three transition regions were identified: amorphous (a_w 0.07–0.437), semicrystalline (a_w 0.437–0.739), and crystalline ($a_w > 0.739$). In the amorphous region, microparticles exhibit a spherical morphology and a fluffy, pale-yellow appearance, with Tg values ranging from 44 to −7 °C. The semicrystalline region shows low-intensity diffraction peaks, merged spherical particles, and agglomerated, intense yellow appearance, with Tg values below 2 °C. The crystalline region is characterized by fully collapsed microstructures and a continuous, solid material with intense yellow color. Optimal storage conditions are within the amorphous region at 25 °C, a_w 0.437, and a water content of 1.98 g H_2O per g of dry powder. Strict moisture control is required at higher storage temperatures (up to 50 °C) to prevent microstructural changes. This research enhances understanding of maltodextrin behavior across diverse dextrose equivalents, aiding the development of stable microencapsulated products.

Keywords: quercetin microencapsulation; maltodextrins; dextrose equivalents; equilibrium state diagram; storage conditions

Citation: Saavedra-Leos, Z.; Carrizales-Loera, A.; Lardizábal-Gutiérrez, D.; López-Martínez, L.A.; Leyva-Porras, C. Exploring the Equilibrium State Diagram of Maltodextrins across Diverse Dextrose Equivalents. *Polymers* **2024**, *16*, 2014. https://doi.org/10.3390/polym16142014

Academic Editor: Patrick Ilg

Received: 1 June 2024
Revised: 28 June 2024
Accepted: 11 July 2024
Published: 14 July 2024

1. Introduction

Food polymers are polymeric materials that can be safely consumed by humans, animals, or microorganisms [1]. These materials are categorized into polysaccharides, proteins, and lipids. The use of food polymers has become increasingly significant in the development of functional food products, including food packaging and nutrient protection. Additionally, they are pivotal in biomedical applications such as controlled drug delivery, tissue engineering, and wound coatings [2,3]. Compared to synthetic polymers, food polymers offer notable advantages, including biodegradability, biocompatibility, and recyclability [4,5].

Currently, the growing interest in nutrition and health has driven the development of foods that not only meet basic dietary needs but also provide additional health benefits [6]. This trend has led to the creation of functional foods, which fulfill nutritional requirements while positively impacting one or more bodily functions. These foods can also play a preventive role by reducing risk factors associated with various diseases [7]. In addition to these health benefits, it is crucial that food products maintain their properties such as nutritional value, taste, texture, and safety under various external environmental conditions. The primary goal of food conservation is to prolong the shelf life of food, ensuring that it remains safe and enjoyable to consume over an extended period [8]. The main factors affecting shelf life are as follows: (i) the storage conditions, such as temperature, humidity, and light exposure; (ii) the type of packing for protecting the product from the environment; (iii) the stability and composition of the ingredients used in the product; (iv) processing methods used for reducing sources of degradation such as pasteurization, sterilization, and dehydration; (v) the use of chemical or natural preservatives can help prolong shelf life. By understanding and optimizing these factors, manufacturers can ensure that their products remain high-quality and safe for consumption over a longer period.

State diagrams are graphical representations that depict the different states of a substance or mixture under varying conditions of temperature, pressure, and composition [9]. In the context of food stability, these diagrams are essential for understanding and predicting the behavior and stability of food materials under different environmental conditions [10]. State diagrams typically include the following components: isotherms, temperature at which they were constructed, water activity, moisture content, and glass transition temperature [11]. By analyzing a state diagram, it is possible to determine the optimal moisture content and temperature for storing dehydrated products to prevent microbial growth [10]. Additionally, the diagram can reveal the conditions under which the product will transition from a glassy to a rubbery state, indicating potential risks for caking or loss of crispness [12]. Understanding the relationships between temperature, moisture content, and other factors through state diagrams allows for the development of more effective preservation, storage, and processing techniques. This ultimately leads to safer and longer-lasting food products.

Maltodextrins are polysaccharides derived from starch through a process called partial hydrolysis, which breaks down the starch into smaller carbohydrate molecules [13]. These molecules are composed of glucose units linked together in chains of varying lengths [14]. Their structure and properties are determined by the degree of polymerization (DP) and the extent of hydrolysis, reflected in the dextrose equivalent (DE). DE is a measure of the amount of reducing sugars present in a sugar product, expressed as a percentage on a dry basis relative to pure dextrose (glucose), which has a DE value of 100. It quantifies the degree of hydrolysis of starch into glucose and other reducing sugars. A higher DE value indicates a greater degree of starch hydrolysis, resulting in shorter chains of glucose units and higher sweetness. Maltodextrin refers to hydrolysates with DE values ranging from 3 to 20, whereas those with DE values above 20 are classified as glucose syrups. Due to their abundant hydroxyl (–OH) groups and branching structures, maltodextrins are highly soluble in water, making them easy to incorporate into different formulations [15,16]. In the food and beverage industry, maltodextrins serve multiple purposes. They enhance texture, act as bulking agents, and stabilize products [17]. Typically, they are white, odorless powders with a neutral to slightly sweet taste, depending on their degree of polymerization. Similar to other carbohydrates, maltodextrins provide about four calories per gram [18]. There is an empirical rule that inversely relates the DP to the DE (DP = 120/DE), indicating that as the DE increases, the DP decreases. For example, maltodextrins with DE values of 10 and 30 have estimated DP values of 12 and 4, respectively. Maltodextrin with a DE of 10 consists predominantly of long-chain oligosaccharides composed of glucose units, with largely linear chains and some degree of branching. In contrast, maltodextrin with a DE of 30 is composed of shorter chains of glucose units, resulting from more extensive hydrolysis, leading to a higher proportion of short oligosaccharides and even monosaccharides. The DE value impacts properties such as solubility, sweetness, and viscosity, with lower DE maltodextrins being less soluble, less sweet, and more viscous, while higher DE maltodextrins are more soluble, sweeter, and less viscous, making them suitable for different industrial applications. Overall, maltodextrins are versatile and valuable additives in various applications due to their solubility, neutral flavor, and functional properties.

Maltodextrins are widely used as microencapsulating agents in the food, pharmaceutical, and cosmetic industries [19–21]. Microencapsulation is a process where active ingredients or sensitive compounds are coated with a protective material to improve their stability, control release, and enhance their handling properties [22,23]. Their use in encapsulating flavors, vitamins, probiotics, colorants, pharmaceuticals, and cosmetic ingredients highlights their importance in improving product quality and functionality [24]. For example, Saavedra-Leos et al. [25] studied a set of four maltodextrin powders with varying DE and DP, all processed via spray drying. Their findings revealed a direct relationship between DE and DP, with the degree of polymerization emerging as a better parameter for describing the microstructure of maltodextrins. Later, Saavedra-Leos et al. [26] compared the performance of maltodextrin and inulin as microencapsulating agents for antioxidants contained in blueberry juice. The study reported that after spray drying, the maltodextrin powder retained a higher concentration of antioxidants, demonstrating superior performance in preserving these valuable compounds. In this context, it was found that for spray drying orange juice with maltodextrin as the carrier agent, the microstructure of the powders obtained using maltodextrins with a DE of less than 30 was stable and did not collapse. This was observed as a white, non-agglomerated powder [27]. Based on the DE, the type of maltodextrin plays a significant role in the yield and antioxidant content of microencapsulated products obtained through spray drying. Maltodextrins with a DE of less than 10 demonstrated higher content and yield values compared to those with DE of 20 and 40 for both quercetin and resveratrol [28,29]. Maltodextrins have been used as carrier agents in the drying of other juices, such as broccoli and strawberry juices, where the properties of these juices, including aroma, color, and composition, were preserved in the resulting dry powders [30,31]. Additionally, the effects of maltodextrin–inulin blends on the preservation of antioxidants and probiotics have been reported [32]. Recently, an equilibrium state diagram for a functional powdered food based on one of these mixtures (25% maltodextrin and 75% inulin) was documented [33]. From this diagram, optimal storage conditions to extend shelf life were identified. These conditions include a monolayer water content of 2.79 g of water per 100 g of dry powder and a storage temperature slightly above 30 °C. However, to the best of our knowledge, an equilibrium state diagram for maltodextrin has not been reported, nor has the effect of DE on this diagram been explored.

Therefore, the aim of this investigation is to develop a functional food with technological application that also provides antioxidant properties, utilizing spray drying as a microencapsulation technique. For the production of the functional powdered food, maltodextrins with varying degrees of dextrose equivalent (DE) were enriched with quercetin as the antioxidant agent. Equilibrium state diagrams for the microencapsulated product (quercetin/maltodextrin) will be employed to understand its behavior under specific environmental conditions (storage temperature and humidity) and to delineate the precise conditions essential for preserving food stability and ensuring the sustained efficacy of the antioxidant properties within the product.

2. Materials and Methods

2.1. Materials

Two commercial maltodextrins with different DE values, DE 10 and DE 30, were used (≥99% purity, Ingredion, Mexico city, Mexico). The antioxidant employed was quercetin (≥95% purity, Merck, Toluca, Mexico). The adsorption isotherms were developed using microenvironments with different relative humidity. This was achieved using the following inorganic salts: sodium hydroxide (NaOH), potassium acetate (CH_3COOK), magnesium chloride ($MgCl_2$), potassium carbonate (K_2CO_3), magnesium nitrate ($Mg(NO_3)_2$), sodium nitrate ($NaNO_3$), potassium chloride (KCl), and potassium sulfate (K_2SO_4) (≥99% purity, Fermont, Monterrey, Mexico).

2.2. Methods

2.2.1. Spray Drying

For the microencapsulation of the antioxidant, spray drying was employed. A 100 mL feed solution was prepared using 20 g of the respective carrier agent (Maltodextrin DE 10 or DE 30), 1 g of quercetin, and distilled water. The microencapsulation process was carried out using a Mini Spray Dryer B290 (BÜCHI, Labortechnik AG, Flawil, Switzerland) under the following conditions: a feed flow rate of 7 cm^3/min, a hot air flow rate of 28 m^3/h, 70% aspiration, and a pressure of 1.5 bar. Based on the findings reported by Saavedra-Leos et al. (2022), an inlet temperature of 210 °C was selected to achieve the highest antioxidant activity in the resulting product [32].

2.2.2. Sorption Isotherms

Sorption isotherms were determined using the static gravimetric method as proposed by Labuza et al. [34]. The moisture content of the obtained products was measured using the oven-drying method [35]. Approximately 2 g of dried food product powder were placed into closed glass containers, comprising different saturated reagents to achieve the desired water activity (a_w) at equilibrium. The reagents used and the corresponding a_w achieved were NaOH (0.070), CH_3COOK (0.225), $MgCl_2$ (0.329), K_2CO_3 (0.437), $Mg(NO_3)_2$ (0.531), KCl (0.739), NaCl (0.84), and K_2SO_4 (0.96). The incubation temperature was set at 25 °C, reflecting the typical room temperature in storage warehouses. Weights were recorded every 24 h until a constant weight was achieved, defined by a difference of ±0.001 g between two consecutive weightings. The isotherms were constructed based on the water activity data and the equilibrium moisture content of the evaluated samples. The mathematical models used to describe sorption behavior are presented in Table 1.

Table 1. Mathematical models used to describe sorption behavior of dry food products.

Model Name	Model and Describing Parameters
GAB	$Xe = \frac{XmCKa_w}{(1-Ka_w)(1-Ka_w+CKa_w)}$ X_e = equilibrium water content (g H_2O/g dry mass) X_m = equilibrium monolayer water content (g H_2O/g dry mass) a_w = water activity K and C = model constants
BET	$Xe = \frac{XmCa_w}{(1-a_w)(1+(C-1)a_w)}$ Xe = equilibrium water content (g H_2O/g dry mass) X_m = equilibrium monolayer water content (g H_2O/g dry mass) a_w = water activity C = model constant

2.2.3. Antioxidant Activity (AA)

The functional powdered food was evaluated for its hydrogen-donating or radical-scavenging activity using 2,2-Diphenyl-1-picrylhydrazyl (DPPH) as the standard. A mixture of 1.7 mL of an alcoholic DPPH solution (0.1 mmol DPPH) and 1.7 mL of a microencapsulated suspension at a concentration of 30 µg/mL was prepared. The mixture was left to stand in the dark for 30 min, and the absorbance was measured at 537 nm using a UV-Vis Evolution 220 spectrophotometer (Thermo Scientific, Waltham, MA, USA). Measurements were performed in triplicate after the spray drying process and during storage under different humidity conditions.

2.2.4. Physicochemical Characterization

To characterize the microstructure of the functional food product, X-ray diffraction (XRD) analysis was performed. This was conducted using a D8 Advance ECO diffractometer (Bruker, Karlsruhe, Germany) equipped with Cu-Kα radiation (λ = 1.5406 Å), operating at 45 kV and 40 mA in Bragg–Brentano geometry. Scans were conducted over a 2θ range of 5° to 50°, with a step size of 0.016° and a scan time of 20 s per step.

Morphological analysis was carried out with a field emission scanning electron microscope (FESEM) (JEOL JSM-7401F, Tokyo, Japan) at an acceleration voltage of 2 kV. Powder samples were spread on double-sided conductive copper tape and coated with a thin gold layer using a sputter coater (Denton Desk II, Denton, TX, USA) to minimize charging effects. Images were captured at various magnifications: 500×, 1000×, 2500×, and 5000×.

The glass transition temperature (Tg) was determined following the procedure outlined by Saavedra-Leos et al. [36]. A Q200 modulated differential scanning calorimeter (MDSC) (TA Instruments, Lukens Drive, New Castle, DE, USA), equipped with an RCS90 cooling system, was utilized. The calorimeter was calibrated for temperature and enthalpy using indium, and for heat capacity using sapphire. Nitrogen gas of HPLC-grade purity was used as the purge gas at a flow rate of 100 mL/min. Approximately 10 mg of each sample was placed in sealed Tzero™ aluminum pans. The thermal program involved heating and cooling cycles with a modulation period of 40 s and an amplitude of 1.5 °C. The temperature range was set from −35 to 240 °C.

3. Results

3.1. Sorption Isotherms at 25 °C

Figure 1A presents the adsorption isotherm for the quercetin–maltodextrin sample with a DE of 10 (Q-MX10) stored at 25 °C. The experimental equilibrium moisture content values (squared data points) for the functional food product were fitted using the GAB and BET mathematical models (solid lines).

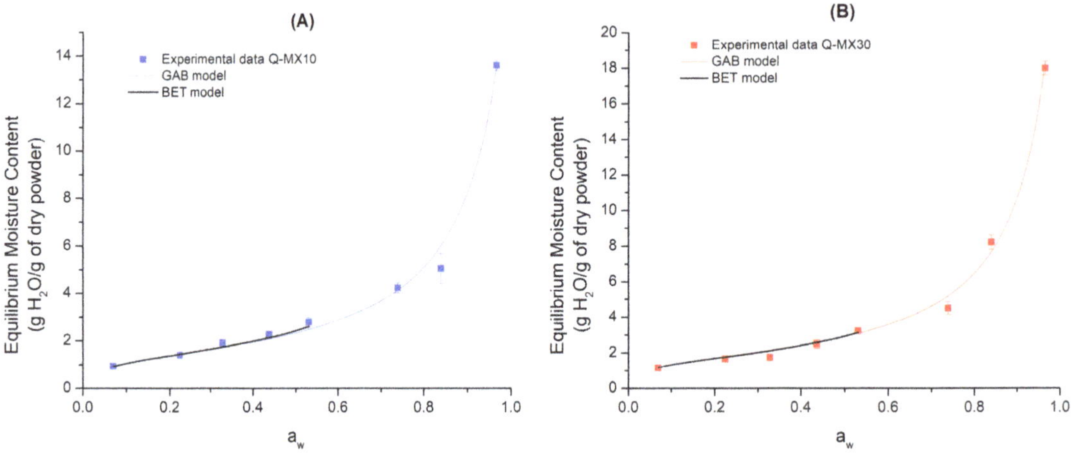

Figure 1. Adsorption isotherms for quercetin-maltodextrin functional food stored at 25 °C: (**A**) Q-MX10 (DE 10); (**B**) Q-MX30 (DE 30).

The GAB model is described by the monolayer moisture content (Xm) and the constants C and K. The first parameter refers to the molecular layer that is bound to free sites which can be occupied by water, creating an aqueous phase in the food. The constants C and K, on the other hand, are related to the amount of energy required to remove water from the monolayer and multilayer, respectively [37]. The mathematical fitting of the GAB model indicates that the Q-MX10 sample exhibits three adsorption zones: (i) a steady increase in moisture content from 1.10 to 2.3 g H_2O per g of dry powder within an a_w range of 0.07 to 0.53; (ii) an increase in moisture content from 2.37 to 5.76 g H_2O per g of dry powder within an a_w range of 0.53 to 0.73; (iii) an accelerated adsorption from 5.76 to 13.53 g H_2O per g of dry powder within an a_w range of 0.73 to 0.967. Regarding the model fitting parameters, the following values were obtained: Xm = 1.29, C = 27.55, K = 0.937, and R^2 = 0.993. The GAB equation is one of the most widely used models for predicting the shelf life of food. It is based on the theory of monolayer and multilayer adsorption, where the food tends to absorb moisture from the environment through the multilayer and subsequently the monolayer. The results obtained in this research align with those reported by Navia et al. [38], who conducted a study on the development of a cassava flour and glycerol biopellicle. They determined the barrier properties of this product by performing sorption isotherms at 15, 25, and 35 °C. These researchers fitted the experimental data to the GAB, Oswin, Smith, and Henderson mathematical models. They reported that the GAB model provided the best fit at all three temperatures. Specifically, at 25 °C, the GAB model parameters were as follows: Xm = 4.52, C = 1.65, and K = 0.84 with an R^2 value of 0.99.

The BET model, on the other hand, is characterized by the monolayer water content (Xm) and the constant C. However, the BET model has a limitation regarding the range of water activity in which it is valid, so the determination of adsorption phenomena was carried out up to an a_w value of 0.531. For the Q-MX10 sample, the parameters obtained from the BET model fit were as follows: Xm = 1.25, C = 28.31, and R^2 = 0.941. Unlike the GAB model, the adsorption isotherm described by the BET model shows only two adsorption zones: (I) the first corresponds to a constant increase in moisture content from 0.9884 to 1.3836 g H_2O per g of dry powder in an a_w range of 0.07–0.225; (II) the second shows a significant increase in equilibrium moisture content from 1.3836 to 2.8666 g H_2O per g of dry powder in an a_w range of 0.225–0.531. These results are consistent with those reported by Talens Oliag et al. [39], who observed moisture adsorption behavior using the BET model for breakfast cereals stored at 20 °C, obtaining Xm = 0.0514 and C = 7.574. Similarly, Cerviño et al. [40] reported that the BET model provided the best fit for describing the stability of sweet potato candies at 30 °C. They also reported an isotherm with two adsorption intervals, where the best BET model fit was observed in the a_w range of 0.0–0.5.

Figure 1B displays the adsorption isotherm for sample Q-MX30 (squared data points) stored at 25 °C. The solid lines represent the fit of the GAB and BET mathematical models.

The GAB model fit revealed three zones of moisture absorption: (I) a constant increase in moisture content from 0.9948 to 3.0609 g H_2O per g of dry powder, corresponding to an a_w range of 0.07–0.53; (II) a moisture increment from 3.0609 to 7.6387 g H_2O per g of dry powder in an a_w range of 0.53–0.84; (III) an accelerated adsorption from 7.6387 to 18.0512 g H_2O per g of dry powder in an a_w range of 0.84–0.967. The parameters obtained for the GAB model were as follows: Xm = 1.58, C = 31.48, K = 0.946, and R^2 = 0.971. These results align with those reported by Domínguez-Domínguez et al. [41], who studied the adsorption isotherm behavior for hibiscus seed (*Hibiscus sabdariffa* L.) at 25, 35, and 45 °C. Specifically, at 25 °C, for the GAB model, they obtained R^2 values of 0.9895, 0.9894, and 0.9872 for the Criolla, China, and Sudan varieties, respectively. According to their findings, the GAB model was the best fit for adsorption isotherms of low-moisture foods. Pascual-Pineda et al. [42], investigated the storage conditions of dehydrated foods using adsorption isotherms at temperatures of 15, 25, and 35 °C, analyzing three low-moisture powdered foods: sucrose–calcium, pineapple, and paprika. They reported that the GAB equation fits powdered products well, with R^2 values exceeding 0.99. Brousse et al. [43] conducted a study on the stability of dehydrated cassava puree (*Manihot esculenta* Crantz) using adsorption isotherms at temperatures of 25, 35, and 45 °C. They concluded that both the GAB and BET mathematical models accurately fit the experimental data, with a relative mean error of less than 10%. Additionally, the primary difference in the results arises from the fact that the BET model accounts only for monolayer adsorption, while the GAB model considers the properties of water adsorbed in multilayers. The calculated BET parameter for sample Q-MX30 were Xm = 1.50, C = 35.62, and R^2 = 0.974.

When comparing the fitting values of the BET and GAB models between the two types of maltodextrins (DE 10 versus DE 30), it is observed that the Q-MXDE30 sample exhibited a higher monolayer moisture content for both models. This suggests that maltodextrin with a higher DE (i.e., DE 30) adsorbs slightly more moisture. This can be explained based on the chemical structure of maltodextrins, which consist of a main carbohydrate chain (amylose) and branches (amylopectin). The arrangement of these components results in differences in dextrose equivalents (DE). According to Saavedra-Leos et al. [25], a higher DE value indicates a greater number of functional groups available for interaction with water molecules, as well as a higher percentage of branching.

3.2. Physicochemical Characterization

Figure 2 presents the X-ray diffraction patterns of samples Q-MX10 and Q-MX30 after being subjected to different moisture adsorption conditions at 25 °C. Generally, the diffractograms exhibited similar behavior, featuring a broad peak around 18° and several low-intensity diffraction peaks at 11°, 12.5°, 16°, and 27°. The broad peak is characteristic of amorphous maltodextrin, while the low-intensity diffraction peaks are associated with the crystallization of quercetin [32,33]. However, for the Q-MX10 sample stored at an a_w of 0.840, there was a notable decrease in the intensity of the broad peak at 18° and the emergence of high-intensity diffraction peaks at 2θ angles of 16°, 17.5°, 18°, and 28°. In contrast, the Q-MX30 sample stored at the same a_w displayed different behavior, showing the appearance of a high-intensity peak at 17°, the disappearance of the broad peak at 18°, and the emergence of relatively low-intensity diffraction peaks between 15° and 20° and at 24°. This reduction in diffraction intensity at 18° is linked to the effect of absorbed water, which promotes crystallization. This suggests that the microstructure of both MD10 and MD30 undergoes changes in stability at high a_w values, transitioning from a glassy solid state to a crystalline solid state. Ballesteros et al. [44] conducted a comparative study on the encapsulation of antioxidant phenolic compounds extracted from coffee using both lyophilization and spray drying methods. Regarding the XRD analysis of samples using maltodextrin as a carrier agent, they found a very low degree of crystallinity. The samples exhibited a broad peak at 18° in 2θ, characteristic of such products, which is attributed to the amorphous nature of the powders obtained through spray drying. Matsuura et al. [45] conducted a study on the effect of dextrose equivalent in maltodextrin on the stability of coconut oil powder obtained through spray drying. The XRD results for all samples showed a characteristic diffraction at a 2θ angle of 20°, observed as a broad peak, which is indicative of amorphous materials. Zhang et al. [46] conducted a study on the use of maltodextrin as an encapsulating agent for xylooligosaccharides during the spray drying process. They concluded that using maltodextrin as a carrier material during spray drying promotes the formation of amorphous solid materials with high dissolution rates.

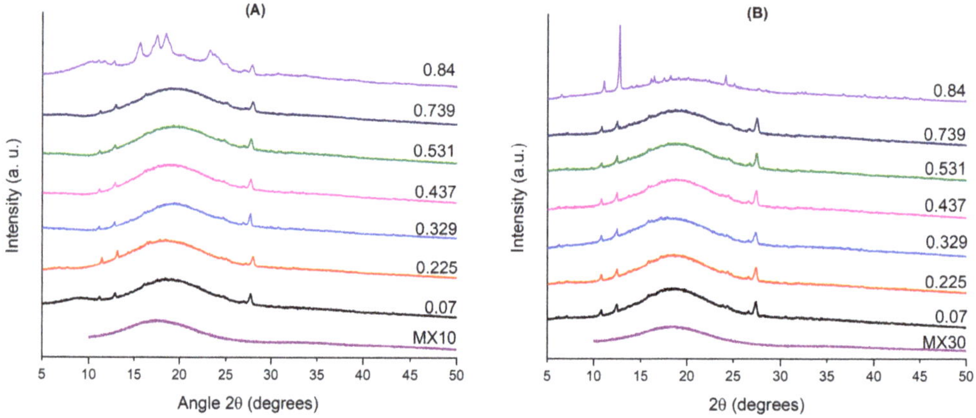

Figure 2. X-ray diffraction patterns of samples Q-MX10 (**A**) and Q-MX30 (**B**) under various moisture adsorption conditions at 25 °C.

The morphological analysis performed using SEM is shown in Figure 3. This figure presents a representative micrograph acquired at 1000X magnification of samples Q-MX10 and Q-MX30 subjected to various storage humidity conditions at 25 °C.

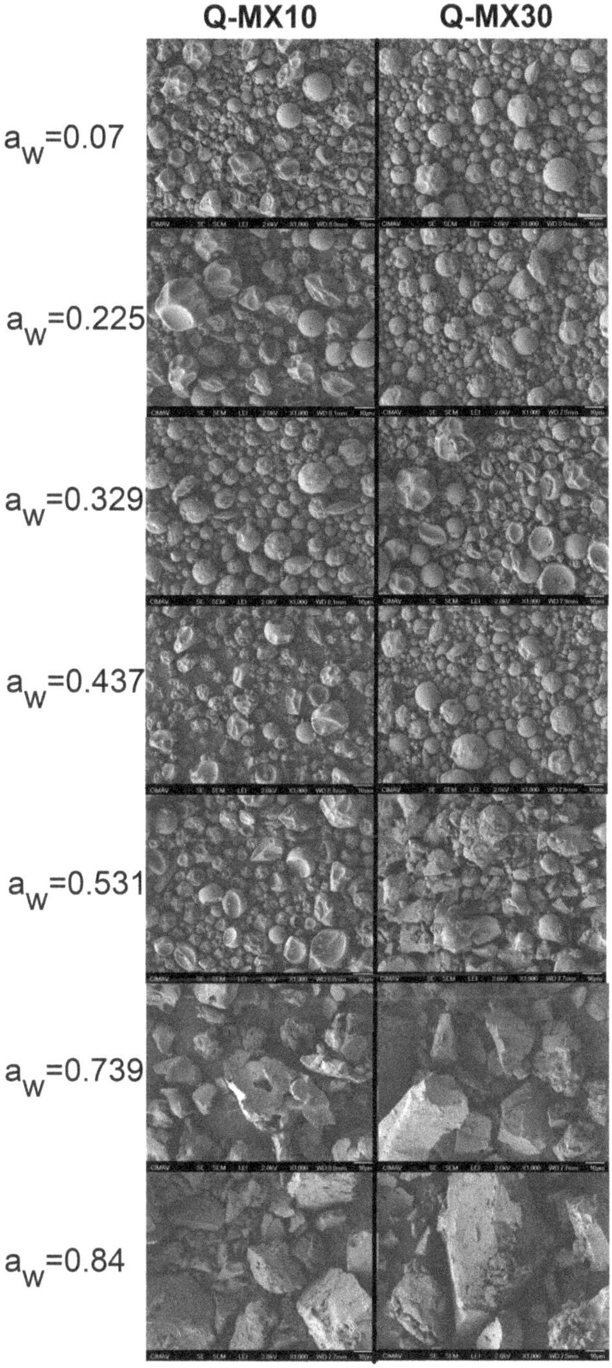

Figure 3. Scanning electron microscopy (SEM) images of samples Q-MX10 and Q-MX30 at 1000× magnification, illustrating the morphological changes under different storage humidity conditions at 25 °C.

For both samples within the a_w range of 0.070–0.329, a very similar behavior was observed, exhibiting a comparable morphology with particle sizes ranging between 2 and 15 μm. The observed morphologies included semi-spherical shapes with smooth surfaces as well as particles with irregular (rough) surfaces. As the water activity increased (a_w = 0.437), a change was noticed in some particles within the samples, displaying irregular spheroidal structures with the appearance of deflated balloons; this phenomenon is associated with the limit of moisture adsorption stability. At an a_w of 0.531, the samples exhibited a different morphology, resembling clusters or agglomerates of irregularly shaped spherical particles larger than 15 μm. Finally, at water activities higher than 0.739, crystallized structures with irregular shapes and smooth surfaces were observed. These observations suggest that the adsorbed water in the samples induced crystallization, leading to changes in particle morphology and size, resulting in a rigid, crystalline solid material. The SEM analysis of MD10 and MD30 provided insights into the morphology of spray-dried samples and established the stability threshold at which microstructural changes occur due to moisture adsorption, altering the properties of the powders. Navarro-Flores et al. [47] conducted a study on the spray drying of a native plant rich in phenolic compounds using maltodextrin as a carrier agent. They reported that the resulting microcapsules exhibited irregular shapes, smooth surfaces with depressions, and particle sizes ranging from 3 to 8 μm. Ferrari et al. [48] reported on the stability of spray-dried blackberry powder using maltodextrin (DE of 20) and gum Arabic as carrier agents. The particles exhibited spherical shapes of various sizes. Specifically, for the samples obtained using maltodextrin, the microcapsules were found to have smooth and wrinkled surfaces, with a size of 43.67 ± 1.76 μm. Additionally, it was noted that during storage for 150 days at 25 and 35 °C, the powders showed a significant tendency to agglomerate.

Thermogravimetric analysis (TGA) quantifies mass loss in relation to temperature increase. Figure 4A,B present the TGA results for the spray-dried Q-MX10 and Q-MX30 samples subjected to different humidity adsorption conditions at 25 °C. Figure 4C,D display the derivative weight curves relative to temperature, providing a clearer view of the onset and completion of each thermal event. In general, four thermal events were identified for both types of maltodextrins, observed under all humidity adsorption conditions: the first event occurs within a temperature range of 50–150 °C, corresponding to an 8% mass loss due to water evaporation; the second thermal event, occurring between 200 and 250 °C, involves a mass loss of approximately 40%, likely related to the degradation of low-molecular-weight carbohydrates; the third event, observed between 250 and 325 °C, is associated with a 40% mass loss, attributed to the thermal degradation of high-molecular-weight carbohydrates; finally, the fourth thermal event appears between 425 and 500 °C, with a 10% mass loss corresponding to the calcination or final degradation of the organic matter in the sample. Costa Ferreira et al. [49] microencapsulated phenolic compounds from the by-product of tucum almonds (*Astrocaryum vulgare* Mart.) using maltodextrin as a carrier agent in spray drying. Their TGA analysis revealed only three thermal events: the first in the temperature range of 50–100 °C, the second between 200 and 239 °C, and the third from 283 to 356 °C. On the other hand, Saavedra-Leos et al. [25] reported a well-defined thermal event at approximately 115 °C, which was observed only in samples with the highest a_w values (i.e., a_w = 0.75). This event was attributed to thermal hydrolysis induced by the large amount of adsorbed water and supplied heat. However, this thermal event was not observed in the TGA analysis presented in this study. In sugar polymers, thermal degradation induces caramelization (Maillard reaction), generating aromas, color changes (yellow and brown tones), and flavor shifts from smooth, caramelized sweetness to bitter, burnt notes. The thermal degradation process involves stages such as enolization, dehydration, carbonyl cleavage, dicarbonyl scission, retroaldolization, aldolization, and radical reactions [50].

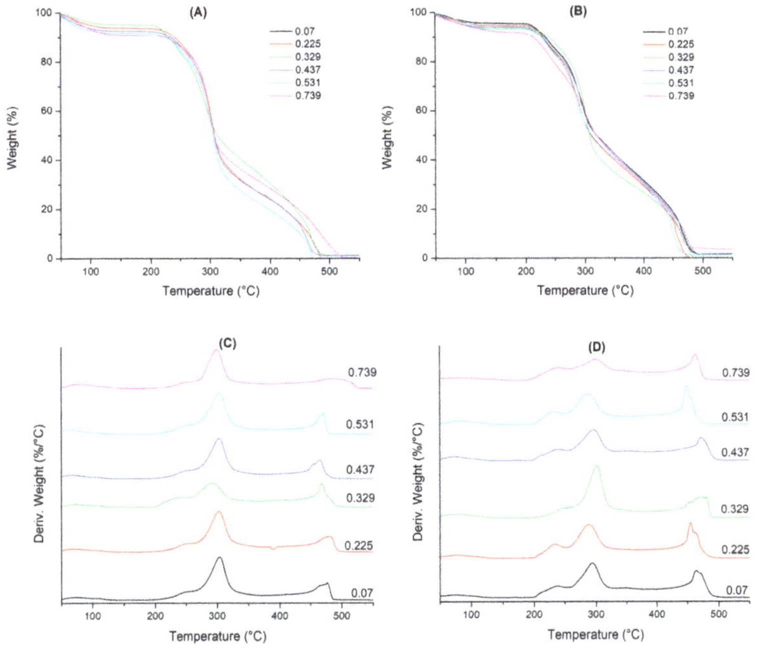

Figure 4. TGA results for spray-dried samples under various humidity adsorption conditions at 25 °C: (**A**,**C**) Q-MX10; (**B**,**D**) Q-MX30.

Figure 5 shows the MDSC thermograms for samples Q-MX10 and Q-MX30 at an a_w of 0.07. Each graph displays three curves: the reversible heat flow (left axis, in red), the modulated heat flow (first right axis, in blue), and the non-reversible heat flow (second right axis, in green). The glass transition temperature (Tg) is observed in the reversible heat flow curve, while other thermal events, such as melting and degradation, are observed in the non-reversible heat flow curve [29]. The Tg appears as a smooth change in the slope of the curve, indicated by arrows in the figures. Table 2 summarizes the Tg values for samples Q-MX10 and Q-MX30 under different humidity conditions. Although both samples exhibited similar behavior, with Tg decreasing monotonically with increasing a_w, sample Q-MX30 presented slightly lower values. The Tg varied from 44.2 to −10.99 °C for sample Q-MX10 and from 44.2 to −9.5 °C for sample Q-MX30. This is a common behavior observed in carbohydrate polymers like maltodextrins, where water molecules act as plasticizers [51]. They increase the free volume between glucose molecules, promoting the transition of the microstructure from a glassy state to a rubbery state.

Table 2. Tg values of samples Q-MX10 and Q-MX30 under different humidity conditions.

Water Activity (a_w)	Sample	
	Q-MX10 Tg (°C)	Q-MX30 Tg (°C)
0.07	44.2 ± 0.04	44.22 ± 0.01
0.225	33.98 ± 0.02	43.43 ± 0.03
0.329	33.05 ± 0.07	−0.34 ± 0.03
0.437	2.89 ± 0.15	−7.35 ± 0.24
0.531	−9.16 ± 0.2	−6 ± 0.02
0.739	−11.2 ± 0.21	−8.83 ± 0.36
0.84	−10.99 ± 0.63	−9.5 ± 0.4

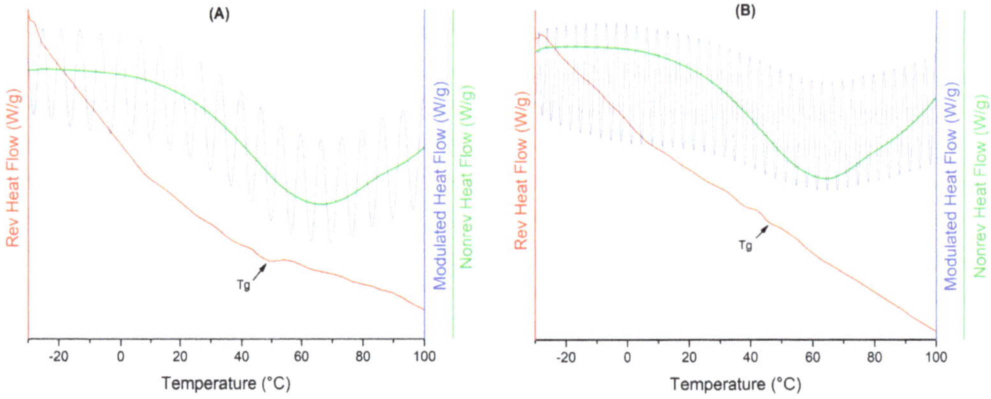

Figure 5. MDSC thermograms for samples at an a_w of 0.07: Q-MX10 (**A**), and Q-MX30 (**B**).

The general appearance of the powders was studied using optical microscopy. Figure 6 shows optical photographs of the Q-MX10 and Q-MX30 samples subjected to different humidity conditions. Both samples exhibited similar behavior, differing only in the water activity level at which the observed changes occurred. Initially, for a_w values of 0.07–0.531, the Q-MX10 sample appeared as a dry, fine, non-agglomerated powder with a light-yellow color. The Q-MX30 sample showed the same appearance in the a_w range of 0.07–0.437. At an a_w of 0.739 for sample Q-MX10, and a_w of 0.531 for sample Q-MX30, a change to a more intense yellow color and increased particle agglomeration was observed. As a_w increased further, the powder crystallized due to moisture adsorption, resulting in a continuous, rigid material with a shiny appearance and intense yellow color. For sample Q-MX10, this occurred at an a_w of 0.739, whereas for sample Q-MX30, it was observed at an a_w of 0.531. Evidently, the differences in water activity observed here are caused by the differences in DE between the two maltodextrins, with the Q-MX30 sample tending to adsorb more moisture than the Q-MX10 sample. Saavedra-Leos et al. [25] observed the macroscopic behavior of a set of four maltodextrins with DE ranging from 10 to 40 using optical photography. They reported a color change in the powder from white to pale yellow up to an a_w range of 0.434–0.532. Beyond this a_w value, the maltodextrins exhibited various physical changes, such as particle agglomeration, volume shrinkage, the appearance of a rubbery state, and moisture saturation. These changes were more pronounced for maltodextrins with higher DE values.

3.3. Antioxidant Activity of the Functional Food

The determination of antioxidant activity (AA) was carried out by scavenging free radicals using the DPPH reagent as a standard. Figure 7 shows the results for the AA of Q-MX10 and Q-MX30 samples subjected to different moisture adsorption conditions at 25 °C. AA is expressed as the percentage inhibition (%) of DPPH radicals due to hydrogen donation for the neutralization of free radicals (OH-). For sample Q-MX10 after the drying process, the initial AA value was 18.02 ± 2.03%, while for sample Q-MX30, the initial AA value was 22.73 ± 1.65%. In order to evaluate the effect of storage conditions (water activity) and the type of carrier agent (DE 10 and DE 30) on the AA value, an ANOVA analysis was conducted with a significance level of $\alpha = 0.05$. The results indicated that there was no significant difference in the mean AA value across the different storage conditions. Additionally, there was no significant effect of using MX10 or MX30 on the AA. This indicates that under these storage conditions, the microencapsulated antioxidant compound in the maltodextrin is preserved, preventing degradation due to the absorbed water. Georgetti et al. [52] evaluated the chemical and biological properties of spray-dried soy extract employing three carrying agents. The results indicated an AA of 59% for concentrated soy extract, 55% for silicon dioxide, 52% for maltodextrin, and 50% for starch. Recently, the antioxidant activity of quercetin microencapsulated in inulin, maltodextrin, and their blends was compared. It was found that in the case of maltodextrin, quercetin exhibits greater interaction, which reduces its bioactivity in interacting with free radicals. As a result, the AA is lower with maltodextrin compared to inulin [32]. For a 25–75% maltodextrin–inulin blend containing quercetin and *B. claussi* as a probiotic, the AA in the a_W range of 0.073–0.856 was statistically similar. This suggests that the humidity conditions did not significantly impact the antioxidant's preservation capacity [33].

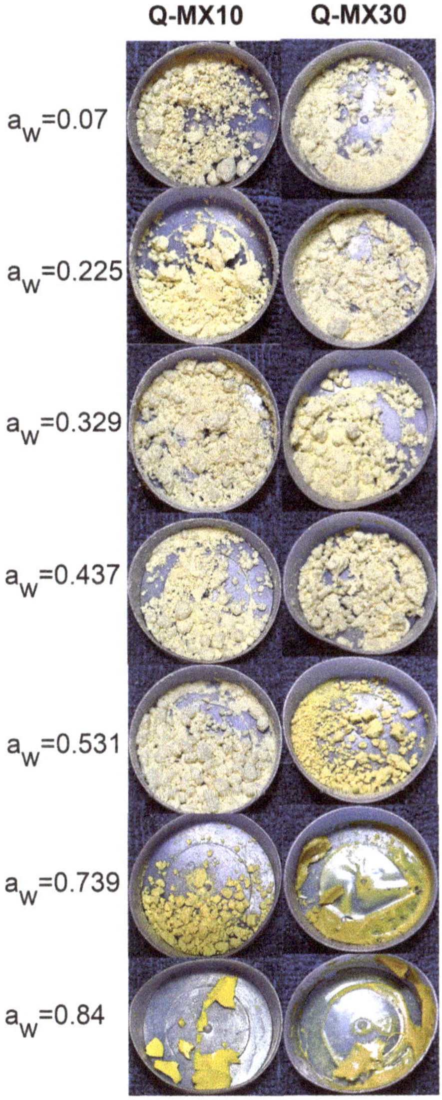

Figure 6. Optical photographs of the Q-MX10 and Q-MX30 samples subjected to different humidity conditions at 25 °C.

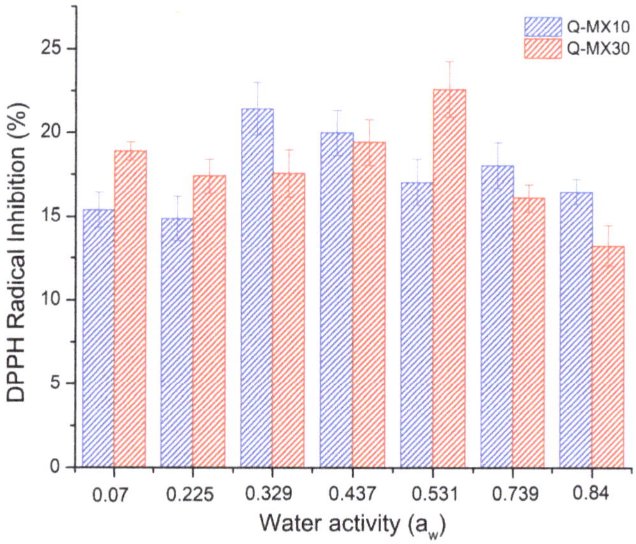

Figure 7. Antioxidant activity, expressed as the percentage inhibition (%) of DPPH radicals, for Q-MX10 and Q-MX30 samples under various moisture adsorption conditions at 25 °C. The values represent the average of three measurements.

4. Discussion

Equilibrium State Diagram

Equilibrium state diagrams can be constructed by combining water sorption isotherms and Tg to determine the critical values for water content and water activity at a specific storage temperature [53]. However, these diagrams can provide additional physicochemical information that helps to understand the behavior of the powder during storage. In this context, Figure 8 presents the equilibrium state diagram for samples Q-MX10 and Q-MX30, constructed based on the physicochemical characterization results. The diagram's main feature includes the adsorption isotherms at 25 °C described by the GAB model, depicted as continuous blue and red lines for samples Q-MX10 and Q-MX30, respectively, which are read on the left *Y*-axis. By comparing the two isotherms, the state diagram reveals that while the adsorption behavior of both maltodextrins is similar in shape, there is a slight difference in the equilibrium moisture content adsorbed. Specifically, sample Q-MX30 adsorbs more water than sample Q-MX10. This behavior is evidently related to the differences in the DE between the two maltodextrins. Additionally, the Tg data are displayed as dashed lines in blue (Q-MX10) and red (Q-MX30), corresponding to the right *Y*-axis. Furthermore, to the individual information obtained from the isotherm and Tg (as previously discussed), combining these data provides parameters related to storage conditions, such as the critical water content (CWC), the critical water activity (CWA), and the amount of moisture adsorbed at a given storage temperature. The CWC and CWA are important thresholds that indicate the limits within which a material can be stored safely without compromising its integrity. Critical parameters can be extrapolated from the intersection of the storage temperature line (dashed green line) with the Tg curve and subsequently with the isotherm curve. These values are read directly from their respective axes. At the storage temperature of 25 °C, the CWC values are nearly identical for both samples, at 1.98 g of H_2O per g of dry powder. However, the CWA values are slightly different, with 0.41 for sample Q-MX10 and 0.27 for sample Q-MX30. In the state diagram is indicated the calculated monolayer water content (M_0) for each isotherm, which were 1.2 and 1.5 for samples Q-MX10 and Q-MX30, respectively. When comparing the M_0 values against the CWC, the latter is several times higher. M_0 represents the amount of water adsorbed in a single molecular layer on the surface of the material, it is often considered the most stable state for a material because it indicates that all available binding sites for water on the surface are occupied, and any additional water will lead to multilayer adsorption, which can be less stable. On the other hand, CWC indicates the water content at which a material transitions from a glassy (solid, stable) state to a rubbery (less stable, more reactive) state. It represents the maximum water content at which the material can maintain its structural integrity and stability [11,54,55]. When the CWC is significantly higher than the M_0, it means the material can tolerate a relatively high amount of water before becoming unstable. This suggests good moisture resistance and a stable glassy state over a wider range of water content.

In addition to the aforementioned data, the equilibrium state diagram provides further insights into the microstructure, morphology, and appearance of the samples. This information is represented by vertical dotted lines, indicating the thresholds at which microstructural transitions occur. These lines help visualize and understand the changes in the material's physical properties under different conditions. Based on the physicochemical characterization results from XRD, SEM, and optical microscopy, three distinct transition regions were identified: amorphous, semicrystalline, and crystalline. In the amorphous region (a_w range: 0.07–0.437), the microparticles exhibit a predominantly spherical morphology. The overall appearance is a fluffy powder with a pale-yellow color. The corresponding Tg values range from 44 to −7 °C. The semicrystalline region (a_w range: 0.437–0.739), or intermediate region, shows a semicrystalline behavior. Here, very-low-intensity diffraction peaks are observed above the broad peak in the XRD pattern. The particles begin to collapse, displaying merged spherical particles with irregular shapes. The overall appearance is an agglomerated powder with an intense yellow color. The Tg values in this region are practically below 2 °C. In the crystalline region (a_w above 0.739) the microstructure of the functional food is completely collapsed. The morphology of the particles becomes fully irregular, and the overall appearance is that of a continuous solid material with an intense yellow color. These transitions help us to understand the material's stability and behavior under varying humidity conditions.

Finally, the optimal storage conditions for these functional foods, which are based on quercetin microencapsulated in maltodextrins with different dextrose equivalents, lie within the amorphous region. These conditions are defined by the CWC and CWA values. The ideal storage parameters are a temperature of 25 °C, an a_w of 0.437, and a maximum water content of 1.98 g of H_2O per gram of dry powder. While these functional foods can tolerate storage temperatures up to 44 °C, the maximum water content is constrained by the monolayer water content value, which is 1.2 g of H_2O per gram of dry powder. In various studies where maltodextrin is used as a microencapsulating agent, storage temperatures close to 50 °C have been reported. However, to prevent microstructural changes, the moisture content at these temperatures must be kept below 0.1 g of H_2O per gram of dry powder. Clearly, in such cases, very strict humidity control is required to avoid the collapse of the microstructure [56,57].

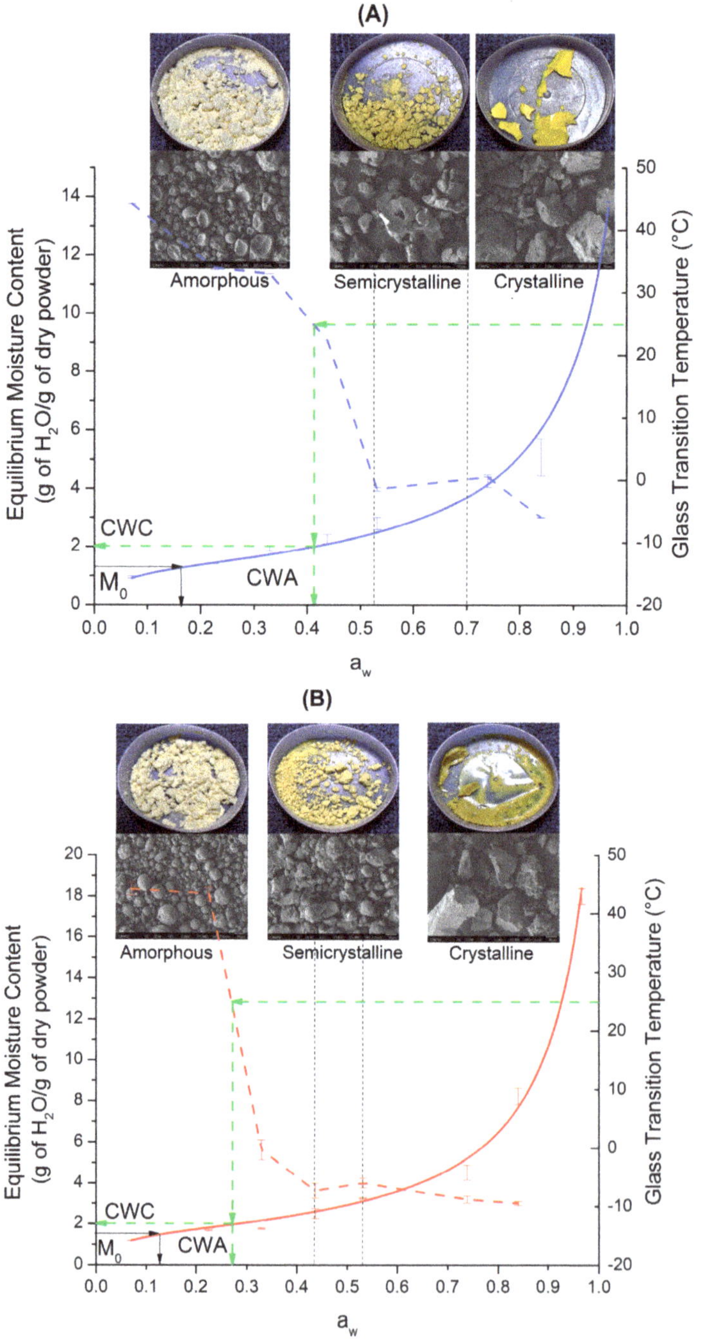

Figure 8. Equilibrium state diagram for samples Q-MX10 (**A**) and Q-MX30 (**B**), constructed from the adsorption isotherms at 25 °C, and the physicochemical characterization results. In both figures, the continuous lines (blue and red) represent the adsorption isotherm, the dotted lines (blue and red) represent the Tg curve, and the dotted green lines the critical storage conditions.

5. Conclusions

This study investigates the equilibrium state diagram of functional foods based on maltodextrins (MX) with varying dextrose equivalents (DE 10 and DE 30) for quercetin microencapsulation via spray drying. Equilibrium adsorption isotherms at 25 °C were constructed to compare the adsorption behavior of these polymeric foods. Among the adsorption models tested, the GAB model presented better fitting results than the BET model. Both samples exhibited similar monolayer water content (M_0) of approximately 1.2–1.5 g of water per g of dry powder. Physicochemical characterization techniques, including XRD, SEM, TGA, MDSC, and optical microscopy, provided comprehensive insights into the microstructural transitions occurring under varying humidity conditions. An equilibrium state diagram was constructed to predict the optimal storage conditions for the functional foods. Three transition regions—amorphous, semicrystalline, and crystalline—were identified based on differences in crystallinity through XRD analysis, changes in particle morphology observed in SEM micrographs, and variations in powder appearance and color noted from optical microscopy. The optimal storage conditions were extrapolated from the isotherms and Tg curves, defined by the CWC and CWA thresholds: a temperature of 25 °C, an a_w of 0.437, and a maximum water content of 1.98 g of water per gram of dry powder. The antioxidant activity showed similar values despite varying water content, indicating the preservation of the antioxidant within the carrying agent across the complete range of water activities. This research enhances understanding of maltodextrin behavior across diverse dextrose equivalents, aiding in the development of stable encapsulated products.

Author Contributions: Conceptualization, Z.S.-L. and C.L.-P.; Data curation, A.C.-L. and D.L.-G.; Formal analysis, Z.S.-L. and C.L.-P.; Funding acquisition, Z.S.-L.; Investigation, A.C.-L. and L.A.L.-M.; Methodology, Z.S.-L., D.L.-G. and C.L.-P.; Project administration, Z.S.-L. and C.L.-P.; Resources, Z.S.-L. and C.L.-P.; Software, A.C.-L. and L.A.L.-M.; Supervision, L.A.L.-M.; Validation, D.L.-G. and L.A.L.-M.; Visualization, A.C.-L. and L.A.L.-M.; Writing—original draft, Z.S.-L. and C.L.-P.; Writing—review and editing, Z.S.-L. and C.L.-P. All authors have read and agreed to the published version of the manuscript.

Funding: This research received no external funding.

Institutional Review Board Statement: Not applicable.

Data Availability Statement: The original contributions presented in the study are included in the article. Further inquiries can be directed to the corresponding author/s.

Acknowledgments: The authors thank to Andrés Isaac González Jáquez for his support with the acquisition of XRD patterns, and the Laboratorio Nacional de Nanotecnología (Nanotech) for the use of the FESEM.

Conflicts of Interest: The authors declare no conflicts of interest.

References

1. Kouhi, M.; Prabhakaran, M.P.; Ramakrishna, S. Edible polymers: An insight into its application in food, biomedicine and cosmetics. *Trends Food Sci. Technol.* **2020**, *103*, 248–263. [CrossRef]
2. Paliwal, R.; Palakurthi, S. Zein in controlled drug delivery and tissue engineering. *J. Control. Release* **2014**, *189*, 108–122. [CrossRef] [PubMed]
3. Shahbaz, A.; Hussain, N.; Basra, M.A.R.; Bilal, M. Polysaccharides-based nano-Hybrid biomaterial platforms for tissue engineering, drug delivery, and food packaging applications. *Starch-Stärke* **2022**, *74*, 2200023. [CrossRef]
4. Shit, S.C.; Shah, P.M. Edible polymers: Challenges and opportunities. *J. Polym.* **2014**, *2014*, 427259. [CrossRef]
5. Ali, A.; Ahmed, S. Recent advances in edible polymer based hydrogels as a sustainable alternative to conventional polymers. *J. Agric. Food Chem.* **2018**, *66*, 6940–6967. [CrossRef] [PubMed]
6. Shi, J.; Mazza, G.; Le Maguer, M. *Biochemical and Processing Aspects. Functional Foods*, 1st ed.; CRC Press: Boca Raton, FL, USA, 2016; Volume 2. [CrossRef]
7. Essa, M.M.; Bishir, M.; Bhat, A.; Chidambaram, S.B.; Al-Balushi, B.; Hamdan, H.; Govindarajan, N.; Freidland, R.P.; Qoronfleh, M.W. Functional foods and their impact on health. *J. Food Sci. Technol.* **2021**, *60*, 820–834. [CrossRef] [PubMed]

8. Granato, D.; Barba, F.J.; Bursać Kovačević, D.; Lorenzo, J.M.; Cruz, A.G.; Putnik, P. Functional foods: Product development, technological trends, efficacy testing, and safety. *Annu. Rev. Food Sci. Technol.* **2020**, *11*, 93–118. [CrossRef]
9. Sablani, S.S.; Syamaladevi, R.M.; Swanson, B.G. A review of methods, data and applications of state diagrams of food systems. *Food Eng. Rev.* **2010**, *2*, 168–203. [CrossRef]
10. Rahman, M.S. State diagram of foods: Its potential use in food processing and product stability. *Trends Food Sci. Technol.* **2006**, *17*, 129–141. [CrossRef]
11. Al-Ghamdi, S.; Hong, Y.K.; Qu, Z.; Sablani, S.S. State diagram, water sorption isotherms and color stability of pumpkin (*Cucurbita pepo* L.). *J. Food Eng.* **2020**, *273*, 109820. [CrossRef]
12. Caballero-Cerón, C.; Guerrero-Beltrán, J.A.; Mújica-Paz, H.; Torres, J.A.; Welti-Chanes, J. Moisture sorption isotherms of foods: Experimental methodology, mathematical analysis, and practical applications. In *Water Stress in Biological, Chemical, Pharmaceutical and Food Systems*; Food Engineering Series (FSES); Springer: New York, NY, USA, 2015; pp. 187–214. [CrossRef]
13. Nurhadi, B.; Roos, Y.H.; Maidannyk, V. Propiedades físicas de la maltodextrina DE 10: Absorción de agua, plastificación del agua y relajación entálpica. *Rev. Ing. Aliment.* **2016**, *174*, 68–74. [CrossRef]
14. Chronakis, I.S. On the molecular characteristics, compositional properties, and structural-functional mechanisms of maltodextrins: A review. *Crit. Rev. Food Sci. Nutr.* **1998**, *38*, 599–637. [CrossRef] [PubMed]
15. Takeiti, C.Y.; Kieckbusch, T.G.; Collares-Queiroz, F.P. Morphological and physicochemical characterization of commercial maltodextrins with different degrees of dextrose-equivalent. *Int. J. Food Prop.* **2010**, *13*, 411–425. [CrossRef]
16. Parikh, A.; Agarwal, S.; Raut, K. A review on applications of maltodextrin in pharmaceutical industry. *Int. J. Pharm. Biol. Sci.* **2014**, *4*, 64–67.
17. Hofman, D.L.; Van Buul, V.J.; Brouns, F.J. Nutrition, health, and regulatory aspects of digestible maltodextrins. *Crit. Rev. Food Sci. Nutr.* **2016**, *56*, 2091–2100. [CrossRef] [PubMed]
18. Buck, A.W. Resistant Maltodextrin Overview. In *Dietary Fiber and Health*, 1st ed.; CRC Press: Boca Raton, FL, USA, 2012; ISBN 9780429184239.
19. Xiao, Z.; Xia, J.; Zhao, Q.; Niu, Y.; Zhao, D. Maltodextrin as wall material for microcapsules: A review. *Carbohydr. Polym.* **2022**, *298*, 120113. [CrossRef] [PubMed]
20. Hendrawati, T.Y.; Sari, A.M.; Rahman MI, S.; Nugrahani, R.A.; Siswahyu, A. Microencapsulation techniques of herbal compounds for raw materials in food industry, cosmetics and pharmaceuticals. In *Microencapsulation-Processes, Technologies and Industrial Applications*; IntechOpen: London, UK, 2019; ISBN 978-1-83881-871-5. [CrossRef]
21. Corrêa-Filho, L.C.; Moldão-Martins, M.; Alves, V.D. Advances in the application of microcapsules as carriers of functional compounds for food products. *Appl. Sci.* **2019**, *9*, 571. [CrossRef]
22. Choudhury, N.; Meghwal, M.; Das, K. Microencapsulation: An overview on concepts, methods, properties and applications in foods. *Food Front.* **2021**, *2*, 426–442. [CrossRef]
23. Piñón-Balderrama, C.I.; Leyva-Porras, C.; Terán-Figueroa, Y.; Espinosa-Solís, V.; Álvarez-Salas, C.; Saavedra-Leos, M.Z. Encapsulation of active ingredients in food industry by spray-drying and nano spray-drying technologies. *Processes* **2020**, *8*, 889. [CrossRef]
24. Mehta, N.; Kumar, P.; Verma, A.K.; Umaraw, P.; Kumar, Y.; Malav, O.P.; Lorenzo, J.M. Microencapsulation as a noble technique for the application of bioactive compounds in the food industry: A comprehensive review. *Appl. Sci.* **2022**, *12*, 1424. [CrossRef]
25. Saavedra-Leos, Z.; Leyva-Porras, C.; Araujo-Díaz, S.B.; Toxqui-Terán, A.; Borrás-Enríquez, A.J. Technological application of maltodextrins according to the degree of polymerization. *Molecules* **2015**, *20*, 21067–21081. [CrossRef]
26. Araujo-Díaz, S.B.; Leyva-Porras, C.; Aguirre-Bañuelos, P.; Álvarez-Salas, C.; Saavedra-Leos, Z. Evaluation of the physical properties and conservation of the antioxidants content, employing inulin and maltodextrin in the spray drying of blueberry juice. *Carbohydr. Polym.* **2017**, *167*, 317–325. [CrossRef]
27. Saavedra–Leos, M.Z.; Leyva-Porras, C.; Alvarez-Salas, C.; Longoria-Rodríguez, F.; López-Pablos, A.L.; González-García, R.; Pérez-Urizar, J.T. Obtaining orange juice–maltodextrin powders without structure collapse based on the glass transition temperature and degree of polymerization. *CyTA-J. Food* **2018**, *16*, 61–69. [CrossRef]
28. Saavedra-Leos, M.Z.; Leyva-Porras, C.; López-Martínez, L.A.; González-García, R.; Martínez, J.O.; Compeán Martínez, I.; Toxqui-Terán, A. Evaluation of the spray drying conditions of blueberry juice-maltodextrin on the yield, content, and retention of quercetin 3-d-galactoside. *Polymers* **2019**, *11*, 312. [CrossRef]
29. Leyva-Porras, C.; Saavedra-Leos, M.Z.; Cervantes-González, E.; Aguirre-Bañuelos, P.; Silva-Cázarez, M.B.; Álvarez-Salas, C. Spray drying of blueberry juice-maltodextrin mixtures: Evaluation of processing conditions on content of resveratrol. *Antioxidants* **2019**, *8*, 437. [CrossRef]
30. Saavedra-Leos, M.Z.; Leyva-Porras, C.; Toxqui-Terán, A.; Espinosa-Solis, V. Physicochemical properties and antioxidant activity of spray-dry broccoli (Brassica oleracea var Italica) stalk and floret juice powders. *Molecules* **2021**, *26*, 1973. [CrossRef]
31. Leyva-Porras, C.; Saavedra-Leos, M.Z.; López-Martinez, L.A.; Espinosa-Solis, V.; Terán-Figueroa, Y.; Toxqui-Terán, A.; Compeán-Martínez, I. Strawberry juice powders: Effect of spray-drying conditions on the microencapsulation of bioactive components and physicochemical properties. *Molecules* **2021**, *26*, 5466. [CrossRef]
32. Saavedra-Leos, M.Z.; Román-Aguirre, M.; Toxqui-Terán, A.; Espinosa-Solís, V.; Franco-Vega, A.; Leyva-Porras, C. Blends of carbohydrate polymers for the co-microencapsulation of Bacillus clausii and quercetin as active ingredients of a functional food. *Polymers* **2022**, *14*, 236. [CrossRef]

33. Leyva-Porras, C.; Saavedra-Leos, Z.; Román-Aguirre, M.; Arzate-Quintana, C.; Castillo-González, A.R.; González-Jácquez, A.I.; Gómez-Loya, F. An equilibrium state diagram for storage stability and conservation of active ingredients in a functional food based on polysaccharides blends. *Polymers* **2023**, *15*, 367. [CrossRef]
34. Labuza, T.; Lomauro, C.; Bakashi, A. Moisture Transfer Properties of Dry and Semimoist Foods. *J. Food Sci.* **1985**, *50*, 397–400. [CrossRef]
35. EUA. AOAC Official Methods of Analysis. Association of Official Analytical Chemists 2003. Available online: https://www.aoac.org/about-aoac-international/ (accessed on 1 October 2023).
36. Saavedra-Leos, M.Z.; Alvarez-Salas, C.; Esneider-Alcalá, M.A.; Toxqui-Terán, A.; Pérez-García, S.A.; Ruiz-Cabrera, M.A. Towards an improved calorimetric methodology for glass transition temperature determination in amorphous sugars. *CyTA-J. Food* **2012**, *10*, 258–267. [CrossRef]
37. Singh, P.C.; Singh, R.K. Application of GAB model for water sorption isotherms of food products 1. *J. Food Process. Preserv.* **1996**, *20*, 203–220. [CrossRef]
38. Navia, D.; Ayala, A.; Villada, H. Modelación Matemática De Las Isotermas De Adsorción En Materiales Bioplásticos De Harina De Yuca. *Vitae* **2012**, *19*, 423–425.
39. Talens Oliag, P. Caracterización del Comportamiento Reológico de un Alimento Fluido Tixotrópico. 2018. Available online: http://hdl.handle.net/10251/103382 (accessed on 15 February 2024).
40. Cerviño, V.F.; Sosa, C.A.; Vergara, L.E.; Schmalko, M.E.; Sgroppo, S.C. Obtención y Modelado de Isotermas de Sorción de Caramelos 'Gummy' de Batatas. In Proceedings of the III Congreso Argentino de Ingeniería—IX Congreso de Enseñanza de La Ingeniería, Resistencia city, Argentina, 7 September 2016.
41. Domínguez-Domínguez, S.; Domínguez-López, A.; González-Huerta, A.; Navarro-Galindo, S. Cinética de Imbibición e Isotermas de Adsorción de Humedad de La Semilla de Jamaica *Hibiscus sabdariffa* L. *Rev. Mex. Ing. Química* **2007**, *6*, 309–316.
42. Pascual-Pineda, L.A.; Alamilla-Beltrán, L.; Gutiérrez-López, G.F.; Azuara, E.; Flores-Andrade, E. Condiciones De Almacenamiento De Alimentos Deshidratados a Partir De Una Isoterma De Adsorción De Vapor De Agua. *Rev. Mex. Ing. Química* **2017**, *16*, 207–220. [CrossRef]
43. Brousse, M.M.; Nieto, A.; Linares, A.; Vergara, M. Cinética de adsorción de agua en purés deshidratados de mandioca (Manihot esculenta Crantz). *Rev. Venez. Cienc. Y Tecnol. Aliment.* **2012**, *3*, 3–096.
44. Ballesteros, L.F.; Ramirez, M.J.; Orrego, C.E.; Teixeira, J.A.; Mussatto, S.I. Encapsulation of antioxidant phenolic compounds extracted from spent coffee grounds by freeze-drying and spray-drying using different coating materials. *Food Chem.* **2017**, *237*, 623–631. [CrossRef]
45. Matsuura, T.; Ogawa, A.; Tomabechi, M.; Matsushita, R.; Gohtani, S.; Neoh, T.L.; Yoshii, H. Effect of dextrose equivalent of maltodextrin on the stability of emulsified coconut-oil in spray-dried powder. *J. Food Eng.* **2015**, *163*, 54–59. [CrossRef]
46. Zhang, L.; Zeng, X.; Fu, N.; Tang, X.; Sun, Y.; Lin, L. Maltodextrin: A consummate carrier for spray-drying of xylooligosaccharides. *Food Res. Int.* **2018**, *106*, 383–393. [CrossRef]
47. Navarro-Flores, M.J.; Ventura-Canseco LM, C.; Meza-Gordillo, R.; Ayora-Talavera, T.D.R.; Abud-Archila, M. Spray drying encapsulation of a native plant extract rich in phenolic compounds with combinations of maltodextrin and non-conventional wall materials. *J. Food Sci. Technol.* **2020**, *57*, 4111–4122. [CrossRef]
48. Ferrari, C.C.; Marconi Germer, S.P.; Alvim, I.D.; de Aguirre, J.M. Storage stability of spray-dried blackberry powder produced with maltodextrin or gum arabic. *Dry. Technol.* **2013**, *31*, 470–478. [CrossRef]
49. Costa-Ferreira LM, D.M.; Pereira, R.R.; Carvalho-Guimarães FB, D.; Remígio MS, D.N.; Barbosa WL, R.; Ribeiro-Costa, R.M.; Silva-Júnior JO, C. Microencapsulation by spray drying and antioxidant activity of phenolic compounds from Tucuma Coproduct (Astrocaryum vulgare Mart.) almonds. *Polymers* **2022**, *14*, 2905. [CrossRef]
50. Kroh, L.W. Caramelisation in food and beverages. *Food Chem.* **1994**, *51*, 373–379. [CrossRef]
51. Kawai, K.; Hagura, Y. Discontinuous and heterogeneous glass transition behavior of carbohydrate polymer–plasticizer systems. *Carbohydr. Polym.* **2012**, *89*, 836–841. [CrossRef]
52. Georgetti, S.R.; Casagrande, R.; Souza, C.R.F.; Oliveira, W.P.; Fonseca MJ, V. Spray drying of the soybean extract: Effects on chemical properties and antioxidant activity. *LWT-Food Sci. Technol.* **2008**, *41*, 1521–1527. [CrossRef]
53. Roos, Y. Thermal analysis, state transitions and food quality. *J. Therm. Anal. Calorim.* **2003**, *71*, 197–203. [CrossRef]
54. Ruiz-Cabrera, M.A.; Rivera-Bautista, C.; Grajales-Lagunes, A.; González-García, R.; Schmidt, S.J. State diagrams for mixtures of low molecular weight carbohydrates. *J. Food Eng.* **2016**, *171*, 185–193. [CrossRef]
55. Espinosa-Andrews, H.; Rodríguez-Rodríguez, R. Water state diagram and thermal properties of fructans powders. *J. Therm. Anal. Calorim.* **2018**, *132*, 197–204. [CrossRef]
56. Braga, M.B.; Rocha SC, D.S.; Hubinger, M.D. Spray-Drying of Milk–Blackberry Pulp Mixture: Effect of Carrier Agent on the Physical Properties of Powder, Water Sorption, and Glass Transition Temperature. *J. Food Sci.* **2018**, *83*, 1650–1659. [CrossRef]
57. Busch, V.M.; Pereyra-Gonzalez, A.; Šegatin, N.; Santagapita, P.R.; Ulrih, N.P.; Buera MD, P. Propolis encapsulation by spray drying: Characterization and stability. *LWT* **2017**, *75*, 227–235. [CrossRef]

MDPI AG
Grosspeteranlage 5
4052 Basel
Switzerland
Tel.: +41 61 683 77 34

Polymers Editorial Office
E-mail: polymers@mdpi.com
www.mdpi.com/journal/polymers